FORM SYMMETRIES AND REDUCTION OF ORDER IN DIFFERENCE EQUATIONS

Advances in
Discrete Mathematics and Applications

Series Editors
Saber Elaydi and Gerry Ladas

ADVANCES IN DISCRETE MATHEMATICS AND ITS APPLICATIONS

FORM SYMMETRIES AND REDUCTION OF ORDER IN DIFFERENCE EQUATIONS

HASSAN SEDAGHAT

VIRGINIA COMMONWEALTH UNIVERSITY
RICHMOND, USA

CRC Press
Taylor & Francis Group
Boca Raton London New York

CRC Press is an imprint of the
Taylor & Francis Group, an **informa** business
A CHAPMAN & HALL BOOK

CRC Press
Taylor & Francis Group
6000 Broken Sound Parkway NW, Suite 300
Boca Raton, FL 33487-2742

First issued in paperback 2019

© 2011 by Taylor & Francis Group, LLC
CRC Press is an imprint of Taylor & Francis Group, an Informa business

No claim to original U.S. Government works

ISBN-13: 978-1-4398-0760-6 (hbk)
ISBN-13: 978-1-138-37412-6 (pbk)

Visit the Taylor & Francis Web site at
http://www.taylorandfrancis.com

and the CRC Press Web site at
http://www.crcpress.com

Contents

Preface

The field of difference equations is old, their study predating calculus and hence differential equations. Since the time of Fibonacci, if not earlier, recursions and finite differences have appeared in a variety of contexts. Further, finite differences and sums naturally lead to differential and integral calculus through the concept of limit. In modern times, difference equations have a symbiotic relationship with their differential cousins: finite differences are routinely used in obtaining numerical solutions of differential equations and insights gained from the study of differential equations in turn inspire new investigations in difference equations.

Increasingly in the last few decades, difference equations have also been studied independently of differential ones, leading to interesting new discoveries and methods of analysis. There are journals currently devoted specifically to difference equations. These sources routinely publish articles of impressive technical quality that reflect the depth and sophistication of present-day researchers in this field.

The majority of articles and books on difference equations currently in print concern topological and convergence issues (e.g., the asymptotic behaviors of solutions) in metric spaces, predominantly finite-dimensional Euclidean spaces due to their importance in scientific models. Although these spaces easily accommodate point sequences (solutions of difference equations) they are primarily and naturally formulated for flows, i.e., solutions of differential equations. There are comparatively few articles and even fewer books that are devoted to the study of difference equations as objects or constructs on other types of sets. Yet other and more general types of sets are needed in order to determine the most natural context for defining difference equations and improve their applicability to solving problems in biology, economics, computer science and other fields.

One may argue that it is entirely natural to formulate and study difference equations in discrete spaces where the topology is trivial but algebraic structure may exist; or even in finite "spaces" such as graphs or networks that often contain a substantial connectivity or adjacency structure but which are typically not amenable to analysis using limit-based techniques. Interesting dynamic behaviors can and do occur in discrete sets but lack of an efficient tool kit like calculus makes studying such dynamics rather difficult. It is necessary to fully utilize existing structures and methods (algebraic and discrete concepts) in addition to the more classical analytical and topological methods where applicable in order to arrive at coherent and potentially useful results.

This book presents a new approach to formulation and study of difference equations in which the underlying space is typically an algebraic group. Topological structure is not shunned; on the contrary, reflecting my past research experience, most of the examples discussed in this book involve equations in finite dimensional Euclidean spaces. In some cases, additional structure beyond that of a group is needed in order to define certain equations and to obtain significant results about them. For instance, in the case of linear equations I take the underlying set to be a field so as to derive complete semiconjugate factorizations into triangular systems of first-order equations based on the eigenvalues (or eigensequences in nonautonomous cases). For certain concepts such as those in the final chapter that do not require a group structure I do not assume it. In short, my aim in this book is to take a fresh and all-inclusive look at difference equations and develop a systematic procedure to examine how the manner in which these equations are constructed from sets, operations, variables, constants, etc. determines their solutions, i.e., sequences that satisfy them.

This book contains many new results some of which are appearing in print for the first time. The easy style of presentation with a generous supply of examples, explanations and exercises all within one cover makes this book an ideal introduction to an exciting new domain in the continuously evolving field of difference equations.

Outline of the book. After some introductory remarks in Chapter 1, I introduce difference equations on groups in Chapter 2, not only to lay the foundation for later chapters but also to illustrate the amazing variety of possible formulations and interpretations of difference equations that occur in concrete contexts. The group formalism includes maps since vector spaces are groups; it includes even some types of partial difference equations that can be formulated as maps of infinite dimensional spaces. Although by themselves too primitive to support the wide variety of function forms needed for dealing with difference equations and their applications, groups can often be found as invariant substructures of more complex constructs such as vector spaces, rings, and fields.

Being able to reduce the order of a difference equation through a change of variables can simplify the analysis of the behaviors of solutions of the equation and in some cases, lead to an explicit formula for the general solution. In Chapters 3–7, I propose a systematic method of decomposition for recursive difference equations that uses a semiconjugate relation between maps. Since this relation may exist between functions with *different* numbers of variables, reductions in order may be formally characterized in terms of the existence of a semiconjugate relation between functions with non-matching numbers of variables. I show how to find the semiconjugate relations, and the accompanying factorizations into two difference equations having strictly lower orders, for large classes of difference equations. In particular, in Chapter 7, I show that nonautonomous, nonhomogeneous linear equations are special members of a class of nonlinear equations that possess a linear form symmetry; this change of variables is one of several that are characterized in terms of the invertibility of a structural mapping of the underlying field.

In Chapter 8, I go beyond semiconjugacy by extending the fundamental ideas based on form symmetries to nonrecursive difference equations. These equations cannot be characterized by maps, thus semiconjugate relations do not exist for them. But such equations include the important class of quadratic difference equations whose solutions collectively exhibit virtually all known types of dynamic behaviors. The reason for this variety is partly because quadratic difference equations include recursive equations such as linear equations (all types) as well as well-known quadratic recursive equations (e.g., the logistic map, the Henon map and related higher-order or delay equations) and rational recursive equations containing first-degree polynomials in their denominators and polynomials of degree at most two in their numerators. But perhaps far more intriguing are nonrecursive quadratic difference equations whose solution sets may span an enormous range of possible sequences. The nontrivial problem of the existence of real solutions is explored in detail for nonrecursive equations on the field of real numbers. It may also be mentioned here that some of the basic results in Chapter 8 may be stated without appealing to algebraic structure and may thus be formulated in arbitrary sets.

Background. The main audiences for this book include not only mathematicians interested in or researching difference equations but also scientists who use difference equations for modeling as well as post-graduate and upper-level undergraduate college students. To help extend the reader's understanding of the material I have included a list of several problems as exercises in Chapters 2 through 8.

College students majoring in mathematics or science may be able to read and understand most of this book since a rudimentary exposure to modern algebra after a calculus sequence should suffice for attaining a basic understanding of the main topics. Having completed a first course in differential equations or difference equations will be additionally beneficial in understanding parts of the text. Certainly the greater his or her mathematical background, the more comfortable the reader will feel reading this book, but perhaps the most essential ingredients of all are an open mind to accept new ideas and a willingness to master them.

H. Sedaghat

1

Introduction

This short chapter offers a gentle introduction to some of the basic ideas in this book. For definitions and additional introductory material see Chapter 2.

Consider the difference equation

$$x_{n+1} = x_n + \frac{a}{x_n - x_{n-1}}, \quad n = 0, 1, 2 \ldots \tag{1.1}$$

where a is a nonzero real number. This equation has order 2, or is second order, because of the difference between the highest index $n+1$ and the lowest $n-1$. Its solution is the sequence of real numbers x_n that may be calculated recursively from a pair of unequal real numbers x_0, x_{-1} as the index n increases through the non-negative integers. Can we find this solution or determine its essential properties *for all values of* n and *all unequal initial pairs* x_0, x_{-1}?

Equation (1.1) has a symmetry in its form that is easy to identify when it is written as

$$x_{n+1} - x_n = \frac{a}{x_n - x_{n-1}}. \tag{1.2}$$

Setting $t_n = x_n - x_{n-1}$ in the right-hand side and $t_{n+1} = x_{n+1} - x_n$ for the left-hand side (to account for the shift in index n) changes Eq. (1.2) to the simpler (and first-order) equation

$$t_{n+1} = \frac{a}{t_n}. \tag{1.3}$$

The expression $x_n - x_{n-1}$ is an example of what we call an order-reducing *form symmetry.* This form symmetry also establishes a link between the second-order (1.1) and the first-order (1.3) in the following sense: Information about each solution $\{t_n\}$ of (1.3) can be translated into information about the corresponding solution of (1.1) using the equation

$$x_n = x_{n-1} + t_n = x_{n-2} + t_1 + t_0 = \cdots = x_0 + \sum_{k=1}^{n} t_k. \tag{1.4}$$

In this particular example, if x_0 and x_{-1} are unequal then each solution of Eq. (1.3) is a sequence of numbers t_n taking only two possible values:

$$t_n = t_0 = x_0 - x_{-1}, \quad \text{if } n \text{ is even}$$
$$t_n = \frac{a}{t_0} = \frac{a}{x_0 - x_{-1}}, \quad \text{if } n \text{ is odd.}$$

Thus from (1.4) we obtain the formula

$$x_n = \begin{cases} x_0 + (t_0 + a/t_0)\, n/2, & \text{if } n \text{ is even} \\ x_0 + a/t_0 + (t_0 + a/t_0)\,(n-1)/2, & \text{if } n \text{ is odd} \end{cases} \qquad (1.5)$$

as giving the explicit solution of (1.1). Note that (1.5) completely determines the numbers x_n for *all values* of n, x_0, x_{-1}.

For more complicated higher order difference equations derivation of explicit solutions is often either infeasible or of little practical value. But the study of qualitative behaviors of solutions can be greatly simplified through reduction of order in the above sense. An example of this latter case is the second-order quadratic equation

$$x_{n+1} = (x_n - x_{n-1})^2 + x_n - a, \quad a > 0 \qquad (1.6)$$

which has the same form symmetry as (1.1) and is reduced by the same change of variables to

$$t_{n+1} = t_n^2 - a. \qquad (1.7)$$

For most positive values of a, the first-order equation (1.7) has no known explicit solutions; however, bifurcations of qualitatively different solutions with changing values of the parameter a are well-known for (1.7) and such information together with (1.4) can be used to study the behaviors of solutions of (1.6) qualitatively.

Most difference equations do not have easily identifiable form symmetries as (1.1) or (1.6) do. For instance, form symmetries of the exponential equation

$$x_{n+1} = x_{n-1} e^{a - x_n - x_{n-1}} \qquad (1.8)$$

are not obvious. Numerical study indicates that the positive solutions of (1.8) exhibit complex behavior *depending on the initial values* x_0, x_{-1} even when with a *fixed* value of the parameter $a > 4.5$. Of course, as seen in (1.6) complexity of behavior is not an indication that reduction of order is difficult or impossible. In fact, we discover a form symmetry for Eq. (1.8) later in this book (Section 6.3) and use it to explain the complexity of behavior seen in its solutions through reduction of order.

As these examples indicate, reducing the order of a difference equation by identifying and substituting for form symmetries like $x_n - x_{n-1}$ can help answer questions about qualitative behaviors of solutions of equations like (1.6) and (1.8) as well as more traditional questions about integrability and the existence of explicit solutions such as (1.5). Of course, simply finding a recurrent expression and substituting a new variable for it is not sufficient unless *all occurrences* of the old variable can be eliminated. Therefore, it is necessary to determine ways in which a particular expression or form symmetry can occur in a difference equation so that substituting for it eliminates all occurrences of the old variable.

This is often a difficult problem. In this book we use the concept of *semi-conjugacy* to formulate the problem in a systematic new way. Then using this new formalism we obtain order-reduction results for large classes of difference equations. Semiconjugacy is a relation that can exist between a given function and another one with *fewer* variables. And being transitive, this relation acquires the character of an ordering relation that formalizes the notion of reduction in order, say from high to low. For difference equations that are defined by functions, i.e., *recursive* equations as in the preceding three examples, the orders of the equations are given by the numbers of variables in the associated mappings. Therefore, for recursive equations the existence of a semiconjugate relation implies the existence of form symmetries and the reducibility of order.

For more general, nonrecursive difference equations semiconjugacy cannot be applied because there is no unfolding map. However, the essential concepts and ideas developed for recursive equations, including a precise definition of form symmetry can be extended to nonrecursive equations by a procedure that does not require the unfolding map. We outline the basics of such a general formalism in a separate chapter.

We work largely with difference equations on groups. An apparent reason for using groups is that they are algebraic generalizations of real and complex numbers. A more subtle and compelling reason is that working with groups (and similar abstract structures) enhances the ability to view a difference equation in different ways. For instance, consider the third-order equation

$$x_{n+1} = x_n + \frac{a(x_n - x_{n-1})^2}{x_{n-1} - x_{n-2}}, \quad a \neq 0, \ x_{-1} \neq x_0, x_{-2} \tag{1.9}$$

on the set of real numbers. First, as in Eq. (1.1) we make the substitution $t_n = x_n - x_{n-1}$, taking account of the extra time delay in n to get the second-order equation

$$t_{n+1} = \frac{at_n^2}{t_{n-1}}. \tag{1.10}$$

Since this new equation can be written as

$$\frac{t_{n+1}}{t_n} = a\frac{t_n}{t_{n-1}}$$

we notice that substituting a new variable s_n for the recurrent expression t_n/t_{n-1} gives the simple first-order equation

$$s_{n+1} = as_n.$$

They may seem different but the two order-reducing substitutions above are closely related. In fact, they are both instances of the *same* form symmetry but in *different groups*. Both expressions are of type $u * v^{-1}$ where u, v are elements in the group, $*$ is the group operation and the power -1 denotes inversion in the group. In the case of $x_n - x_{n-1}$ the underlying group is the set $\mathbb{R}$ of real numbers *under addition* while for t_n/t_{n-1} we refer to the group of all nonzero real numbers *under multiplication*. We see later that the form symmetry $u * v^{-1}$ characterizes a sizable class of homogeneous difference equations to which both (1.9) and (1.10) belong.

In the remainder of this book we make the aforementioned concepts precise. But our treatment is far from being comprehensive. Reducing the order of an arbitrary difference equation is a nontrivial task and finding order-reducing form symmetries or even just showing that one must exist for a given difference equation is a generally difficult thing to do. This book aims to show its committed readers that in spite of the inherent difficulties, there is a great deal that can be done. With a reasonable amount of patience and due diligence, the resulting work has its rewards not only by offering a set of ideas and methods to tackle problems in theory and applications but also by providing a deeper and more comprehensive understanding of difference equations.

2

Difference Equations on Groups

The two difference equations

$$x_{n+1} = x_n + x_{n-1} \qquad (2.1)$$

and

$$x_{n+1} = x_n x_{n-1} \qquad (2.2)$$

are different by some accounts; e.g., the first is linear and the second is not. Thus an explicit formula for the solutions of (2.1) may be found using classical methods that are not applicable to (2.2). Yet there are also some similarities: the positive solutions of (2.2) are the images of solutions of (2.1) under the exponential function. Indeed, the exponential function is both a group *isomorphism* between all real numbers under addition and all positive real numbers under multiplication and also a *homeomorphism* between all real numbers and all positive real numbers under the usual topology. It follows that the two equations (2.1) and (2.2) are dynamically equivalent (i.e., they have essentially the same solutions) if (2.2) is restricted to the multiplicative group of positive real numbers.

These similarities and differences are perhaps best explained by the fact that (2.1) and (2.2) are both special cases of the *same* equation being defined on *different* groups. Eq. (2.2) may be defined on still other groups; for example, on the group of all nonzero real numbers under multiplication where (2.2) and (2.1) are *not* dynamically equivalent (there are solutions that are essentially different). We explore the idea of difference equations on groups in some detail in this chapter and illustrate the influence of algebraic structure by examples in which equations are defined in exactly the same way but generate different dynamics because the underlying group structures are different (nonisomorphic).

In this and the next five chapters, we discuss *explicit* or *recursive* difference equations on algebraic groups; i.e., difference equations of type

$$x_{n+1} = f_n(x_n, x_{n-1}, \dots, x_{n-k}), \quad n = 0, 1, 2, \dots \qquad (2.3)$$

where k is a fixed non-negative integer and the functions $f_n : D \to G$ with G a group and $D \subset G^{k+1}$ a nonempty common domain for all f_n. These equations are not the most general that may occur but they are important in modeling applications and have the desirable feature of unfolding to self-maps of G^{k+1} that relate order to dimensionality. As a result, solutions of

(2.3) can be generated by iteration, a feature that substantially simplifies the discussions of solutions while also guaranteeing their existence in most cases of interest.

In the final chapter of the book we study general scalar difference equations for which there may be no unique explicit or recursive forms globally and the existence of solutions is not guaranteed by direct iteration because the equations do not unfold to self-maps of a higher dimensional space. Such equations may be called *implicit* or *nonrecursive*. In spite of difficulties in obtaining information about the solutions of such equations, we discover that essential ideas and methods from Chapters 3–7 on decompositions (or factorizations) of equations and reductions of their orders can be extended to nonrecursive equations in many cases.

2.1 Basic definitions

The next few definitions introduce basic concepts and terminology pertaining to Eq. (2.3).

DEFINITION 2.1   *The **order** of the difference equation (2.3) is the integer $k + 1$, i.e., the difference between the highest order $n + 1$ and the lowest order $n - k$ in the equation. The independent **variable** n may also be called the **time index** of (2.3). If f_n is time independent, i.e., $f_n = f$ for all $n \geq 0$ then (2.3) is **autonomous**; otherwise, (2.3) is **nonautonomous**.*

REMARK 2.1   (Classical operators) In the classical theory of difference equations on real or complex numbers two basic operators appear frequently. They are the forward shift E and the backward difference Δ, which are defined as follows: If $\{s_n\}$ is a sequence of real or complex numbers then

$$Es_n = s_{n+1}, \quad \Delta s_n = s_n - s_{n-1}.$$

These concepts readily extend to an arbitrary group G since E is independent of any structure on the underlying set and Δ takes the form

$$\Delta x_n = x_n * x_{n-1}^{-1}$$

for every sequence $\{x_n\}$ in G. In an algebraic field $\mathcal{F}$ we may define Δ using the additive notation, i.e., $\Delta x_n = x_n - x_{n-1}$ which is closer to the classical meaning. We occasionally refer to these operators in this book, but they do not play a fundamental role in our discussions. ∎

DEFINITION 2.2 *A (**forward**) solution of (2.3) is a sequence $\{x_n\}_{n=-k}^{\infty}$ in G that is generated recursively from a given set of **initial values** (i.v.)*

$$x_0, x_{-1}, \ldots, x_{-k} \in G$$

*by setting $n = 0, 1, 2, \ldots$ one step at a time. The plot of points x_n versus n is sometimes called a **time-series** plot of the orbit. The solution $\{x_n\}_{n=-k}^{\infty}$ is also a **base-space orbit** or **trajectory** of (2.3) in G, the underlying **base-space**.*

A more standard definition of orbit follows Definition 2.3 below.

REMARK 2.2 The preceding definition gives a *forward* solution since the index n only increases. Other concepts of solution are possible. In analogy with ordinary differential equations where solutions are typically defined for all time (both forward and backward) we may also define solutions of difference equations as functions on integers $\mathbb{Z}$. In this case, a solution is a *"doubly infinite sequence"* $\{x_n\}_{n=-\infty}^{\infty}$ in G. The initial values simply define a given point through which the solution passes.

In most applications of difference equations only forward solutions are of interest. In order to keep things as simple as possible without loss of significance, in this book we limit attention to forward solutions and unless stated otherwise, by a "solution" we mean a forward solution. ∎

DEFINITION 2.3 *For $k \geq 1$ and each $n \geq 0$ the **unfolding** (or **vectoriziation** or **associated vector map**) $F_n : D \to G^{k+1}$ of f_n is defined as*

$$F_n(u_0, \ldots, u_k) = [\, f_n(u_0, \ldots, u_k), u_0, \ldots, u_{k-1}].$$

*We refer to the collection of maps $\{F_n\}_{n=0}^{\infty}$ as the unfolding of Eq. (2.3). The vector $(x_0, x_{-1}, \ldots, x_{-k})$ of initial values is the **initial state** and for each $n \geq 0$ the vector $(x_n, x_{n-1}, \ldots, x_{n-k})$ is the **state at time** n or the n-**th state**. The space G^{k+1} is called the **state-space** of (2.3).*

The concept of state-space in Definition 2.3 is analogous to the "phase space" in differential equations.

Evidently, each solution $\{x_n\}_{n=-k}^{\infty}$ of (2.3) in G corresponds uniquely to a solution $\{X_n\}$ of the first-order equation

$$X_{n+1} = F_n(X_n), \quad X_0 = (x_0, x_{-1}, \ldots, x_{-k}) \in D \subset G^{k+1}. \qquad (2.4)$$

Starting from an initial state X_0 iteration yields a sequence of states $X_n = (x_n, x_{n-1}, \ldots, x_{n-k})$ for $n = 0, 1, 2, \ldots$ The sequence $\{X_n\}_{n=0}^{\infty}$ is an *orbit* (or *trajectory*) of (2.4) in the state-space G^{k+1}. These are the standard defintions of orbit (or trajectory) which when $k = 1$ coincide with the earlier versions in the base-space G defined above.

A remark about the vector concept may be in order at this stage. Eq. (2.3) is traditionally defined on the real or complex number systems and therefore it is often called a *scalar equation* to distinguish it from its vector (or system) version (2.4). In the context of this book this distinction remains important but acquires a technical rather than a conceptual flavor since G may itself be a group of finite or infinite dimensional vectors of real or complex numbers.

DEFINITION 2.4   *A solution* $\{x_n\}_{n=-k}^{\infty}$ *of (2.3) is* **eventually periodic** *with period* $p \geq 1$ *(or* **eventually** p**-periodic***) if there is an integer* $n_0 \geq 0$ *such that* $x_{n+p} = x_n$ *for all* $n \geq n_0$*. If* $n_0 = 0$ *then the solution is* **periodic** *with period* p *(or* p**-periodic***). If* $p = 1$ *then the solution is a* **constant solution** *(eventually if* $n_0 > 0$*) or a* **fixed point** *of (2.3). A fixed point* $\bar{x}$ *of (2.3), also sometimes called a point of* **equilibrium**, *is a solution of the equation*

$$u = f(u, \ldots, u).$$

A solution that is not eventually periodic is a **nonperiodic** *(or* **aperiodic***) solution.*

From this definition and the vector form (2.4) it follows that if a particular state vector X_q is revisited after p iterations then the sequence of states $\{X_n\}$ is eventually p-periodic. An immediate consequence of this observation is that *if G is a finite group with m elements then every solution of (2.3) is eventually periodic with period at most m^{k+1}.* This is true because there are m^{k+1} points in the state-space G^{k+1} and in at most that many iterations of the vector equation (2.4) any initial point in G^{k+1} must be revisited.

REMARK 2.3   If a topological structure is present then solutions that are not periodic but converge to periodic solutions are generally not considered aperiodic or nonperiodic. They are in fact identified with their **limit cycles**, a terminology that we use when appropriate.   ∎

A function may be defined on a group but its iterations may not be. Hence, there is no guarantee that every initial value leads to a solution of the difference equation in its group. As an extreme example, consider the autonomous first-order difference equation

$$x_{n+1} = \sqrt{x_n} - \frac{1}{\sqrt{x_n}}. \tag{2.5}$$

Here $f(u) = (u-1)/\sqrt{u} = F(u)$ is defined on $G = (0, \infty)$ which is a group under ordinary multiplication. However, note that if $x_0 \in (0, 1]$ then $x_1 \notin G$; since $f(u) < u$ for all $u > 0$ it follows that $x_{n+1} < x_n$ for all n. Therefore, every solution of (2.5) is strictly decreasing and eventually negative, then complex (or possibly undefined if passing through 1). It follows that Eq. (2.5) has *no solutions in the group* $G = (0, \infty)$ on which it is defined.

The next definition rules out equations like (2.5) and ensures the existence of solutions in a relevant group.

DEFINITION 2.5   *(Invariance) If there is a nonempty subset $A \subset G^{k+1}$ such that for all $n \geq 0$,*

$$F_n(A) \subset A$$

*where F_n is the unfolding of f_n then A is an **invariant subset** of G^{k+1}.*

Eq. (2.5) with $k = 0$ has no invariant sets in $G = (0, \infty)$. Although without invariance the existence of solutions for (2.3) is not guaranteed, difference equations of type (2.3) with non-invariant domains do arise in applications. These equations are generally not as bad as Eq. (2.5) and often have nontrivial solutions in their base-spaces. When considering such equations a key concept is the following.

DEFINITION 2.6   *(Singularity set) Let M be a nonempty set and for functions $f_n : D \to M$ where $D \subset M^{k+1}$ let F_n be the unfolding of f_n for each n. The **singularity set** $S \subset M^{k+1}$ of Eq. (2.3) is the set of all initial states $X_0 \in D$ such that for some $n_0 \geq 1$ (depending on X_0)*

$$F_{n_0} \circ F_{n_0-1} \circ \cdots \circ F_0(X_0) \notin D. \tag{2.6}$$

*The initial states in S may be called **singular states** (or **singular points**) in the sense that such points do not generate complete orbits in M.*

In the autonomous case, i.e., $F_n = F$ for all n, (2.6) takes the simpler form

$$F^{n_0}(X_0) \notin D$$

where F^{n_0} is the composition of F with itself n_0 times. Note that if $D = M^{k+1}$ then $F_n(M^{k+1}) \subset M^{k+1}$. Therefore, M^{k+1} is an invariant domain in this case and S is empty. Thus a necessary condition for S to be nonempty is that $D \neq M^{k+1}$. In the extreme case of the autonomous Eq. (2.5) where $k = 0$, we have $f = F : (0, \infty) \to \mathbb{R}$. Letting $D = (0, \infty)$ and $M = \mathbb{R}$, it is evident that for every $x_0 \in D$ there is $n_0 \geq 1$ such that the iterate $F^{n_0}(x_0) \notin D$; for example, if $x_0 \in (0, 1]$ then $n_0 = 1$. It follows that every point of D is singular, i.e., $S = D$.

In Definition 2.6 the set M may have a group structure defined on it that may or may not be relevant to the context of a model or a problem. Therefore, it is not necessary that M have any algebraic structure in that definition or that it retain the algebraic structure of G.

Recall that a binary relation $\leq$ is a *total ordering* on a nonempty set T if $\leq$ satisfies the conditions:

(i) $u \leq u$ for all $u \in T$;

(ii) $u \leq v$ and $v \leq u$ imply that $u = v$ for every $u, v \in T$;

(iii) $u \leq v$ and $v \leq w$ imply that $u \leq w$ for every $u, v, w \in T$.

DEFINITION 2.7   *Let T be an invariant set for Eq. (2.3) that is totally ordered by a relation $\leq$. A solution $\{x_n\}_{n=-k}^{\infty}$ of (2.3) is **eventually nondecreasing** (respectively, **eventually nonincreasing**) if there is an integer $n_0 \geq 1$ such that $n > m \geq n_0$ implies $x_m \leq x_n$ (respectively, $x_n \leq x_m$). If $\{x_n\}_{n=-k}^{\infty}$ is either eventually nondecreasing or eventually nonincreasing then $\{x_n\}_{n=-k}^{\infty}$ is **eventually monotonic**. If $\{x_n\}_{n=-k}^{\infty}$ is not eventually monotonic then it is an **oscillatory** solution of (2.3).*

In particular, constant solutions of (2.3) are eventually monotonic as they are both nondecreasing and nonincreasing while periodic solutions having periods at least two are oscillatory.

2.2   One equation, many interpretations

Let $(G, *)$ represent a nontrivial group with its binary operation $*$ and define the simple, second-order difference equation

$$x_{n+1} = x_n * x_{n-1}. \tag{2.7}$$

The unique solution of this equation in G starts out as follows:

$$\{x_{-1}, x_0, x_0 * x_{-1}, x_0 * x_{-1} * x_0, x_0 * x_{-1} * x_0 * x_0 * x_{-1}, \cdots\}$$

The nature of this sequence depends crucially on G and its binary operation. In special cases, the above solution of (2.7) may have simple, explicit formulas: For instance, if $x_{-1} = x_0$ then by inspection of the above sequence $x_n = x_0^{\varphi_n}$ where φ_n is the n-th term of the Fibonacci sequence (see Example 2.1 below). However, a little reflection should convince the reader that a general formula for expressing all solutions of (2.7) on an arbitrary group does not exist. Indeed, even the special solution $x_n = x_0^{\varphi_n}$ that starts on the diagonal in the state-space may have a complicated nature depending on the group G; see Example 2.4 below. Thus, *the role of G in shaping the solutions of (2.7) is important* and needs to be carefully examined.

To gain further insight into this matter, in this section we examine (2.7) on a few *specific* groups. Changing the group structure while keeping the equation's *form* fixed leads to a new and deeper understanding of difference equations and their solutions on groups.

Example 2.1
Let G be the set of all real numbers $\mathbb{R}$ and let $*$ be the ordinary addition of real numbers. Then (2.7) takes the form

$$x_{n+1} = x_n + x_{n-1}. \tag{2.8}$$

We may refer to Eq. (2.8) as the Fibonacci difference equation since the famous "Fibonacci sequence"

$$\{\varphi_n\}_{n=-1}^{\infty} = \{1, 1, 2, 3, 5, 8, 13, 21, \ldots\} \tag{2.9}$$

is, by its definition, a solution of (2.8) with initial values $x_0 = x_{-1} = 1$. An explicit formula for the general solution of the linear equation (2.8) is easily obtained by familiar, classical methods as

$$x_n = C_1 \left(\frac{1 + \sqrt{5}}{2} \right)^n + C_2 \left(\frac{1 - \sqrt{5}}{2} \right)^n \tag{2.10}$$

where the constants C_1, C_2 are determined by the initial values x_0, x_1 in the customary way (linear difference equations are discussed in detail later in this book). Using (2.10) we can quickly determine the nature of all solutions of (2.8) in $\mathbb{R}$. In particular, since $-1 < (1 - \sqrt{5})/2 < 0$ it is evident that for sufficiently large n, the terms x_n are essentially proportional to the dominant exponential term whose base $(1 + \sqrt{5})/2$ (known as the "golden ratio") is greater than 1. In particular, the solutions of (2.8) are all eventually monotonic.

An interesting feature of Eq. (2.8) is that if the initial values are in a subgroup of $(\mathbb{R}, +)$, e.g., the rationals $(\mathbb{Q}, +)$ or the integers $(\mathbb{Z}, +)$, then so are x_n for all $n \geq 1$. For example, all terms of the Fibonacci sequence above are integers since $x_0 = x_{-1} = 1$. Therefore, formula (2.10) which involves irrational real numbers such as the golden ratio produces only integers or rationals depending on the initial values. Later in this book, by studying general linear difference equations on arbitrary fields in sufficient depth we discover a general formula for solutions of (2.8) that does not involve irrational numbers; i.e., a formula that exists in $\mathbb{Q}$, itself a field under ordinary addition and multiplication; see Example 7.4 in Chapter 7.

Finally, we point out that Eq. (2.8) has invariant sets in $\mathbb{R}$ that are not subgroups; e.g., intervals of type $[r, \infty)$ for all $r \geq 0$ or intervals $(-\infty, s]$ for $s \leq 0$. $\square$

A slight generalization of the preceding example is the following.

Example 2.2
Let G be the vector space $\mathbb{R}^m$ of all vectors with m real components under the usual component-wise vector addition. The difference equation (2.7) takes the form

$$v_{n+1} = v_n + v_{n-1}. \tag{2.11}$$

This equation can also be written equivalently as a system of m Fibonacci component equations

$$v_{i,n+1} = v_{i,n} + v_{i,n-1}, \quad i = 1, 2, \ldots, m.$$

Since each component equation is independent of the others, the solutions of (2.11) consist of vectors whose components are determined by (2.10). □

In the next example, we examine (2.7) on a different group and discover a different behavior.

Example 2.3

Consider the group of all nonzero real numbers $\mathbb{R}_0$ under ordinary multiplication. Then (2.7) takes the form

$$x_{n+1} = x_n x_{n-1}. \tag{2.12}$$

A formula for the general solution of (2.12) on $\mathbb{R}_0$ is obtained by straightforward iteration as

$$x_n = x_0^{\varphi_{n-1}} x_{-1}^{\varphi_{n-2}} \tag{2.13}$$

where φ_n is the Fibonacci sequence (2.9) for $n \geq 1$ and its value for every n may be calculated from (2.10). Note that formula (2.13) remains valid with the inclusion of zero; therefore, it gives all solutions of (2.12) in $\mathbb{R}$ not just $\mathbb{R}_0$.

It is clear from (2.13) that the behavior of solutions of (2.12), namely, Eq. (2.7) on $\mathbb{R}_0$ under multiplication is qualitatively different from that of (2.8) on $\mathbb{R}$ under addition. For example, if $x_{-1} = 1$ and $x_0 = -a$ where $a > 0$ then by (2.13) the corresponding solution of (2.12):

$$x_n = (-a)^{\varphi_{n-1}}$$

is oscillatory; further, $\{x_n\}$ is bounded if $a \leq 1$ and unbounded if $a > 1$. These patterns are not exhibited by the eventually monotonic solutions of Eq. (2.8). As in Example 2.1, there are invariant *sets* that are not groups. For example, the interval $[-1, 1]$ is not a subgroup of $\mathbb{R}_0$ but it is an invariant subset of $\mathbb{R}$ under Eq. (2.12) because $|x_0 x_{-1}| \leq 1$ implies $|x_n| \leq 1$ for all n. Since both (2.12) and its general solution (2.13) are defined at zero, it follows that the behaviors of solutions in $[-1, 1]$ can be determined using arguments similar to the preceding discussion. For instance, if $|x_0 x_{-1}| < 1$ then it is evident from (2.13) that $x_n \to 0$ as $n \to \infty$. If furthermore, $x_0, x_{-1} > 0$ then the solution decreases monotonically to zero. □

It may be worth mentioning that the nonlinear equation (2.12) in Example 2.3 may behave like the linear equation (2.8) in Example 2.1 under certain restrictions. In the multiplicative subgroup $(0, \infty)$ of $\mathbb{R}_0$, $x_0, x_{-1} > 0$ so

$x_n > 0$ for all n; i.e., the subgroup $(0, \infty)$ is invariant under (2.12). The change of variables $x_n = e^{y_n}$ under the real exponential function transforms (2.12) into the equivalent additive form (2.8) from which an explicit formula for x_n can be easily obtained using (2.10). This similarity of behavior is a reflection of the fact that the the additive group $\mathbb{R}$ and the multiplicative group $(0, \infty)$ are algebraically isomorphic under the real exponential function and the unfoldings of (2.8) and (2.12) are topologically equivalent (or "conjugate") via the homeomorphism $H : \mathbb{R}^2 \to (0, \infty)^2$ defined as $H(u, v) = [e^u, e^v]$.

In the next example we discuss equation (2.7) in yet another group and discover a new solution set that is different from those in the preceding two examples.

Example 2.4

(The circle group) Consider the unit circle $\mathbb{T} = \{z \in \mathbb{C} : |z| = 1\}$ in the complex plane. Since $|zw| = |z||w| = 1$ for every $z, w \in \mathbb{T}$, it follows that $\mathbb{T}$ is closed under the ordinary multiplication of complex numbers. Therefore, $\mathbb{T}$ forms a multiplicative subgroup of $\mathbb{C}$. Now we consider Eq. (2.12) on $\mathbb{T}$ i.e.,

$$z_{n+1} = z_n z_{n-1}, \quad z_0, z_{-1} \in \mathbb{T}. \tag{2.14}$$

The solutions of (2.14) satisfy a formula of type (2.13) upon iteration, yet they are qualitatively different from solutions of either (2.8) or (2.12). For instance, all solutions of (2.14) are bounded because they are in $\mathbb{T}$; in fact, every term of each solution has modulus 1. Clearly, this is not the case for all solutions of (2.12) or (2.8). More precisely, given that $\mathbb{T}$ can be represented in polar form as

$$\mathbb{T} = \{e^{i\theta} = \cos\theta + i\sin\theta : 0 \le \theta < 2\pi\},$$

formula (2.13) for $\mathbb{T}$ may be written in the following more descriptive form

$$\begin{aligned} z_n &= z_0^{\varphi_{n-1}} z_{-1}^{\varphi_{n-2}} \\ &= e^{i(\theta_0\varphi_{n-1} + \theta_{-1}\varphi_{n-2})} \\ &= \cos(\theta_0\varphi_{n-1} + \theta_{-1}\varphi_{n-2}) + i\sin(\theta_0\varphi_{n-1} + \theta_{-1}\varphi_{n-2}) \end{aligned}$$

with $e^{i\theta_0} = z_0$ and $e^{i\theta_{-1}} = z_{-1}$. In the above form we see that the behaviors of solutions of (2.14) are clearly different (and more complicated) than those of (2.8) or (2.12). To illustrate further, let $z_{-1} = z_0 = e^{i\theta_0}$ and recall that $\varphi_{n-1} + \varphi_{n-2} = \varphi_n$ for all n to obtain

$$z_n = \cos(\theta_0\varphi_n) + i\sin(\theta_0\varphi_n). \tag{2.15}$$

If $\theta_0 = j\pi/2$ where j is any given integer then using the regular occurrence of even numbers among φ_n we can show that $\{z_n\}$ is periodic with period 6; this can be also verified by iterating (2.14) with all possible initial values

$$z_0 = e^{i\theta_0} = e^{i\pi j/2} = i^j \in \{\pm 1, \pm i\}.$$

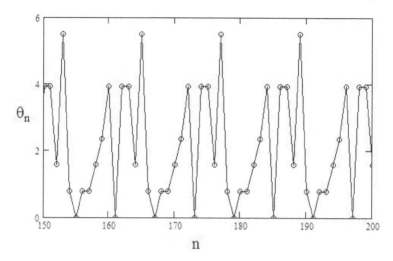

FIGURE 2.1

Values of the arguments θ_n with $\theta_0 = \theta_{-1} = \pi/4$ showing a periodic solution.

Figure 2.1 shows a plot of arguments of the state-space orbit with $z_0 = (1+i)/\sqrt{2}$ or $\theta_0 = \pi/4$.

Examination of Figure 2.1 indicates that the trajectory has period 12. A plot of the trajectory in state-space is shown in Figure 2.2. Some states in this figure (seen as dots on the unit circle) are visited more than once so there are fewer than 12 *distinct* states. The manner in which the states in Figure 2.2 are traversed can be inferred from Figure 2.1.

Other values of θ_0, including irrational multiples of π may generate solutions $\{z_n\}$ in (2.15) that are oscillatory but nonperiodic, or aperiodic, in $\mathbb{T}$. See Figures 2.3 and 2.4 in which a numerically generated solution with slightly different initial values is nonperiodic. Such a behavior is clearly more complicated than what occurs in Example 2.1 or in Example 2.3. $\square$

We may write Eq. (2.14) in polar form as

$$e^{i\theta_{n+1}} = e^{i\theta_n + i\theta_{n-1}}, \quad e^{i\theta_{n+j}} = z_j, \ j = -1, 0, 1.$$

In this form an additive equation is apparent in exponents that is modulo 2π; i.e.,

$$\theta_{n+1} = (\theta_n + \theta_{n-1}) \bmod 2\pi. \tag{2.16}$$

Refer to the Problems for this chapter for a look at the solutions of (2.14) in the set of all complex numbers. In this context we find interesting relationships between ideas discussed in Examples 2.1, 2.3 and 2.4.

Addition modulo a positive number also appears in the next example which gives yet another incarnation of Eq. (2.7).

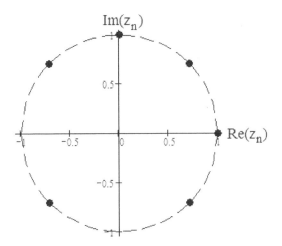

FIGURE 2.2

The periodic state-space orbit z_n on the unit circle with
$z_0 = z_{-1} = (1+i)/\sqrt{2}$ or $\theta_0 = \theta_{-1} = \pi/4$.

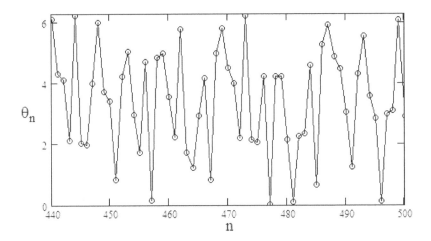

FIGURE 2.3

Values of the arguments θ_n with $\theta_0 = \theta_{-1} = \pi/3$ showing a nonperiodic
solution.

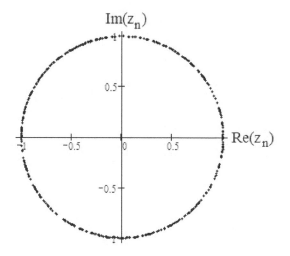

FIGURE 2.4

The nonperiodic state-space orbit z_n on the unit circle with $z_0 = z_{-1} = (1+i)/\sqrt{2}$ or $\theta_0 = \theta_{-1} = \pi/3$.

Example 2.5

Consider the group $\mathbb{Z}_m = \{0, 1, \ldots, m-1\}$ for each fixed integer $m \geq 2$ where the operation in $\mathbb{Z}_m$ is addition modulo m, i.e.,

$$i * j = (i + j) \bmod m.$$

Here $k \bmod m$ is the remainder of the quotient k/m for all $k = 0, 1, 2, \ldots$ Clearly $\mathbb{Z}_m$ is commutative for each m and the inverse of 1 in $\mathbb{Z}_m$ is $m - 1$, the inverse of 2 is $m - 2$, etc. If m is even then the element $m/2$ is its own inverse. Eq. (2.7) on $\mathbb{Z}_m$ takes the form

$$x_{n+1} = x_n * x_{n-1} = (x_n + x_{n-1}) \bmod m. \tag{2.17}$$

Given that $\mathbb{Z}_m$ is finite and the state-space of (2.17) has m^2 elements, it follows that all solutions of (2.17) are periodic with period at most m^2. The solutions of (2.17) are therefore qualitatively different from the perviously seen versions of (2.7) in Examples 2.1, 2.3 and 2.4. ☐

Despite their simple natures, the solution sets of (2.17) are not easy to characterize for arbitrary m. To illustrate, consider the analog of the Fibonacci sequence in $\mathbb{Z}_m$. With the usual initial values $x_0 = x_{-1} = 1$ but with different values of m, "Fibonacci sequences" of different periods are generated by truncating the actual Fibonacci sequence, modulo m. The following table lists the periods of Fibonacci sequences modulo m for the first 25 values of m. Note the significant variations in periods as m changes:

m	2	3	4	5	6	7	8	9	10	11	12	13
Period	3	8	6	20	24	16	12	24	60	10	24	28
m	14	15	16	17	18	19	20	21	22	23	24	25
Period	48	40	24	36	24	18	60	16	30	48	24	100

In all of the preceding equations, the groups involved are commutative. In the next example we consider (2.7) on a noncommutative group.

Example 2.6
(Matrix groups) The collection M of all $m \times m$ matrices A with real entries and determinants $\det A = 1$ is a group under the usual multiplication of matrices. On M (2.7) is the matrix difference equation

$$A_{n+1} = A_n A_{n-1}, \quad A_0, A_{-1} \in M. \tag{2.18}$$

If $m \geq 2$ then M is noncommutative and therefore, solutions of (2.18) are again qualitatively different from those seen in previous examples. For instance, if $A_0 A_{-1} \neq A_{-1} A_0$ then switching the values of A_0 and A_{-1} generates two different solutions of (2.18), a situation that did not occur in the commutative settings of the previous examples. ⬚

The next two examples look at functional interpretations of (2.7).

Example 2.7
(Groups of self-maps of a group) Let G_0 be a nontrivial group.

(a) If G_0 is commutative then certain sets of self-maps $\phi : G_0 \to G_0$ form commutative groups under the usual pointwise addition of functions

$$(\phi_1 + \phi_2)(u) = \phi_1(u) + \phi_2(u) \quad \text{for all } u \in G_0$$

where as long as there is no confusion we use the same symbol $+$ for both the operation in G_0 and that defined among functions on G_0. Now let G be any nontrivial group of self-maps of G_0. Then Eq. (2.7) takes the following form in G

$$\phi_{n+1} = \phi_n + \phi_{n-1}$$

which is a generalization of the Fibonacci difference equation (2.8). This functional equation can be extended as follows: For every $\phi \in G$ and positive integer m define

$$m\phi = \phi + \cdots + \phi, \quad \phi \text{ is added } m \text{ times.}$$

We may also define $m\phi = (-m)(-\phi)$ for $m < 0$ and $m\phi = 0$ (i.e., the zero function, which is the identity of G) if $m = 0$. So the following equation makes sense on G:

$$\phi_{n+1} = p\phi_n + q\phi_{n-1}, \quad p, q \in \mathbb{Z}, \ \phi_0, \phi_{-1} \in G. \tag{2.19}$$

◻

Example 2.8

(Groups of bijections of a set) Let M be a nonempty set. It is not hard to verify that the collection $\mathcal{G}$ of all bijections of M is a group under function composition. If G is any nontrivial subgroup of $\mathcal{G}$ then Eq. (2.7) takes the following form as a *functional* difference equation

$$\phi_{n+1} = \phi_n \circ \phi_{n-1}. \tag{2.20}$$

Well-known examples of groups of bijections include groups of permutations of a finite set. For instance, If $M = \{a, b, c\}$ is a set of three distinct elements then the noncommutative group G of all permutations (or bijections) of M consists of the following six self-maps

$$\sigma_0 : a \to a, b \to b, c \to c \quad \sigma_1 : a \to a, b \to c, c \to b$$
$$\sigma_2 : a \to c, b \to b, c \to a \quad \sigma_3 : a \to b, b \to a, c \to c$$
$$\sigma_4 : a \to b, b \to c, c \to a \quad \sigma_5 : a \to c, b \to a, c \to b.$$

Note that σ_0 is the identity of G and $\sigma_i^{-1} = \sigma_i$ for $i = 1, 2, 3$ while σ_4 and σ_5 are inverses of each other. Since G is finite we expect that all solutions of (2.20) are periodic with period at most $6^2 = 36$. Solutions with different periods occur; for instance, if $\phi_{-1} = \sigma_0$ and $\phi_0 = \sigma_4$ then we obtain the following solution

$$\{\sigma_0, \sigma_4, \sigma_4, \sigma_5, \sigma_0, \sigma_5, \sigma_5, \sigma_4, \sigma_0, \sigma_4, \ldots\}$$

which has period 8; we leave the details of calculation to the reader in the Problems for this chapter. Similarly, $\phi_{-1} = \sigma_1$ and $\phi_0 = \sigma_2$ give the solution

$$\{\sigma_1, \sigma_2, \sigma_5, \sigma_3, \sigma_2, \sigma_4, \sigma_1, \sigma_2, \ldots\}$$

which has period 6.   ◻

Examples 2.1–2.8 show that when defining a difference equation by algebraic operations, the nature of solution set of the equation depends critically not only on the specific manner in which the equation is defined but *also on the underlying algebraic structure.* If a result is obtained for a difference equation that is defined on a general group then such a result is valid for all interpretations of the group. In this book we often obtain reductions of orders for such generally defined difference equations, usually on groups and fields, by looking for symmetries in forms that make up a difference equation such as (2.7).

2.3 Examples of difference equations on groups

Having discussed different interpretations of a single difference equation on various groups, we now turn to a discussion of different types of difference equations on groups in order to illustrate more of the basic definitions and concepts introduced in Section 2.1.

The first example discusses a difference equation whose solution on an arbitrary commutative group is easy to obtain. The utility of having such a general solution is indicated in the remarks after the example.

Example 2.9
Let $(G, *)$ be an arbitrary commutative group and consider the difference equation

$$x_{n+1} = a * x_n * x_{n-1}^{-1} \tag{2.21}$$

where a is a fixed element of G and the power -1 denotes group inversion. For arbitrary parameter values a, x_0, x_{-1} let us calculate the first few terms of the general solution:

$$x_1 = a * x_0 * x_{-1}^{-1}, \quad x_2 = a * x_1 * x_0^{-1} = a^2 * x_{-1}^{-1},$$
$$x_3 = a * x_2 * x_1^{-1} = a^2 * x_0^{-1}, \quad x_4 = a * x_3 * x_2^{-1} = a * x_0^{-1} * x_{-1},$$
$$x_5 = a * x_4 * x_3^{-1} = x_{-1}, \quad x_6 = a * x_5 * x_4^{-1} = x_0.$$

From the above calculation it is evident that the 6-step pattern repeats and results in a solution of period 6. This result is valid on any commutative group.

In particular, the above calculation is valid in the specific group of all real numbers under addition. In this group (2.21) is the nonhomogeneous linear equation

$$x_{n+1} = a + x_n - x_{n-1}, \quad a, x_0, x_{-1} \in \mathbb{R} \tag{2.22}$$

whose solutions are sequences of period 6 as follows

$$\{x_{-1}, x_0, a + x_0 - x_{-1}, 2a - x_{-1}, 2a - x_0, a - x_0 + x_{-1}, x_{-1}, x_0, \cdots\}.$$

In certain additive groups the solution may have a smaller minimal period. For example, in the group $\mathbb{Z}_2 = \{0, 1\}$ with addition modulo 2 the solution reduces to

$$\{x_{-1}, x_0, a + x_0 - x_{-1}, x_{-1}, x_0, \ldots\}$$

whose minimal period is at most three (note that $2 = 0$ and $\pm 1 = 1$ in $\mathbb{Z}_2$).

Next, let G be any nontrivial subgroup of nonzero complex numbers under ordinary multiplication (e.g., $\mathbb{R}_0$ or $\mathbb{T}$) and obtain the equation

$$x_{n+1} = \frac{a x_n}{x_{n-1}}, \quad a, x_0, x_{-1} \in G. \tag{2.23}$$

Solutions of (2.23) are the following sequences of period 6

$$\left\{ x_{-1}, x_0, \frac{ax_0}{x_{-1}}, \frac{a^2}{x_{-1}}, \frac{a^2}{x_0}, \frac{ax_{-1}}{x_0}, x_{-1}, x_0, \cdots \right\}.$$

◻

REMARK 2.4

1. While the approach in Example 2.9 is the quickest way of obtaining the general solution of the nonhomogeneous linear equation (2.22), we can also obtain this solution using the classical theory of linear equations. The eigenvalues of the homogeneous part of Eq. (2.22) are complex

$$\lambda_\pm = \frac{1}{2} \pm i\frac{\sqrt{3}}{2} = \cos\frac{\pi}{3} \pm i\sin\frac{\pi}{3} = e^{\pm i\pi/3}.$$

Since $\lambda_\pm^6 = 1$ the complementary solution has period 6 and the general solution can be obtained using standard methods, e.g., undetermined coefficients.

2. When $G = \mathbb{T}$, the circle group, then Eq. (2.23) can be written in polar form as

$$\theta_{n+1} = (\theta_n - \theta_{n-1} + \alpha) \bmod 2\pi, \quad e^{i\alpha} = a. \tag{2.24}$$

If a sequence of real numbers $\{\theta_n\}_{n=-1}^\infty$ satisfies (2.24) then the sequence $\{z_n\}_{n=-1}^\infty$ in $\mathbb{T}$ defined by $z_n = e^{i\theta_n}$ evidently satisfies Eq. (2.23) in $\mathbb{T}$ (recall that $e^{2\pi mi} = 1$ for all integers m). Since by Example 2.9 every solution of (2.23) has period 6, it follows that all solutions of Eq. (2.24) are also periodic with period 6.

The above relationship between equations on $\mathbb{R}$ modulo 2π and equations on $\mathbb{T}$ can be extended to the following more general

$$\theta_{n+1} = (a\theta_n + b\theta_{n-1} + c) \bmod 2\pi, \quad a, b, c, \theta_{-1}, \theta_0 \in \mathbb{R} \tag{2.25}$$

which can be viewed as the polar form of the following equation on $\mathbb{T}$

$$z_{n+1} = \gamma z_n^a z_{n-1}^b, \quad \gamma = e^{ic}, \ z_n = e^{i\theta_n}. \tag{2.26}$$

Note that Eq. (2.16) following Example 2.4 is another special case of (2.25). The Problems for this chapter further explore the relationship between (2.25) and (2.26). Eq. (2.26) is a special case of a general class of equations on $\mathbb{C}$ that we study in some detail in Section 6.3 below. ∎

Other simple equations similar to (2.21) exist on groups that can be solved generally; see the Problems for this chapter.

If G is not commutative then the calculations of the preceding example are not valid and different solutions may exist. Solutions of period 6 may still occur if a, x_0, x_{-1} are all in a nontrivial commutative subgroup of G. See the Problems for this chapter.

Often when using difference equations in modeling applications, a given equation is proposed on a suitable set without regard to any algebraic structure that may exist on the set. In such cases, it may be necessary to account for singularity sets, if any. The next example provides a simple illustration.

Example 2.10

Consider the difference equation of order $k + 1$ defined as

$$x_{n+1} = \frac{a_n x_n}{x_{n-k}}, \quad a_n, x_0, x_{-1}, \ldots, x_{-k} \in \mathbb{R}, \ a_n \neq 0 \text{ for all } n. \quad (2.27)$$

If all of the initial values are nonzero then $x_1 \neq 0$ and thus by induction $x_n \neq 0$ for all n. Conversely, if $x_m = 0$ for some least integer m then either $m \leq 0$ or $a_{m-1} x_{m-1} = 0$. But this implies $x_{m-1} = 0$ which contradicts the assumption that m is the least index for a zero term. Thus $m \leq 0$. These observations show that the only singular values or states of (2.27) occur on the coordinate hyperplanes, i.e.,

$$S = \{0\} \times \mathbb{R}^k \cup \mathbb{R} \times \{0\} \times \mathbb{R}^{k-1} \cup \cdots \cup \mathbb{R}^k \times \{0\} \subset \mathbb{R}^{k+1}$$

which form an invariant subset of $\mathbb{R}^{k+1}$. With regard to algebraic structures, although the set $\mathbb{R}$ has a natural additive group structure, in this example Eq. (2.27) is multiplicative in nature so the relevant algebraic structure is the multiplicative group $\mathbb{R}_0$. Since $\mathbb{R}_0^{k+1} \cup S = \mathbb{R}^{k+1}$ the nonsingular solutions are precisely those that are in the multiplicative group $\mathbb{R}_0$. We also note that if

$$F_n(u_0, \ldots, u_k) = \left[\frac{a_n u_0}{u_k}, u_0, u_1, \ldots, u_{k-1} \right]$$

is the unfolding for Eq. (2.27) then $F_n(\mathbb{R}_0^{k+1}) = \mathbb{R}_0^{k+1}$ so in particular, $\mathbb{R}_0^{k+1}$ is invariant under every F_n. ⬚

Often an invariant set of interest for a difference equation is a subset of some relevant group for the equation as seen in the next example.

Example 2.11

The difference equation

$$x_{n+1} = x_{n-1} e^{a_n - x_n - x_{n-1}}, \quad a_n, x_0, x_{-1} \in \mathbb{R} \quad (2.28)$$

is used in biological models; see the Notes section of this chapter. Although this equation is well defined on $\mathbb{R}$ the invariant subset of interest for modeling is clearly $(0, \infty)$, a subgroup of the group $\mathbb{R}_0$ of nonzero real numbers under ordinary multiplication (which is also invariant). Indeed, all occurrences of addition can be removed from the right-hand side using the exponent properties:

$$x_{n+1} = x_{n-1} e^{a_n} (1/e^{x_n})(1/e^{x_{n-1}})$$

i.e., Eq. (2.28) can be defined using the group multiplication only. Because of their boundedness and complexity, the behavior of solutions of (2.28) on the invariant subgroup $(0, \infty)$ is much more interesting than solutions starting outside this group. A detailed analysis of this equation is given in Section 6.3 below.   ☐

Although multiplicative groups in $\mathbb{R}$ were natural choices to consider as algebraic structures in the preceding example, it should be mentioned that the definition of the exponential function requires the usual (Euclidean) topological structure of $\mathbb{R}$ as well as its algebraic field structure, features that are implicitly assumed in defining the difference equation and can play subtle roles. For instance, *use of the exponential function effectively prevents groups of rational numbers from being relevant or invariant in this example.*

In many applications a difference equation is defined using more than a single group operation as we saw in Example 2.11. The examples in this section involve algebraic fields on which certain important types of difference equations can be defined.

Example 2.12

Let $\mathcal{F}$ be an algebraic field with its two operations denoted as usual by the addition and multiplication symbols. If

$$a_{j,n}, b_n \in \mathcal{F} \text{ for } j = 0, 1, \ldots, k \text{ and all } n \geq 0$$

then the difference equation

$$x_{n+1} = a_{0,n} x_n + a_{1,n} x_{n-1} + \cdots + a_{k,n} x_{n-k} + b_n \tag{2.29}$$

is the general linear nonhomogeneous difference equation of order $k + 1$ in $\mathcal{F}$ provided that $a_{k,n} \neq 0$ for all n. Eq. (2.29) has of course been extensively studied on the fields $\mathbb{R}$ and $\mathbb{C}$. However, there are other algebraic fields such as finite fields $\mathbb{Z}_p$ of integers modulo a prime $p \geq 2$ on which linear equations have not been studied extensively. Our study of decompositions (or factorizations) and reductions of orders of linear equations in this book takes all algebraic fields into account.   ☐

A linear form such as (2.29) can be defined on weaker algebraic structures such as rings but noncommutativity and the existence of zero divisors add complications that are not characteristically associated with linear equations. For example, even in a division ring (or a noncommutative field) we need to distinguish between $a_{j,n} x_{n-j}$ and $x_{n-j} a_{j,n}$ when defining a linear equation. In this first book on the topic, we limit attention to fields so as to avoid such complications.

Example 2.13

Extending the idea in Example 2.12, polynomial and rational difference equations can also be defined on fields. In particular, consider the following types of rational equations

$$x_{n+1} = \frac{\alpha_{0,n}x_n + \alpha_{1,n}x_{n-1} + \cdots + \alpha_{k,n}x_{n-k} + \beta_n}{A_{0,n}x_n + A_{1,n}x_{n-1} + \cdots + A_{k,n}x_{n-k} + B_n} \qquad (2.30)$$

and

$$x_{n+1} = \frac{\alpha_{0,n}x_n + \cdots + \alpha_{k,n}x_{n-k} + \beta_n}{A_{0,n}x_n + \cdots + A_{k,n}x_{n-k} + B_n} + a_{0,n}x_n + \cdots + a_{k,n}x_{n-k} + b_n \quad (2.31)$$

where $a_{j,n}, \alpha_{j,n}, A_{j,n}, b_n, \beta_n, B_n \in \mathcal{F}$ for $j = 0, 1, \ldots, k$ and $n \geq 0$. For nontriviality, it may be assumed in these equations that $\alpha_{k,n}A_{j,n} \neq 0$ or $\alpha_{j,n}A_{k,n} \neq 0$ for some $j \leq k$ and all n. Eq. (2.27) in Example 2.10 is a special case of (2.30).

Eqs.(2.30) and (2.31) with all coefficients in the field $\mathbb{R}$ of all real numbers appear often in pure and applied literature in difference equations; see the Notes section of this section. We refer to Eq. (2.30) as the *Ladas rational difference equation* in recognition of the pioneering work on these types of equations by G. Ladas and colleagues. Both Ladas difference equations and equations such as (2.31) are examples of quadratic difference equations; see Section 8.4. We refer to Eq. (2.31) as a *quadratic-linear rational difference equation*. This type of equation is capable of generating complex (including chaotic) solutions even in the autonomous case, i.e., with all coefficients being constants. We discuss one such example later in this book; see Section 5.6.5.

In studying rational equations it may be necessary to determine the singularity sets; i.e., sets of initial values that after a finite number of iterations lead to the occurrence of a zero in the denominator. Singularity sets of arbitrary rational equations are difficult to determine; however, in many applications they can be avoided without loss of significance. For example, if all the coefficients and parameters $\alpha_{j,n}, A_{j,n}, b_n, B_n$ are positive then all solutions of (2.30) with positive initial values are nonsingular. Positive solutions are often what are sought in modeling applications so that we can stay safely away from singularities. Further, positive solutions are in the interval $(0, \infty)$, which has a natural multiplicative group structure inherited from the field structure of $\mathbb{R}$.

In implicit form (if both sides of each is multiplied by the expression in the denominator) both of the equations (2.30) and (2.31) are examples of *quadratic equations*. Later in this book we study quadratic difference equations on fields as natural extensions of linear equations. ∎

Example 2.14
Let $\mathcal{F}$ be a nontrivial field. The following equation

$$x_{n+1} = \frac{x_n + a}{x_{n-1}}, \qquad a, x_0, x_{-1} \in \mathcal{F} \qquad (2.32)$$

is a second-order special case of (2.30) that is commonly known as the "Lyness equation" in the case $\mathcal{F} = \mathbb{R}$. In this case, the properties of real solutions of (2.32) have been determined in some detail. If $a, x_0, x_{-1} > 0$ then the singularity set of (2.32) is clearly avoided. Further, it can be checked by direct calculation that every solution of (2.32) lies on the curve

$$(a + x_n + x_{n-1})\left(1 + \frac{1}{x_n}\right)\left(1 + \frac{1}{x_{n-1}}\right) = c(x_0, x_{-1})$$

that is known as an "invariant" for the Lyness equation (the constant $c(x_0, x_{-1})$ depends on the initial values and of course, on a). Invariants are discussed in Section 3.1 below.

We note that if $a = 0$ then (2.32) reduces to a special case of Eq. (2.23); see Example 2.9. In this example, we focus on the oldest known nondegenerate case $a = 1$, i.e., the equation

$$x_{n+1} = \frac{x_n + 1}{x_{n-1}}. \tag{2.33}$$

In this case, every nonconstant real solution of (2.32) with positive initial values has period 5 as a quick calculation demonstrates

$$\left\{ x_{-1}, x_0, \frac{x_0 + 1}{x_{-1}}, \frac{x_0 + x_{-1} + 1}{x_0 x_{-1}}, \frac{x_{-1} + 1}{x_0}, x_{-1}, x_0, \dots \right\}.$$

Eq. (2.32) has a positive fixed point in $\mathbb{R}$ given by none other than the golden ratio

$$\bar{x} = \frac{1 + \sqrt{5}}{2}.$$

As usual this is calculated by setting $x_{n+1} = x_n = x_{n-1} = \bar{x}$ in (2.33) and solving the resulting quadratic equation for its positive solution. Thus if $x_0 = x_{-1} = \bar{x}$ then the corresponding solution of (2.32) is constant at $\bar{x}$. It is noteworthy that if x_0, x_{-1} are both rational (and positive) then every solution of (2.33) is rational so that in the field $\mathbb{Q}$ the solutions behave in essentially the same way as they do in $\mathbb{R}$ with one interesting exception: There are no constant solutions in $\mathbb{Q}$ since $\bar{x}$ is irrational. Therefore, in $\mathbb{Q}$ *every* positive solution of (2.33) has period 5, as calculated above.

In very different fields, however, the nature of solutions changes significantly. For example, it is easy to see by direct calculation that (2.33) has no solutions in the finite fields $\mathbb{Z}_2$ and $\mathbb{Z}_3$ because in each of these the solutions become zero in a few steps; i.e., the singularity sets of (2.33) of these fields are $\mathbb{Z}_2^2$ and $\mathbb{Z}_3^2$, respectively. The field $\mathbb{Z}_5$ is more interesting; the solutions of (2.33) for various initial points (x_0, x_{-1}) in $\mathbb{Z}_5$ fall into two main categories that are listed below:

5-cycle	$(1, 1), (2, 1), (1, 2), (3, 2), (2, 3), (3, 3)$
Singularity	$(2, 2), (3, 1), (1, 3), (x_0, 0), (0, x_{-1}), (x_0, 4), (4, x_{-1})$

We give the calculations for some of these and leave the verifications of the rest to the reader. If $(x_0, x_{-1}) = (2, 1)$ then

$$x_1 = \frac{2+1}{1} = 3, \ x_2 = \frac{3+1}{2} = 2, \ x_3 = 1, \ x_4 = 1, \ x_5 = 2$$

so that the solution has period 5 or is the 5-cycle

$$\{1, 2, 3, 2, 1, 1, 2, \ldots\}.$$

If $(x_0, x_{-1}) = (2, 2)$ then

$$x_1 = \frac{2+1}{2} = \frac{3}{2}, \ x_2 = \frac{3/2+1}{2} = \frac{5}{4} = 0$$

so that (2,2) is in the singularity set. The actual value of $3/2$ in $\mathbb{Z}_5$ is found as follows

$$\frac{3}{2} = \frac{3}{2}(1) = \frac{3}{2}(6) = 9 = 4 + 5 = 4.$$

Changing the rationals to actual values is not always necessary as the preceding calculation for (2,2) indicates but in some cases it is important to know the actual values. For instance, if $(x_0, x_{-1}) = (3, 3)$ then

$$x_1 = \frac{3+1}{3} = \frac{4}{3} = \frac{4}{3}(6) = 8 = 5 + 3 = 3, \ x_2 = \frac{3+1}{3} = 3, \ldots$$

i.e., *3 is in fact a fixed point of (2.33) in $\mathbb{Z}_5$!* This 5-cycle has minimal period 1 (constant solution) unlike the other 5-cycles listed in the above table. Thus the solutions of (2.33) in $\mathbb{Z}_5$ to some extent resemble the positive solutions in $\mathbb{R}$ in their variety. However, a different conculsion holds if $a \neq 1$; for a further exploration of the Lyness equation on the fields $\mathbb{Z}_p$ see the problems below.
⬜

The next example shows that by using matrices as operators on finite or infinite dimensional vector spaces we may represent certain types of partial difference equations as difference equations on groups. Partial difference equations are encountered in many different contexts, including numerical solutions of differential equations and cellular automata.

Example 2.15
(Vector spaces and partial difference equations on groups) Let $\mathcal{F}$ be a nontrivial field and let $\mathcal{F}_0 = \mathcal{F}\backslash\{0\}$ be the group of units of $\mathcal{F}$ under multiplication. Then the set of all sequences in $\mathcal{F}_0$

$$\mathcal{F}_0^{\mathbb{N}} = \{(u_1, u_2, u_3, \ldots)^{\top} : u_j \in \mathcal{F}_0, \ j = 1, 2, 3, \ldots\}$$

is again a group (the direct product) under pointwise multiplication (we have stretched the vector notation and terminology to include sequences; the symbol $\top$ denotes transposition). Let G be a nontrivial subgroup of $\mathcal{F}_0^{\mathbb{N}}$. Define the matrix $A = (a_{ij})_{i,j=1}^{\infty}$ where $a_{j,j+1} = 1$ (the identity of $\mathcal{F}_0$) and $a_{ij} = 0$ otherwise; i.e.,

$$A = \begin{bmatrix} 0 & 1 & 0 & 0 & 0 & \\ 0 & 0 & 1 & 0 & 0 & \cdots \\ 0 & 0 & 0 & 1 & 0 & \\ & & \vdots & & \ddots \end{bmatrix}.$$

Given two initial vector values $x_i = (x_{i,1}, x_{i,2}, \ldots)^{\top}$ in G for $i = -1, 0$ we define for each integer $n \geq 0$

$$x_{n+1} = (Ax_n)x_{n-1}^{-1} \tag{2.34}$$

where for convenience we used ordinary multiplication symbol and where the inverse of each sequence $(u_1, u_2, \ldots)^{\top} \in G$ is simply the sequence of inverse terms $(u_1^{-1}, u_2^{-1}, \ldots)^{\top}$. Notice that the parentheses are required in (2.34) as $A(x_n x_{n-1}^{-1})$ has a different meaning.

We now consider writing the vector equation (2.34) in component form. First, for every n

$$Ax_n = \begin{bmatrix} x_{2,n} \\ x_{3,n} \\ \vdots \end{bmatrix}$$

and by the definition of multiplication in $\mathcal{F}_0^{\mathbb{N}}$ we obtain

$$x_{j,n+1} = x_{j+1,n} x_{j,n-1}^{-1}, \quad j = 1, 2, 3, \ldots \tag{2.35}$$

Unlike the system of equations in Example 2.2, Eqs.(2.35) are not independent of each other and there are infinitely many of them. They may be viewed as having two independent variables, j and n and as such, (2.35) are considered to be a *single partial difference equation*. The variable j is analogous to the space variable in a partial differential equation. We see that the ordinary, autonomous vector difference equation (2.34) on the infinite-dimensional product group $\mathcal{F}_0^{\mathbb{N}}$ is equivalent to the partial difference equation (2.35).

A solution of (2.35) may be listed in matrix form as follows:

$$\begin{array}{cccc} x_{1,1} = x_{2,0}x_{1,-1}^{-1} & x_{2,1} = x_{3,0}x_{2,-1}^{-1} & x_{3,1} = x_{4,0}x_{3,-1}^{-1} & \cdots \\ x_{1,2} = x_{2,1}x_{1,0}^{-1} & x_{2,2} = x_{3,1}x_{2,0}^{-1} & x_{3,2} = x_{4,1}x_{3,0}^{-1} & \cdots \\ x_{1,1} = x_{2,2}x_{1,1}^{-1} & x_{2,3} = x_{3,2}x_{2,1}^{-1} & x_{3,3} = x_{4,2}x_{3,1}^{-1} & \cdots \\ \vdots & \vdots & \vdots & \ddots \end{array}$$

We leave it to the reader to verify that n-th row above represents the term $x_n^{\top}$ in the solution of (2.34) corresponding to the given vector initial values. $\square$

In the next example we determine the vector analog of a linear partial difference equation.

Example 2.16

(A linear partial difference equation) Let $\mathcal{F}$ be a nontrivial field and let

$$\alpha_{j,n}, \ \beta_{j,n}, \ \gamma_{j,n}, \ \delta_{j,n}, \ n \geq 0, \ j \geq 1$$

be given sequences in $\mathcal{F}$. The partial difference equation

$$x_{j,n+1} = \alpha_{j,n} x_{j-1,n} + \beta_{j,n} x_{j,n} + \gamma_{j,n} x_{j,n-1} + \delta_{j,n} x_{j+1,n-1} \tag{2.36}$$

is an example of a linear partial difference equations with variable coefficients that depend on both the "time" variable n and the "space" variable j. We now determine the vector difference equation corresponding to (2.36). As in Example 2.15 denote the set of all sequences in $\mathcal{F}$ by $\mathcal{F}^{\mathbb{N}}$. Since two time levels are involved in (2.36) we seek an equation of type

$$x_{n+1} = A_n x_n + B_n x_{n-1}, \quad x_n \in \mathcal{F}^{\mathbb{N}} \text{ for all } n \geq 0.$$

To find the matrix sequence A_n we pair the first two terms on the right hand side of (2.36) together since they both have the time index n. Similarly, B_n is found by pairing the terms with time index $n-1$. By direct examination of the pattern of indices in (2.36) we determine that

$A_n = (a_{i\,jn})_{i,j=1}^{\infty}$ where $a_{j\,jn} = \beta_{j,n}$, $a_{j+1,j,n} = \alpha_{j,n}$, $a_{j\,jn} = 0$ otherwise;

$B_n = (b_{i\,jn})_{i,j=1}^{\infty}$ where $b_{j\,jn} = \gamma_{j,n}$, $b_{j+1,j,n} = \delta_{j,n}$, $b_{j\,jn} = 0$ otherwise;

i.e.,

$$A_n = \begin{bmatrix} \beta_{1,n} & 0 & 0 & 0 \\ \alpha_{2,n} & \beta_{2,n} & 0 & 0 & \cdots \\ 0 & \alpha_{3,n} & \beta_{3,n} & 0 \\ & \vdots & & \ddots \end{bmatrix}$$

and

$$B_n = \begin{bmatrix} \gamma_{1,n} & \delta_{1,n} & 0 & 0 \\ 0 & \gamma_{2,n} & \delta_{2,n} & 0 & \cdots \\ 0 & 0 & \gamma_{3,n} & \delta_{3,n} \\ & \vdots & & \ddots \end{bmatrix}.$$

☐

Certain types of partial difference equations known as "cellular automata" are often used in modeling discrete dynamical systems. The next example features this type of equation and the set-up behind it.

Example 2.17

Let $\mathcal{F}$ be the binary field $\mathbb{Z}_2$ and similarly to preceding examples, denote by $\mathbb{Z}_2^{\mathbb{Z}}$ the set of all doubly infinite sequences of 0's and 1's, i.e.,

$$\mathbb{Z}_2^{\mathbb{Z}} = \{(\ldots, u_{-1}, u_0, u_1, \ldots)^{\top} : u_j \in \mathbb{Z}_2 \text{ for all } j \in \mathbb{Z}\}.$$

The partial difference equation

$$x_{j,n+1} = \psi(x_{j-1,n}, x_{j,n}, x_{j+1,n})$$

where $\psi : \mathbb{Z}_2^3 \to \mathbb{Z}_2$ is a given function defines a "one-dimensional, 3-point cellular automata (CA)." The function ψ defines a 3-point or 3-input "neighborhood rule," which determines whether cell j is active at time $n+1$ (i.e., $x_{j,n+1} = 1$) or inactive ($x_{j,n+1} = 0$) depending on the activity status of cell j and its immediate neighbors at time n. There are eight points in $\mathbb{Z}_2^3$ and thus $2^8 = 256$ possible definitions for ψ. Clearly this idea can be extended to m-point CA for all integers $m \geq 1$ by defining functions on $\mathbb{Z}_2^m$.

Certain choices of ψ can be represented by vector difference equations on $\mathbb{Z}_2^{\mathbb{Z}}$. For instance, it is readily verified that the linear partial difference equation

$$x_{j,n+1} = x_{j-1,n} + x_{j,n} + x_{j+1,n} \qquad (2.37)$$

is represented by the linear, autonomous first-order vector difference equation

$$x_{n+1} = Ax_n$$

where $A = (a_{ij})_{i,j=-\infty}^{\infty}$ with $a_{j-1,j} = a_{jj} = a_{j+1,j} = 1$ and $a_{ij} = 0$ otherwise; i.e.,

$$A = \begin{bmatrix} \ddots & \vdots & & & \vdots & \\ \cdots & 0\ 1\ 1\ 1\ 0\ 0 & \cdots \\ \cdots & 0\ 0\ 1\ 1\ 1\ 0 & \cdots \\ & \vdots & & & \vdots & \ddots \end{bmatrix}, \quad x_n = \begin{bmatrix} \vdots \\ x_{0,n} \\ x_{1,n} \\ \vdots \end{bmatrix}.$$

Eq. (2.37) can be more pictorially (and conventionally in CA terms) stated as a CA "rule"

000	001	010	011	100	101	110	111
0	1	1	0	1	0	0	1

In the above table the numbers in the bottom row are sums of the three numbers immediately above them, modulo 2 as indicated by (2.37). In CA terms, each cell j is active at time $n+1$ if either exactly one or all three of the cells $j-1$, j, $j+1$ are active at time n. Otherwise, cell j is inactive at time $n+1$. ▯

2.4   Notes

For an interesting discussion of difference and differential equations and the relationship between them on general algebraic structures see Bertram (2007).

Textbooks such as Hungerford (1974) or Fraleigh (1976) have all the required information on algebra for this section. Books such as Jordan (1965), Mickens (1991) and Elaydi (1999) offer additional introductory background material on difference equations and their applications should the reader need further clarification of some material in this book; for instance, such basic topics as exploring local stability by linearization or solving linear difference equations using classical methods are discussed in these books.

Information about the Fibonacci sequence is widely available both in print and on the Internet. Fibonacci sequences modulo m are discussed in several publications; see, e.g., Wall (1960), Andreassian (1974), and Renault (1996). Fibonacci sequences have also been considered modulo some noninteger real number, e.g., modulo 2π as in Eq. (2.16), or the modulo π case, which arises as an open problem in Kocic and Ladas (1993), p. 175.

Ladas rational difference equations have been objects of interest to many researchers in difference equations for at least the past 15 years. References are too numerous to be listed individually here but we refer to the textbooks Kocic and Ladas (1993), Kulenovic and Ladas (2002), Grove and Ladas (2004), and Camouzis and Ladas (2008), which collectively contain extensive bibliographies.

The more general quadratic-linear rational difference equations have been studied in Dehghan, et al. (2008a, 2008b) and Sedaghat (2009b).

Additional books that offer a wide range of topics in theory and applications of difference equations of higher order include Agarwal (2000) and Sedaghat (2003). For additional material on partial difference equations and cellular automata mentioned in Example 2.15 and later examples see Wolfram (1984), Cheng (2003), Gil (2007), and Ceccherini-Silberstein and Coornaert (2010).

2.5 Problems

2.1 Show that for Eq. (2.5) there are no invariant sets A in $G = (0, \infty)$.

2.2 Verify by induction that formula (2.13) gives the general solution of Eq. (2.12) in all of $\mathbb{R}$.

2.3 Let $z_0 = \rho_0 e^{i\theta_0}$ and $z_{-1} = \rho_{-1} e^{i\theta_{-1}}$ be complex numbers. Prove the following statements about Eq. (2.14) and its solutions $\{z_n\}$:

(a) Eq. (2.14) divides the complex plane $\mathbb{C}$ into three disjoint invariant sets: the interior of the unit disk, the exterior of the unit disk and the boundary $\mathbb{T}$ of the unit disk.

(b) If $\rho_0, \rho_{-1} < 1$ then $\lim_{n\to\infty} z_n = 0$; i.e., all solutions of (2.14) with initial points in the interior of the unit disk converge to zero.

(c) If $\rho_0, \rho_{-1} > 1$ then $\lim_{n\to\infty} z_n = \infty$; i.e., all solutions of (2.14) with initial points in the exterior of the unit disk converge to infinity.

2.4 Prove the following stronger results about a solution $\{z_n\}$ of Eq. (2.14) in the preceding problem:

(a) If $\rho_0 \leq 1$ and $\rho_{-1} < 1/\rho_0$ then $\lim_{n\to\infty} z_n = 0$.

(b) If $\rho_0 \geq 1$ and $\rho_{-1} > 1/\rho_0$ then $\lim_{n\to\infty} z_n = \infty$.

2.5 This problem brings together ideas in three examples 2.1, 2.3 and 2.4. We prove in a few steps that the difference equation

$$z_{n+1} = z_n z_{n-1} \tag{2.38}$$

has solutions in $\mathbb{C}$ that converge neither to 0 nor to ∞; instead, the circle $\mathbb{T}$ is the limit set of such solutions. Therefore, the limiting (or asymptotic) behaviors of these solutions may be periodic or nonperiodic depending on the initial values.

(a) Let φ_n be the Fibonacci sequence $1,1,2,3,5,8,13,\ldots$ discussed in Example 2.1 and let $\gamma = (1+\sqrt{5})/2$ be the "golden ratio". Use formula (2.10) to prove that for all $m = 0, 1, 2, \ldots$

$$\varphi_{2m} < \gamma\varphi_{2m-1}, \quad \varphi_{2m+1} > \gamma\varphi_{2m}, \quad \lim_{n\to\infty} \frac{\varphi_n}{\varphi_{n-1}} = \gamma.$$

Note that as it approaches its limit γ the sequence φ_n/φ_{n-1} oscillates about γ for all n. Specifically, $\varphi_n/\varphi_{n-1} < \gamma$ if n is even and $\varphi_n/\varphi_{n-1} > \gamma$ if n is odd.

(b) Show that each of the quantities $\varphi_{2m+1} - \gamma\varphi_{2m}$ and $\gamma\varphi_{2m-1} - \varphi_{2m}$ decreases to zero monotonically as $m \to \infty$.

(c) Let z_0 be a nonzero complex number and define the complex number z_{-1} so that

$$|z_{-1}| = \frac{1}{|z_0|^\gamma}.$$

If $\{z_n\}$ is the corresponding solution of (2.38) then prove that for all $m = 0, 1, 2, \ldots$

$$|z_{2m+1}| = \frac{1}{|z_0|^{\gamma\varphi_{2m-1}-\varphi_{2m}}}, \quad |z_{2m+2}| = \frac{1}{|z_0|^{\varphi_{2m+1}-\gamma\varphi_{2m}}}.$$

(d) If $|z_0| < 1$ then $|z_n| < 1$ for all even integers n and $|z_n| > 1$ for all odd integers n, and this statement remains true if all three inequalities are reversed. Further,

$$\lim_{n\to\infty} |z_n| = 1.$$

Therefore, the limit set of the solution $\{z_n\}$ of (2.38) is the circle $\mathbb{T}$.

(e) If the initial values z_0, z_{-1} are both real and nonzero then $\{z_n\}$ either converges to 1 (a fixed point of (2.38)) or to the 3-cycle

$$\{\ldots, 1, -1, -1, 1, -1, -1, \ldots\}.$$

To prove this statement, let $a > 0$ and

$$z_0 = \pm a, \quad z_{-1} = \pm \frac{1}{a^\gamma}.$$

Then show that

$$z_n = \begin{cases} a^{\varphi_{n-1} - \gamma \varphi_{n-2}}, & \text{if } z_0, z_{-1} > 0 \\ (-1)^{\varphi_{n-1}} a^{\varphi_{n-1} - \gamma \varphi_{n-2}}, & \text{if } z_0 z_{-1} < 0 \\ (-1)^{\varphi_n} a^{\varphi_{n-1} - \gamma \varphi_{n-2}}, & \text{if } z_0, z_{-1} < 0 \end{cases}$$

from which the asymptotic behavior follows. This is another type of behavior that may occur in Example 2.3.

2.6 Let ϕ_0, ϕ_{-1} be self-maps of an arbitrary commutative group G_0 as in Example 2.7. Find the solutions of Eq. (2.19) if (a) $q = 0$, (b) $p = 0$, (c) $p = q = 1$.

2.7 Referring to Example 2.8, verify that the solution sequences of permutations have periods 8 and 6, respectively as stated.

2.8 Show that the general solution of $x_{n+1} = x_n^{-1} * x_{n-1}$ on an arbitrary commutative group $(G, *)$ is

$$x_n = x_0^{(-1)^n \varphi_{n-1}} * x_{-1}^{(-1)^{n-1} \varphi_{n-2}} \tag{2.39}$$

where φ_n is the n-th term of the Fibonacci sequence and for each $a \in G$ the negativer integer power a^{-j} is interpreted as $(a^{-1})^j$. Examine the difference equation and its solutions for the two specific groups $\mathbb{R}$ under addition and $\mathbb{R}_0$ under multiplication. In the case of $\mathbb{R}$ compare the results with what is obtained using classical calculations.

2.9 Establish the following stronger form of Example 2.9: Let G be a nontrivial group (not necessarily commutative). If the parameter values a, x_0, x_{-1} in Eq. (2.21) commute with each other (i.e., they are contained in a commutative subgroup of G) then the corresponding solution of (2.21) has period 6.

2.10 Referring to Remark 2.4, consider the equation $\mathbb{R}$

$$\theta_{n+1} = (-\theta_n + \theta_{n-1}) \bmod 2\pi \tag{2.40}$$

which corresponds to the following equation on $\mathbb{T}$ under the substitution $z_n = e^{i\theta_n}$

$$z_{n+1} = \frac{z_{n-1}}{z_n}. \tag{2.41}$$

(a) The general solution of (2.41) is given by formula (2.39) above. If $z_{-1} = z_0$ then verify that the formula reduces to

$$z_n = z_0^{(-1)^n \varphi_{n-3}} \text{ if } n \geq 2 \text{ and } z_1 = 1. \tag{2.42}$$

For the three values $z_0 = 1, -1, i$ show that the solutions given by (2.42) are periodic with periods 1, 3 and 6, respectively.

(b) Let $\theta_{-1} = \theta_0$ in (2.40). For the three values $\theta_0 = 0, \pi, \pi/2$ calculate the corresponding solutions of (2.40). Show that these solutions are periodic with periods 1, 3, and 6, respectively, and correspond term by term to the solutions obtained in Part (a).

2.11 Consider the following version of Eq. (2.21) over the group of 2×2 matrices with unit determinants:

$$A_{n+1} = A_n A_{n-1}^{-1}. \tag{2.43}$$

Let

$$A_0 = \begin{bmatrix} 1 & 1 \\ 0 & 1 \end{bmatrix}, \quad A_{-1} = \begin{bmatrix} 0 & 1 \\ 1 & 1 \end{bmatrix}.$$

By calculating A_1 through A_6 show that the corresponding solution of (2.43) does *not* have period 6. Note that A_0 and A_{-1} above do not commute, i.e., $A_0 A_{-1} \neq A_{-1} A_0$.

Note: If we choose A_{-1} and A_0 so that they commute with each other (say, pick one to be the identity matrix) then by Problem 2.9 the corresponding solution of (2.43) in fact has period 6. Therefore, Eq. (2.21) in Example 2.9 has a greater variety of solutions if G is not commutative.

2.12 Find all initial points (or states) that lead to 5-cycles for the Lyness equation (2.33) in the field $\mathbb{Z}_7$. Does (2.33) have any fixed points in $\mathbb{Z}_7$?

13 In this exercise we explore the behaviors of solutions of the following case of the Lyness equation in finite fields $\mathbb{Z}_p$

$$x_{n+1} = \frac{x_n + 2}{x_{n-1}}. \tag{2.44}$$

(a) Show that Eq. (2.44) has a fixed point in $\mathbb{Z}_p$ for all $p \geq 2$. What is this fixed point for each value of p? Consider the cases $p = 2$ and $p > 2$ separately.

(b) For $p \geq 5$ show that $p - 1$ is a distinct, additional fixed point for (2.44). It may help to recall that $p - 1 = -1$ in $\mathbb{Z}_p$.

(c) For $p = 2, 3, 5$ determine the singularity sets and the behaviors of all solutions of (2.44).

2.14 Refer to Example 2.15.

(a) Establish that the n-th row of the solution list of the partial difference equation (2.35) represents the term x_n in the solution of (2.34).

(b) Let x_{-1} be the constant sequence $(1, 1, 1, \ldots)$ in a field $\mathcal{F}$ with multiplicative identity 1. If $c \neq 0, 1$ is in $\mathcal{F}$ then for each of the following choices of x_0 determine the corresponding solution of (2.34) (which also gives the corresponding solution of (2.35)):

$$x_0 = \begin{bmatrix} c \\ c \\ c \\ c \\ c \\ c \\ \vdots \end{bmatrix}, \quad x_0 = \begin{bmatrix} c \\ 1 \\ c \\ 1 \\ c \\ 1 \\ \vdots \end{bmatrix}, \quad x_0 = \begin{bmatrix} c \\ 1 \\ c \\ c \\ 1 \\ c \\ \vdots \end{bmatrix}.$$

(c) Determine the partial difference equation for each of the following vector difference equations:

(a) $\quad x_{n+1} = A(x_n x_{n-1}^{-1})$; (b) $\quad x_{n+1} = x_n(Ax_{n-1}^{-1})$;

(c) $\quad x_{n+1} = x_n^{-1}(Ax_{n-1})$; (d) $\quad x_{n+1} = (Ax_n)(Ax_{n-1}^{-1})$.

2.15 Find the vector difference equations corresponding to each of the following nonlinear partial difference equations:

(a) $\quad x_{j,n+1} = \alpha_n x_{j,n} + \beta_n x_{j-1,n} x_{j,n-1}, \quad n \geq 0, \ j \geq 1$

(b) $\quad x_{j,n+1} = \alpha_j x_{j,n} + \beta_j x_{j-1,n} x_{j,n-1}, \quad n \geq 0, \ j \geq 1$

where $\{\alpha_i\}, \{\beta_i\}, \ i = 0, 1, 2, \ldots$ are given sequences of real numbers.

3

Semiconjugate Factorization and Reduction of Order

This chapter begins our study of the main topic of this book, namely, decompositions (or factorizations) and reductions of order in difference equations. We introduce the basic concept of semiconjugacy on which a systematic framework for decomposition and reduction of order can be build for recursive (or explicit) equations of type (2.3), i.e.,

$$x_{n+1} = f_n(x_n, x_{n-1}, \ldots, x_{n-k}). \qquad (3.1)$$

Despite its abstract-sounding name, we see below that the concept of semiconjugacy can be used to formalize the highly concrete ideas of change of variables and reduction of order in recursive difference equations.

Patience with the developments in this chapter will be rewarding as the reader proceeds to later chapters. In Chapters 4, 5, and 6 we discuss applications and particular cases of the ideas and methods introduced in this chapter. Then, in Chapters 7 and 8 we extend the methods and concepts of this chapter to more general difference equations that include broader classes of nonautonomous equations as well as general nonrecursive equations.

3.1 Semiconjugacy and ordering of maps

3.1.1 Basic concepts

DEFINITION 3.1 *Let S and M be arbitrary nonempty sets and let F, Φ be self-maps of S and M, respectively. If there is a mapping $H : S \to M$ such that H is not constant and*

$$H \circ F = \Phi \circ H \qquad (3.2)$$

*then we say that the mapping F is **semiconjugate** to Φ and refer to Φ as a semiconjugate (SC) **factor** of F. The function H may be called a link map. We refer to the equality (3.2) is the **semiconjugate relation**.*

The semiconjugate relation (3.2) is illustrated by the following diagram

$$
\begin{array}{ccc}
S & \xrightarrow{\;F\;} & F(S) \\
\downarrow{\scriptstyle H} & & \downarrow{\scriptstyle H} \\
H(S) \subset M & \xrightarrow{\;\Phi\;} & \Phi(H(S)) = H(F(S))
\end{array}
$$

If H is a bijection (one-to-one and onto) then we call F and Φ **conjugates** and alternatively write

$$\Phi = H \circ F \circ H^{-1}. \tag{3.3}$$

We use the notation $F \simeq \Phi$ when F and Φ are conjugates. It is easy to show that $\simeq$ is an equivalence relation for self-maps of a set S onto itself.

Example 3.1
Consider the following self-maps of the plane $\mathbb{R}^2$:

$$F_1(u,v) = [u^2 + v^2, 2uv], \quad F_2(u,v) = [u^2 - v^2, 2uv].$$

Note that $u^2 + v^2 + 2uv = (u+v)^2$ so let $H_1(u,v) = u + v$ to obtain

$$H_1(F_1(u,v)) = (u+v)^2 = [H_1(u,v)]^2.$$

It follows that F_1 is semiconjugate to $\Phi(t) = t^2$ on $\mathbb{R}$. For F_2 define $H_2(u,v) = (u^2 + v^2)^{1/2}$ to get

$$H_2(F_2(u,v)) = \sqrt{(u^2 - v^2)^2 + 4u^2v^2} = u^2 + v^2 = [H_2(u,v)]^2.$$

Therefore, F_2 is also semiconjugate to $\Phi(t) = t^2$ though this time on $[0, \infty)$ since $H_2(u,v) \geq 0$ for all $(u,v) \in \mathbb{R}^2$. □

REMARK 3.1   (nonuniqueness of semiconjugates)
   The maps F, Φ and H are not uniquely determined by each other. In particular, the link map H is not uniquely defined by F and Φ. In Example 3.1 defining the function H_1 as $u - v$ would give the same Φ.
   The motivation for H_2 came from interpreting F_2 as the $\mathbb{R}^2$ version of the complex function z^2 whose modulus satisfies $|z^2| = |z|^2$. But we also notice that the same H_2 would work if the components of F_2 were switched or if one of its component functions were replaced by its negative so that there could be no direct relationship to the complex square function. Hence, different F may correspond to the same H and Φ. In none of the above cases the semiconjugate relation is a conjugacy since neither H_1 nor H_2 are bijections.   ∎

3.1.2   Coordinate transformations vs semiconjugacy

Coordinate transformations have classically been the means by which variables are changed in difference equations. One application of coordinate transformations is to reduce the order of a difference equation by "uncoupling" it, i.e., splitting it into two or more lower order equations. However, while in some cases coordinate transformations work very well, in other cases they may not be the best way of achieving a reduction in order.

To illustrate both cases, consider the two-dimensional or planar difference equations

$$(x_{n+1}, y_{n+1}) = F_i(x_n, y_n), \quad i = 1, 2$$

where F_i are given as in Example 3.1. We can change variables in each of these difference equations on $\mathbb{R}^2$ by transforming the coordinates from the given rectangular one to something else. For this problem we switch to polar coordinates:

$$x_n = r_n \cos \theta_n, \ y_n = r_n \sin \theta_n, \quad r_n^2 = x_n^2 + y_n^2, \ \tan \theta_n = \frac{y_n}{x_n}.$$

Then for F_2 we obtain

$$x_{n+1} = x_n^2 - y_n^2 = r_n^2 \cos 2\theta_n,$$
$$y_{n+1} = 2x_n y_n = r_n^2 \sin 2\theta_n.$$

Eliminating the rectangular coordinates gives

$$r_{n+1} = \sqrt{x_{n+1}^2 + y_{n+1}^2} = r_n^2 \sqrt{\cos^2 2\theta_n + \sin^2 2\theta_n} \Rightarrow r_{n+1} = r_n^2$$

$$\tan \theta_{n+1} = \frac{y_{n+1}}{x_{n+1}} = \frac{r_n^2 \sin 2\theta_n}{r_n^2 \cos 2\theta_n} = \tan 2\theta_n \Rightarrow \theta_{n+1} = 2\theta_n \bmod \pi.$$

The important thing to notice is that the system of polar equations is *uncoupled* because r_n and θ_n are determined independently by means of first-order equations: The first equation above shows that the radius is squared in each iteration while the second one indicates a doubling of the polar angle. A similar uncoupling via polar coordinates does *not* occur for F_1; in fact, a similar calculation yields the polar system of equations in this case as

$$r_{n+1} = r_n^2 \sqrt{1 + \sin^2 2\theta_n}$$
$$\theta_{n+1} = \tan^{-1} \sin 2\theta_n,$$

which is not uncoupled into first-order equations that are independent of each other.

The semiconjugate relations in Example 3.1 reduce the number of variables from 2 to 1 with equal ease in both F_1 and F_2 *through loss of information.* For example, in the case of F_2 the SC relation accounts for the squaring of radius through the one-dimensional map Φ but it ignores the doubling of angles. Since coordinate transformations retain more information about a map, they can be more difficult to obtain. On the other hand, if the "lost" information in the SC relation can be recovered somehow, then the SC relations may be easier to obtain as in the case of the mapping F_1. Later in this chapter we discuss a method not only of obtaining a semiconjugate mapping but also of recovering the lost information.

3.1.3   Semiconjugacy as a map-ordering relation

In Definition 3.1 the function H is not constant but otherwise unrestricted. This does not rule out some trivial or improper situations. For instance, if $S \subset M$ and Φ is any extension of F to M, i.e., $\Phi|_S = F$ then the inclusion map $H : S \rightarrow M$ defined as $H(x) = x$ for all $x \in S$ is a semiconjugate link between F and Φ since for all

$$\Phi(H(x)) = \Phi|_S(x) = F(x) = H(F(x)) \quad \text{for all } x \in S.$$

More generally, suppose that S and M are such that there is an injective (one-to-one) function $G : S \rightarrow M$ with $S' = G(S) \neq M$. If $F' : S' \rightarrow S'$ is the mapping

$$F' = G \circ F \circ G^{-1}$$

and we define $H : S' \rightarrow M$ as the inclusion map and Φ' as any extension of F' to M then as in the above we find that Φ' is semiconjugate to F', a conjugate equivalent of F on S'.

To avoid uninteresting situations like these, the following restriction is introduced.

DEFINITION 3.2   *F is **surjectively (or properly) semiconjugate** to Φ in Definition 3.1 if the link map H is surjective (onto). In this case we write $F \trianglerighteq \Phi$. From now on we implicitly assume that the link map is surjective except when stated otherwise.*

The surjective semiconjugate relation is illustrated by the following diagram

$$\begin{array}{ccc} S & \xrightarrow{F} & F(S) \\ \downarrow H & & \downarrow H \\ M & \xrightarrow{\Phi} & \Phi(M) = H(F(S)) \end{array}$$

Note that since by Definition 3.1 the link map H is not a constant function, M cannot be a singleton set. Further, the cases cited before Definition 3.2 are no longer problematic since the inclusion maps in those examples are not onto M. In a precise sense, the surjectivity requirement ensures that all elements of M are "interesting."

The binary relation in Definition 3.2 is not generally an equivalence relation but the next result shows that $\trianglerighteq$ is transitive like an *ordering relation* among three or more semiconjugate maps.

PROPOSITION 3.1
If $F \trianglerighteq \Phi$ and $\Phi \trianglerighteq \Psi$ then $F \trianglerighteq \Psi$.

PROOF Let the mappings H and H' be surjective and

$$H \circ F = \Phi \circ H \quad \text{and} \quad H' \circ \Phi = \Psi \circ H'$$

relative to appropriate domain and range sets. Then

$$H' \circ H \circ F = H' \circ \Phi \circ H = \Psi \circ H' \circ H.$$

Therefore, $F \trianglerighteq \Psi$ with the link mapping $H' \circ H$. Note that $H' \circ H$ is surjective. ∎

In Example 3.1, the function $\Phi(t) = t^2$, which is simpler than F_1 may be considered "below" or "lower than" F_1 in the ordering $\trianglerighteq$ since $F_1 \trianglerighteq \Phi$.

3.1.4 The semiconjugacy problem

Suppose that the sets S and M are given together with a self-map F of S. We may ask if there are maps H and Φ that satisfy Definition 3.2. The answer to this question may be trivial without further narrowing of our focus. For example, if $M = S$ then choosing H as the identity map and $\Phi = F$ would answer our question in the affirmative. But clearly, this is not a satisfactory answer and in cases of interest involving reduction of order $M \neq S$. The following refines our question.

The Semiconjugacy (SC) Problem. *Suppose that the sets S and M are given together with a self-map F of S. If $\mathfrak{M}$ is a given nonempty class of self-maps of M then is there a surjective link map H relative to which F is semiconjugate to some $\Phi \in \mathfrak{M}$?*

This is a nontrivial problem; in fact, there are no general results that guarantee the existence of H for every class $\mathfrak{M}$ of mappings Φ. For instance, if we modify Example 3.1 slightly by replacing the coefficient 2 in F_1 with an arbitrary real number a to get $F_1^a(u, v) = [u^2 + v^2, auv]$ then it is not clear whether (or not) there is a function Φ_a *on the real line* $\mathbb{R}$ such that $F_1^a \trianglerighteq \Phi_a$ for $a \neq 2$. Note that the requirement that Φ_a be restricted to the real line is the reason for increased difficulty of the problem. If we allowed Φ_a to be a self-map of $\mathbb{R}^2$ then trivially $F_1^a \trianglerighteq F_1^a$ relative to the identity function H for all $a \in \mathbb{R}$.

Questions about the existence of semiconjugates can sometimes be more easily answered in finite settings, as the next example shows; also see the Problems for this chapter and Examples 3.4 and 3.5.

Example 3.2
Consider the following autonomous difference equation of order 2 on the finite field $\mathbb{Z}_2$:

$$x_{n+1} = x_n x_{n-1} + 1, \quad x_0, x_{-1} \in \{0, 1\}. \tag{3.4}$$

Eq. (3.4) is represented by its unfolding $F(u, v) = [uv + 1, u]$ on $\mathbb{Z}_2^2$. We determine all self-maps of $\mathbb{Z}_2$ that are SC factors of F. There are four self-maps of $\mathbb{Z}_2$; they are the constant maps $\xi_0 \equiv 0$, $\xi_1 \equiv 1$, the identity map ι and the mapping $\psi(t) = t + 1$ which maps 0 to 1 and conversely. Also there are 16 link maps $H : \mathbb{Z}_2^2 \to \mathbb{Z}_2$ of which the 14 non-constant ones are surjective. We list these 14 maps succinctly as follows:

$$H_1 = (0,0,0,1) \quad H_2 = (0,0,1,0) \quad H_3 = (0,0,1,1) \quad H_4 = (0,1,0,0)$$
$$H_5 = (0,1,0,1) \quad H_6 = (0,1,1,0) \quad H_7 = (0,1,1,1) \quad H_8 = (1,0,0,0)$$
$$H_9 = (1,0,0,1) \quad H_{10} = (1,0,1,0) \quad H_{11} = (1,0,1,1) \quad H_{12} = (1,1,0,0)$$
$$H_{13} = (1,1,0,1) \quad H_{14} = (1,1,1,0).$$

In the above list, e.g., H_5 is defined by the rule

$$H_5 : (0,0) \to 0, \ (0,1) \to 1, \ (1,0) \to 0, \ (1,1) \to 1$$

The other link maps are defined by the same rule. Writing F as

$$F : (0,0) \to (1,0), \ (0,1) \to (1,0), \ (1,0) \to (1,1), \ (1,1) \to (0,1).$$

we calculate the compositions $H_j \circ F$ for $j = 1, \ldots, 14$:

$$H_1 \circ F = (0,0,1,0) \quad H_2 \circ F = (1,1,0,0) \quad H_3 \circ F = (1,1,1,0)$$
$$H_4 \circ F = (0,0,0,1) \quad H_5 \circ F = (0,0,1,1) \quad H_6 \circ F = (1,1,0,1)$$
$$H_7 \circ F = (1,1,1,1) \quad H_8 \circ F = (0,0,0,0) \quad H_9 \circ F = (0,0,1,0)$$
$$H_{10} \circ F = (1,1,0,0) \quad H_{11} \circ F = (1,1,1,0) \quad H_{12} \circ F = (0,0,0,1)$$
$$H_{13} \circ F = (0,0,1,1) \quad H_{14} \circ F = (1,1,0,1).$$

Next we check the compositions $\xi_0 \circ H_j$, $\xi_1 \circ H_j$, $\iota \circ H_j$, $\psi \circ H_j$ against each entry in the above table. For instance, $H_1 \circ F$ is not constant, not equal to H_1 and not equal to the "negation" of H_1, i.e., $\psi \circ H_1 = (1,1,1,0)$. Therefore, $H_1 \circ F$ is different from $\xi_0 \circ H_j$, $\xi_1 \circ H_j$, $\iota \circ H_j$, $\psi \circ H_j$ and therefore, H_1 is not a SC link map for F. Repeating this calculation shows that H_j is a SC link map only for $j = 7, 8$. For these two cases we obtain

$$H_7 \circ F = (1,1,1,1) = \xi_1 \circ H_7, \quad H_8 \circ F = (0,0,0,0) = \xi_0 \circ H_8$$

so the constant maps ξ_0 and ξ_1 are both SC factors of F. The above calculations also indicate that these constant maps are the only semiconjugate factors of F. ▯

As the preceding example shows, if S and M are both finite then for each given self-map $F : S \to S$ one can, in principle, calculate all pairs $H \circ F$ and $\Phi \circ H$ for all surjective (onto) link maps H and all self-maps $\Phi : M \to M$ and determine which pairs, if any, are equal. Thus with adequate computing resources one can solve the SC Problem for relatively small finite sets. However, since the number of pairs that are required to check increases exponentially with the sizes of sets S and M the brute-force computational approach to studying the SC Problem becomes impractical even for moderately-sized finite sets.

3.1.5 Semiconjugacy and difference equations

We now apply the notion of semiconjugacy to decomposing difference equations into pairs of lower order ones. Not surprisingly, we find that the difficulties in reducing the order of a difference equation in this way can often be traced to the Semiconjugacy Problem; see Section 3.1.4; also see Section 3.2 below. To make the transition to reduction of order in difference equations, let k be a nonnegative integer and G a nonempty set (not necessarily a group). Let $\{F_n\}$ be a family of functions $F_n : D_n \to G^{k+1}$ where $D_n \subset G^{k+1}$ for all $n = 0, 1, 2, \ldots$ We assume that

$$D = \bigcap_{n=0}^{\infty} D_n$$

is a nonempty subset of G^{k+1} and write $F_n = [f_{1,n}, \ldots, f_{k+1,n}]$ where $f_{j,n} : D \to G$ are the component functions of F_n restricted to D for all j and all n.

Let m be an integer, $1 \le m \le k+1$ and assume that each F_n is semiconjugate to a map $\Phi_n : D_n' \to G^m$ where $D_n' \subset G^m$. Let $H : G^{k+1} \to G^m$ be the link map such that for every n,

$$H \circ F_n = \Phi_n \circ H. \tag{3.5}$$

The following diagram illustrates 3.5:

$$
\begin{array}{ccc}
G^{k+1} & \xrightarrow{F_n} & F_n(G^{k+1}) \\
\downarrow{H} & & \downarrow{H} \\
G^m & \xrightarrow{\Phi_n} & \Phi_n(G^m) = H(F_n(G^{k+1}))
\end{array}
$$

We assume that $D_n' \cap H(D_n)$ is nonempty for all n. By keeping only the portion of D_n' that overlaps $H(D_n)$ we may assume without loss of generality that $D_n' \supset H(D_n)$ for all n. Then

$$\bigcap_{n=0}^{\infty} D_n' \supset \bigcap_{n=0}^{\infty} H(D_n) \supset H(D).$$

Thus for every n, if F_n is restricted to D then we may assume that Φ_n is restricted to $H(D)$. Next, let us write

$$H(u_0, \ldots, u_k) = [h_1(u_0, \ldots, u_k), \ldots, h_m(u_0, \ldots, u_k)]$$
$$\Phi_n(t_1, \ldots, t_m) = [\phi_{1,n}(t_1, \ldots, t_m), \ldots, \phi_{m,n}(t_1, \ldots, t_m)]$$

where $h_j : G^{k+1} \to G$ and $\phi_{j,n} : H(D) \to G$ are the corresponding component functions for $j = 1, 2, \ldots, m$. Then identity (3.5) is equivalent to the system of equations

$$h_j(f_{1,n}(u_0, \ldots, u_k), \ldots, f_{k+1,n}(u_0, \ldots, u_k)) =$$
$$\phi_{j,n}(h_1(u_0, \ldots, u_k), \ldots, h_m(u_0, \ldots, u_k)), \quad j = 1, 2, \ldots, m. \tag{3.6}$$

If the functions $f_{j,n}$ are given then (3.6) is a system of *functional equations* whose solutions $h_j, \phi_{j,n}$ give the maps H and Φ_n. Note that if $m < k + 1$ then the functions Φ_n on G^m define a system with lower dimension than that defined by the functions F_n on G^{k+1}. The system of equations (3.6) forms the basis for our work in the next section.

Now, suppose that D is an invariant subset of G^{k+1} to avoid the occurrence of singularities and related technical problems in the following discussion.

PROPOSITION 3.2

Assume that D is invariant under all F_n, i.e., $F_n(D) \subset D$ for all n and (3.5) holds. Then $H(D)$ is invariant under Φ_n for all n.

PROOF It is evident that $H(F_n(D)) = \Phi_n(H(D))$ for all n. Therefore,

$$F_n(D) \subset D \Rightarrow H(F_n(D)) \subset H(D) \Rightarrow \Phi_n(H(D)) \subset H(D).$$

Therefore, $H(D)$ is invariant under Φ_n for every n. ∎

The following equation in the $k + 1$ dimensional space G^{k+1} generalizes difference equations of order $k + 1$ on G:

$$X_{n+1} = F_n(X_n), \quad X_0 \in D \subset G^{k+1}. \tag{3.7}$$

For a given solution $\{X_n\}$ of (3.7) let $Y_n = H(X_n)$ for $n = 0, 1, 2, \ldots$ Then

$$Y_{n+1} = H(X_{n+1}) = H(F_n(X_n)) = \Phi_n(H(X_n)) = \Phi_n(Y_n)$$

so that $\{Y_n\}$ satisfies the difference equation

$$Y_{n+1} = \Phi_n(Y_n), \quad Y_0 = H(X_0) \in H(D) \subset G^m. \tag{3.8}$$

If $k \geq 1$ and $1 \leq m \leq k$ then (3.8) is a system with lower dimension m than the original $k + 1$ dimensional system. We will see later that this apparent loss of structure and information actually leads to reduction of order in difference equations. We also discuss a way of recovering the (temporarily) lost information.

3.1.6 Invariants

In (3.5) let $F_n = F$ and $\Phi_n = \iota$, the identity map for all n. Then for each initial state X_0 we obtain

$$H(X_1) = H(F(X_0)) = \iota(H(X_0)) = H(X_0).$$

Repeating the preceding argument gives

$$H(F^n(X_0)) = H(X_n) = H(X_0) \quad \text{for all } n \geq 1. \tag{3.9}$$

From (3.9) it is evident that the value of H does not change, or is constant on the orbits of F. This observation justifies the following definition.

DEFINITION 3.3 *If F is semiconjugate to the identity map then the link map H is an **invariant** for F and it satisfies (3.9).*

Example 3.3
A well-known example of a map with an invariant is the Lyness map of the plane $\mathbb{R}^2$ that is defined as

$$F_L(u, v) = \left[\frac{a + u}{v}, u\right].$$

It is easy to check that the function

$$H_L(u, v) = \left(1 + \frac{1}{u}\right)\left(1 + \frac{1}{v}\right)(a + u + v)$$

is an invariant for F_L. Also see Example 2.14 above. ⬜

The question of whether an invariant exists or how to compute it for a given map F is as difficult to answer as the SC problem itself. Again, in finite settings this question is easier to answer. For instance, the map F in Example 3.2 clearly has no invariants. However, in the next example, we do find an invariant.

Example 3.4
Let $\mathbb{Z}_2$ be as in Example 3.2 but now consider the following difference equation of order 2 in $\mathbb{Z}_2$,

$$x_{n+1} = (x_n + 1)x_{n-1}, \quad x_0, x_{-1} \in \{0, 1\}.$$

With $F(u, v) = [(u + 1)v, u]$, proceeding as in Example 3.2 we obtain the following results:

$$H_1 \circ F = (0, 0, 0, 0) = \xi_0 \circ H_1, \quad H_{14} \circ F = (1, 1, 1, 1) = \xi_1 \circ H_{14}$$
$$H_7 \circ F = (0, 1, 1, 1) = H_7, \quad H_8 \circ F = (1, 0, 0, 0) = H_8.$$

In particular, H_7 and H_8 are now invariants, unlike the situation encountered in Example 3.2. It is easy to check by inspection that they can be written in algebraic form as

$$H_7(u, v) = uv + u + v, \quad H_8(u, v) = uv + u + v + 1.$$

⬜

3.2   Form symmetries and SC factorizations

Let $k \geq 1$ and $1 \leq m \leq k$ so that Eq. (3.1), i.e.,

$$x_{n+1} = f_n(x_n, x_{n-1}, \ldots, x_{n-k})$$

has order at least 2. Let F_n be the associated map (or unfolding) of the function f_n as in Definition 2.3, i.e.,

$$F_n(u_0, \ldots, u_k) = [f_n(u_0, \ldots, u_k), u_0, \ldots, u_{k-1}].$$

Even with each such F_n semiconjugate to an m-dimensional map Φ_n as in (3.5) the preceding discussion only gives the system (3.8) in which the maps Φ_n are not necessarily of scalar type similar to F_n. To ensure that each Φ_n is also of scalar type and hence unfolds a difference equation we define the first component function of H as

$$h_1(u_0, \ldots, u_k) = u_0 * h_0(u_1, \ldots, u_k) \tag{3.10}$$

where $h_0 : G^k \to G$ is a function to be determined and $*$ denotes the group operation. This restriction on H makes sense for Eq. (3.1) which is of recursive type; i.e., x_{n+1} given explicitly by functions f_n that depend on states before x_{n+1}. With these restrictions on H and F_n the first equation in (3.6) is given by

$$f_n(u_0, \ldots, u_k) * h_0(u_0, \ldots, u_{k-1}) = \tag{3.11}$$
$$g_n(u_0 * h(u_1, \ldots, u_k), h_2(u_0, \ldots, u_k), \ldots, h_m(u_0, \ldots, u_k))$$

where for notational convenience we have set

$$g_n \doteq \phi_{1,n} : G^m \to G.$$

3.2.1   Order-reducing form symmetries

Eq. (3.11) is a functional equation in which the functions h_j, g_n may be determined in terms of the given functions f_n. Our aim is to extract a scalar equation of order m such as

$$t_{n+1} = g_n(t_n, \ldots, t_{n-m+1}) \tag{3.12}$$

from (3.11) in such a way that the maps Φ_n will be of scalar type. The basic framework for carrying out this process is already in place; let $\{x_n\}$ be a solution of Eq. (2.3) and define

$$t_n = x_n * h_0(x_{n-1}, \ldots, x_{n-k}).$$

Then the left-hand side of (3.11) for the solution $\{x_n\}$ is

$$x_{n+1} * h_0(x_n, \ldots, x_{n-k+1}) = t_{n+1}, \tag{3.13}$$

which gives the initial part of the difference equation (3.12). In order that the right-hand side of (3.11) coincide with that in (3.12) it is necessary to define

$$h_j(x_n, \ldots, x_{n-k}) = t_{n-j+1} = x_{n-j+1} * h_0(x_{n-j}, \ldots, x_{n-k+1-j}), \tag{3.14}$$
$$\text{for } j = 2, \ldots, m.$$

Since the left-hand side of (3.14) does not depend on the terms

$$x_{n-k-1}, \ldots, x_{n-k-j+1}$$

it follows that the function h_j must be constant in its last few coordinates. Since h_0 does not depend on j it must be constant in some of its coordinates. The number of its constant coordinates is found from the last function h_m. Specifically, we have

$$h_m(x_n, \ldots, x_{n-k}) = x_{n-m+1} * h_0(\underbrace{x_{n-m}, \ldots, x_{n-k}}_{k-m+1 \text{ variables}}, \underbrace{x_{n-k-1}, \ldots, x_{n-k-m+1}}_{m-1 \text{ terms } h_0 \text{ is constant at}}).$$
$$\tag{3.15}$$

The preceding condition leads to the necessary restrictions on h_0 and every h_j for a consistent derivation of (3.12) from (3.11). Therefore, (3.15) is a consistency condition. Now from (3.14) and (3.15) for $(u_0, \ldots, u_k) \in G^{k+1}$ we obtain

$$h_j(u_0, \ldots, u_k) = u_{j-1} * h(u_j, u_{j+1} \ldots, u_{j+k-m}), \quad j = 1, \ldots, m. \tag{3.16}$$

In (3.16) the function $h : G^{k+1-m} \to G$ represents the variable part of h_0; i.e., it takes precisely the same values as h_0. In fact, if G is a group then we may define

$$h(u_0, \ldots, u_{k-m}) = h_0(u_0, \ldots, u_{k-m}, 1, \ldots, 1), \quad (u_0, \ldots, u_{k-m}) \in G^{k+1-m}$$

where 1 is the identity of G. Using (3.16) in (3.11) we obtain for all $n \geq 0$

$$f_n(u_0, \ldots, u_k) * h(u_0, \ldots, u_{k-m}) = g_n(u_0 * h(u_1, u_2 \ldots, u_{k-m+1}),$$
$$u_1 * h(u_2, u_3 \ldots, u_{k-m+2}), \ldots$$
$$u_{m-1} * h(u_m, u_{m+1} \ldots, u_k)). \tag{3.17}$$

Eq. (3.17) is a special case of the semiconjugate relation (3.5); with the definitions of F_n, Φ_n, g_n and H given above, the left-hand side of (3.17) is recognizable as $H \circ F_n$ and the right-hand side as $\Phi_n \circ H$. The next definition formalizes the role of the somewhat special H in this context.

DEFINITION 3.4   *The function $H = [h_1, \ldots, h_m]$ whose components h_j are defined by (3.16) is a (recursive)* **form symmetry** *for Eq. (3.1). Since the range of H has a lower dimension m than the dimension $k + 1$ of its domain, we say that H is an* **order-reducing** *form symmetry.*

In the next section we show how to use form symmetries to change variables, decompose and obtain a reduction of order in a difference equation.

3.2.2   The semiconjugate factorization theorem

Assume that G denotes a nontrivial group and let $\{x_n\}$ be a solution of (3.1) in G. As in the previous section, we use (3.13), (3.16), and (3.17) to obtain the following pair of lower-order equations

$$t_{n+1} = g_n(t_n, \ldots, t_{n-m+1}), \tag{3.18}$$

$$x_{n+1} = t_{n+1} * h(x_n, \ldots, x_{n-k+m})^{-1} \tag{3.19}$$

where -1 denotes group inversion in G. Before stating our main decomposition theorem we give names to the above equations for easy later reference.

DEFINITION 3.5   *Eq. (3.18) is a* **factor** *of Eq. (3.1) since it is distilled from the semiconjugate factor Φ_n. Eq. (3.19) that links the factor to the original equation is a* **cofactor** *of Eq. (3.1). We call the system of equations (3.18) and (3.19) a* **semiconjugate (SC) factorization** *of Eq. (3.1).*

If $\{t_n\}$ is a given solution of Eq. (3.18) then using it in (3.19) produces a solution $\{x_n\}$ of Eq. (3.1). Conversely, if $\{x_n\}$ is a solution of (3.1) then the sequence $t_n = x_n * h(x_{n-1}, \ldots, x_{n-k+m-1})$ is a solution of (3.18) with initial values

$$t_{-j} = x_{-j} * h(x_{-j-1}, \ldots, x_{-j-k+m-1}), \quad j = 0, \ldots, m-1.$$

Since solutions of the pair of equations (3.18) and (3.19) coincide with the solutions of (3.1), we say that the pair (3.18) and (3.19) constitute a system that is equivalent to the higher-order equation (3.1).

We now present the fundamental theorem on semiconjugate decomposition of difference equations.

THEOREM 3.1

Let $k \geq 1$, $1 \leq m \leq k$ and suppose that there are functions $h : G^{k+1-m} \to G$ and $g_n : G^m \to G$ that satisfy equations (3.16) and (3.17).

(a) With the order-reducing form symmetry

$$H(u_0, \ldots, u_k) = [u_0 * h(u_1, \ldots, u_{k+1-m}), \ldots, u_{m-1} * h(u_m, \ldots, u_k)] \tag{3.20}$$

Eq. (3.1) is equivalent to the SC factorization consisting of the system of equations (3.18) and (3.19) whose orders m and k + 1 − m respectively, sum to the order of (3.1).

(b) The function $H : G^{k+1} \to G^m$ defined by (3.20) is surjective (onto).

(c) For each n, the SC factor function $\Phi_n : G^m \to G^m$ in (3.5) is the unfolding of Eq. (3.18). In particular, each Φ_n is of scalar type.

PROOF   (a) To show that the SC factorization system consisting of equations (3.18) and (3.19) is equivalent to Eq. (3.1) we show that: (i) each solution $\{x_n\}$ of (3.1) uniquely generates a solution of (3.18) and (3.19) and conversely (ii) each solution $\{(t_n, y_n)\}$ of the system (3.18) and (3.19) correseponds uniquely to a solution $\{x_n\}$ of (3.1). To establish (i) let $\{x_n\}$ be the unique solution of (3.1) corresponding to a given set of initial values $x_0, \ldots x_{-k} \in G$. Define the sequence

$$t_n = x_n * h(x_{n-1}, \ldots, x_{n-k+m-1}) \tag{3.21}$$

for $n \geq -m + 1$. Then for each $n \geq 0$ using (3.17)

$$
\begin{aligned}
x_{n+1} &= f_n(x_n, \ldots, x_{n-k}) \\
&= g_n(x_n * h(x_{n-1}, \ldots, x_{n-k+m-1}), \ldots, x_{n-m+1} * h(x_{n-m}, \ldots, x_{n-k})) \\
&\qquad * [h(x_n, \ldots, x_{n-k+m})]^{-1} \\
&= g_n(t_n, \ldots, t_{n-m+1}) * [h(x_n, \ldots, x_{n-k+m})]^{-1}.
\end{aligned}
$$

Therefore, $g_n(t_n, \ldots, t_{n-m+1}) = x_{n+1} * h(x_n, \ldots, x_{n-k+m}) = t_{n+1}$ so that $\{t_n\}$ is the unique solution of the factor equation (3.18) with initial values

$$t_{-j} = x_{-j} * h(x_{-j-1}, \ldots, x_{-j-k+m-1}), \quad j = 0, \ldots, m - 1.$$

Further, by (3.21) for $n \geq 0$ we have $x_{n+1} = t_{n+1} * [h(x_n, \ldots, x_{n-k+m})]^{-1}$ so that $\{x_n\}$ is the unique solution of the cofactor equation (3.19) with initial values $y_{-i} = x_{-i}$ for $i = 0, 1, \ldots, k - m$ and t_n obtained above.

To establish (ii) let $\{(t_n, y_n)\}$ be a solution of the factor-cofactor system with given initial values

$$t_0, \ldots, t_{-m+1}, y_{-m}, \ldots y_{-k} \in G.$$

Note that these numbers determine $y_{-m+1}, \ldots, y_0$ through the cofactor equation

$$y_{-j} = t_{-j} * [h(y_{-j-1}, \ldots, y_{-j-1-k+m})]^{-1}, \quad j = 0, \ldots, m - 1. \tag{3.22}$$

Now for $n \geq 0$,

$$
\begin{aligned}
y_{n+1} &= t_{n+1} * [h(y_n, \ldots, y_{n-k+m})]^{-1} \\
&= g_n(t_n, \ldots, t_{n-m+1}) * [h(y_n, \ldots, y_{n-k+m})]^{-1} \\
&= g_n(y_n * h(y_{n-1}, \ldots, y_{n-k+m-1}), \ldots, y_{n-m+1} * h(y_{n-m}, \ldots, y_{n-k})) \\
&\qquad\qquad\qquad\qquad\qquad\qquad\qquad * [h(y_n, \ldots, y_{n-k+m})]^{-1} \\
&= f_n(y_n, \ldots, y_{n-k})
\end{aligned}
$$

where the last step is justified by (3.17). Thus $\{y_n\}$ is the unique solution of Eq. (3.1) that is generated by the initial values (3.22) and $y_{-m}, \ldots y_{-k}$. This completes the proof of (a).

(b) Choose an arbitrary point $[v_1, \ldots, v_m] \in G^m$ and set

$$
u_{m-1} = v_m * h(u_m, u_{m+1} \ldots, u_k)^{-1}
$$

where $u_m = u_{m+1} = \ldots u_k = \bar{u}$ where $\bar{u}$ is a fixed element of G, e.g., the identity. Then

$$
\begin{aligned}
v_m &= u_{m-1} * h(\bar{u}, \bar{u} \ldots, \bar{u}) \\
&= u_{m-1} * h(u_m, u_{m+1} \ldots, u_k) \\
&= h_m(u_0, \ldots, u_k) \\
&= h_m(u_0, \ldots, u_{m-2}, v_m * h(\bar{u}, \bar{u} \ldots, \bar{u})^{-1}, \bar{u} \ldots, \bar{u})
\end{aligned}
$$

for any choice of $u_0, \ldots, u_{m-2} \in G$. Similarly, define

$$
u_{m-2} = v_{m-1} * h(u_{m-1}, u_m \ldots, u_{k-1})^{-1}
$$

so as to get

$$
\begin{aligned}
v_{m-1} &= u_{m-2} * h(u_{m-1}, u_m \ldots, u_{k-1}) \\
&= h_{m-1}(u_0, \ldots, u_k) \\
&= h_{m-1}(u_0, \ldots, u_{m-3}, v_{m-1} * h(u_{m-1}, \bar{u} \ldots, \bar{u})^{-1}, u_{m-1}, \bar{u} \ldots, \bar{u})
\end{aligned}
$$

for any choice of $u_0, \ldots, u_{m-3} \in G$. Continuing in this way, induction leads to selection of $u_{m-1}, \ldots, u_0$ such that

$$
v_j = h_j(u_0, \ldots, u_{m-1}, \bar{u} \ldots, \bar{u}), \quad j = 1, \ldots, m.
$$

Therefore, $H(u_0, \ldots, u_{m-1}, \bar{u} \ldots, \bar{u}) = [v_1, \ldots, v_m]$ and it follows that H is onto G^m.

(c) It is necessary to prove that each coordinate function $\phi_{j,n}$ is the projection into coordinate $j - 1$ for $j > 1$. Suppose that the maps h_j are given by (3.16). For $j = 2$ (3.6) gives

$$
\begin{aligned}
\phi_{2,n}(h_1(u_0, \ldots, u_k), \ldots, h_m(u_0, \ldots, u_k)) &= h_2(f_n(u_0, \ldots, u_k), u_0, \ldots, u_{k-1}) \\
&= u_0 * h(u_1, u_2 \ldots, u_{k-m+1}) \\
&= h_1(u_0, \ldots, u_k).
\end{aligned}
$$

Therefore, $\phi_{2,n}$ projects into coordinate 1. Generally, for $j \geq 2$ we have

$$\phi_{j,n}(h_1(u_0, \ldots, u_k), \ldots, h_m(u_0, \ldots, u_k)) = h_j(f_n(u_0, \ldots, u_k), u_0, \ldots, u_{k-1})$$
$$= u_{j-2} * h(u_{j-1}, u_j \ldots, u_{j+k-m-1})$$
$$= h_{j-1}(u_0, \ldots, u_k).$$

Therefore, for each n and for every $(t_1, \ldots, t_m) \in H(G^{k+1})$ we have

$$\Phi_n(t_1, \ldots, t_m) = [g_n(t_1, \ldots, t_m), t_1, \ldots, t_{m-1}]$$

i.e., $\Phi_n|_{H(G^{k+1})}$ is of scalar type. Since by Part (b) $H(G^{k+1}) = G^m$ it follows that Φ_n is of scalar type. ∎

We point out that the SC factorization in Theorem 3.1(a) does not require the determination of $\phi_{j,n}$ for $j \geq 2$. However, as seen in Parts (b) and (c) of the theorem the rest of the picture fits together properly. In particular, in Part (b) we see that the order-reducing form symmetry H is a surjective semiconjugate link.

REMARK 3.2 (The semiconjugacy problem for higher-order difference equations)

Let G be a nontrivial group. In Section 3.1.4 assume that $S = G^{k+1}$, $M = G^m$ and let F_n be the sequence of unfoldings of the difference equation (3.1) of order $k + 1$. Then the existence of a form symmetry H is equivalent to the existence of a solution to the semiconjugacy problem with $\mathfrak{M}$ being the collection of all maps that are unfoldings of difference equations of order m. By the Factorization Theorem above, $\mathfrak{M}$ may be assumed to be the class of all self-maps of M, since if there a form symmetry H relative to which $F_n \unrhd \Phi_n$ for each n then Φ_n must be the unfolding of the factor difference equation of order m. In either case, this is a highly nontrivial problem for the values $1 \leq m \leq k$. ∎

We discuss many applications of Theorem 3.1 in various sections and chapters that follow. The following example gives a flavor.

Example 3.5

Let G be any nontrivial commutative or Abelian group and consider the following difference equation of order two

$$x_{n+1} = x_{n-1} + a, \quad a, x_0, x_{-1} \in G, \ a \neq 0. \tag{3.23}$$

The unfolding of Eq. (3.23) is the map $F(u, v) = [v + a, u]$. Define

$$H(u, v) = u + v \text{ and } \phi(t) = t + a \tag{3.24}$$

and note that

$$H(F(u, v)) = u + v + a = \phi(u + v) = \phi(H(u, v)).$$

Therefore, $F \unrhd \phi$ with H being a form symmetry of type $u + h(v)$ where h is the identity function on G. Now using Theorem 3.1 we obtain the SC factorization of Eq. (3.23) as

$$t_{n+1} = \phi(t_n) = t_n + a, \quad t_0 = H(x_0, x_{-1}) = x_0 + x_{-1}$$
$$x_{n+1} = t_{n+1} - x_n.$$

Note that if $G = \mathbb{Z}_2$ under addition then $H = H_6$ and $\phi = \psi$ in Example 3.2. This fact motivated the more general example here.   ☐

3.3 Order-reduction types

Theorem 3.1 provides a natural classification scheme for order reduction that we discuss in this section. In order to obtain all solutions of Eq. (3.1) the lower-order factor equation is not sufficient. The cofactor equation is also required and it is generally as important as the factor in understanding the properties of (3.1). The next definition introduces the appropriate designations.

3.3.1 The basic concept

DEFINITION 3.6   *The SC factorization of Eq. (3.1) into (3.18) and (3.19) gives a* **type-$(m, k + 1 - m)$** **order reduction** *(or just* **type-$(m, k + 1 - m)$** **reduction***) for (3.1). We also say that (3.1) is a type-$(m, k + 1 - m)$ equation in this case.*

A second-order difference equation ($k = 1$) can have only the type-(1,1) order reduction into two first-order equations. A third-order equation has two order reduction types, namely, (2,1) and (1,2), a fourth-order equation has three order reduction types (2,1), (2,2) and (1,2) and so on. We shall encounter many examples of the two extreme types $(k, 1)$ and $(1, k)$ in the rest of the book. The following example furnishes a type-$(m, k + 1 - m)$ equation for any $1 \le m \le k$.

Example 3.6
Consider the rational difference equation

$$x_{n+1} = \frac{a_n x_{n-m+1} x_{n-k} + b_n}{x_{n-k+m}} \tag{3.25}$$

where $1 \leq m \leq k$ and $\{a_n\}$, $\{b_n\}$ are sequences in some field (e.g., the real numbers) with $a_n \neq 0$ for all n. Multiplying the above equation by x_{n-k+m} and substituting

$$t_n = x_n x_{n-k+m-1}$$

for all n in it gives

$$t_{n+1} = a_n t_{n-m+1} + b_n$$

as a factor equation of order m with a cofactor of order $k + 1 - m$ that is derived from the above substitution as

$$x_{n+1} = \frac{t_{n+1}}{x_{n-k+m}}.$$

In particular, if $k = 2m - 1$ in Eq. (3.25) then for every $m \geq 1$ we obtain a type-(m, m) reduction for

$$x_{n+1} = a_n x_{n-2m+1} + \frac{b_n}{x_{n-m+1}}$$

via the SC factorization

$$t_{n+1} = a_n t_{n-m+1} + b_n,$$
$$x_{n+1} = \frac{t_{n+1}}{x_{n-m+1}}.$$

⬚

3.3.2 Nonuniqueness of factorizations and reduction types

In general, a given difference equation may have different order reduction types or SC factorizations. The next two examples illustrate these facts; more cases are encountered later on.

Example 3.7
(Different SC factorizations for the same reduction type) Consider the second-order difference equation

$$x_{n+1} = a x_n + b x_{n-1} \quad \text{with } a + b = 1. \tag{3.26}$$

The only possible order reduction type for this equation is $(1,1)$. However, this equation has at least three distinct form symmetries with their corresponding SC factorizations. First, setting $a = 1 - b$ and rearranging terms in (3.26) reveals the form symmetry

$$H(x_n, x_{n-1}) = x_n - x_{n-1}$$

and the corresponding SC factorization on the additive group of real numbers:

$$x_{n+1} - x_n = -b(x_n - x_{n-1}) \Rightarrow t_{n+1} = -bt_n,$$
$$x_{n+1} = t_{n+1} + x_n.$$

On the other hand,

$$H(x_n, x_{n-1}) = x_n + bx_{n-1}$$

is also a form symmetry of (3.26) with its SC factorization again on the additive reals because

$$x_{n+1} + bx_n = (a+b)x_n + bx_{n-1} = x_n + bx_{n-1} \Rightarrow t_{n+1} = t_n,$$
$$x_{n+1} = t_{n+1} - bx_n.$$

Finally, if $a, b \geq 0$ then there is also the form symmetry

$$H(x_n, x_{n-1}) = x_n / x_{n-1}$$

with its SC factorization on the group of positive real numbers under ordinary multiplication:

$$\frac{x_{n+1}}{x_n} = a + \frac{bx_{n-1}}{x_n} \Rightarrow t_{n+1} = a + \frac{b}{t_n},$$
$$x_{n+1} = t_{n+1}x_n.$$

▯

The third factorization above is valid also on the multiplicative group of all nonzero real numbers with $a, b \in \mathbb{R}$ although on this larger group there is also a nonempty singularity set for the factor equation which limits the equivalence of the SC factorization to the original equation (the solutions of the latter that contain a zero must be excluded to keep the ratios well defined). See Section 5.4 for further discussion of this issue.

REMARK 3.3   Of the three form symmetries and SC factorizations of Eq. (3.26) discussed in Example 3.7 we discover in Chapter 4 that the first and the third belong to the same general class even though they look different and are expressed on different groups. Note that the third SC factorization is valid for $a + b > 0$ and does not require that $a + b$ add up to unity. The first form symmety and its SC factorization also belongs to two different classes that contains the second form symmetry; these classes of form symmetries and the corresponding SC factorizations are broad enough to contain all linear difference equations in their intersection without any restrictions on the coefficients. We discuss these important classes in Chapters 5 and 6.   ∎

The next example uses the ideas in Example 3.7 to show that a given difference equation can have different order reduction types.

Example 3.8

(Different reduction types for the same equation) Consider the third-order difference equation

$$x_{n+1} = ax_n + bx_{n-1} + cx_{n-2}, \quad a, b, c \geq 0, \ a + b + c = 1.$$

Since

$$x_{n+1} + (b+c)x_n + cx_{n-1} = x_n + (b+c)x_{n-1} + cx_{n-2}$$

there is the form symmetry

$$H(x_n, x_{n-1}, x_{n-2}) = x_n + (b+c)x_{n-1} + cx_{n-2}$$

which gives a type-(1,2)reduction with SC factorization

$$t_{n+1} = t_n,$$
$$x_{n+1} = t_{n+1} - (b+c)x_n - cx_{n-1}.$$

on the additive group of real numbers. In addition, we have

$$\frac{x_{n+1}}{x_n} = a + b\frac{x_{n-1}}{x_n} + c\frac{x_{n-2}}{x_n} = a + b\frac{x_{n-1}}{x_n} + c\frac{x_{n-1}}{x_n}\frac{x_{n-2}}{x_{n-1}}$$

which gives the form symmetry

$$H(x_n, x_{n-1}, x_{n-2}) = \left[\frac{x_n}{x_{n-1}}, \frac{x_{n-1}}{x_{n-2}}\right]$$

and a type-(2,1) reduction with SC factorization

$$t_{n+1} = a + \frac{b}{t_n} + \frac{c}{t_n t_{n-1}}$$
$$x_{n+1} = t_{n+1}x_n.$$

on the multiplicative group of positive real numbers. ◻

As noted in Remark 3.3 above, both of the conditions $a, b, c \geq 0$ and $a + b + c = 1$ may be dropped after introducing the general methods in either Chapter 5 or Chapter 6.

3.3.3 Reduction types $(k, 1)$ and $(1, k)$

Of the k possible order reduction types

$$(k, 1), \ (k-1, 2), \cdots, (2, k-1), \ (1, k)$$

for an equation of order $k + 1$ the two extreme ones, namely, $(k, 1)$ and $(1, k)$ have the extra appeal of having an equation of order 1 as either a factor or a cofactor.

For a type-$(k,1)$ reduction, $m = k$. Therefore, the function $h : G \to G$ in (3.16) is of one variable and yields the form symmetry

$$H(u_0, \ldots, u_k) = [u_0 * h(u_1), u_1 * h(u_2) \ldots, u_{k-1} * h(u_k)] \tag{3.27}$$

and SC factorization

$$t_{n+1} = g_n(t_n, t_{n-1}, \ldots, t_{n-k+1}) \tag{3.28}$$

$$x_{n+1} = t_{n+1} * h(x_n)^{-1} \tag{3.29}$$

where the functions $g_n : G^k \to G$ are determined by the given functions f_n in (3.1) as in Section 3.2. Example 3.8 above gives a type-(2,1) reduction for a third-order difference equation. In Chapters 4 and 5 we discuss substantial classes of difference equations that have type-$(k,1)$ reductions together with their SC factorizations.

For a type-$(1,k)$ reduction, $m = 1$. Therefore, $h : G^k \to G$ and the form symmetry has the scalar form

$$H(u_0, \ldots, u_k) = u_0 * h(u_1, \ldots, u_k). \tag{3.30}$$

This form symmetry gives the SC factorization

$$t_{n+1} = g_n(t_n) \tag{3.31}$$

$$x_{n+1} = t_{n+1} * h(x_n, \ldots, x_{n-k+1})^{-1}. \tag{3.32}$$

Example 3.8 above gives a type-(1,2) reduction for a third-order difference equation. In Chapter 6 we discuss substantial classes of higher-order difference equations having type-$(1,k)$ reductions together with their SC factorizations.

3.3.4   Factor and cofactor chains

A difference equation such as (3.1) may admit repeated reductions of order through its factor equation, its cofactor equation or both as follows:

$$\text{Eq. (3.1)} \to \begin{cases} \underbrace{t_{n+1} = g_n(t_n, \ldots, t_{n-m+1})}_{\text{factor equation}} \quad \to \quad \begin{cases} \text{factor} \ \to \cdots \\ \text{cofactor} \ \to \cdots \end{cases} \\[2em] \underbrace{x_{n+1} = t_{n+1} * h(x_n, \ldots, x_{n-k+m})^{-1}}_{\text{cofactor equation}} \to \begin{cases} \text{factor} \ \to \cdots \\ \text{cofactor} \ \to \cdots \end{cases} \end{cases}$$

In the above binary tree structure, we call each branch a *reduction chain*. If a reduction chain consists only of factor (or cofactor) equations then it is a *factor* (or *cofactor*) *chain*. As an example, consider the following rational difference equation of order three with its tree structure over a given nontrivial field $\mathcal{F}$:

$$x_{n+1} = x_n + \frac{a(x_n - x_{n-1})^2}{x_{n-1} - x_{n-2}} \to \begin{cases} t_{n+1} = \dfrac{at_n^2}{t_{n-1}} \\ x_{n+1} = t_{n+1} + x_n \end{cases} \to \begin{cases} s_{n+1} = as_n \\ t_{n+1} = s_{n+1}t_n \end{cases}$$

$$\tag{3.33}$$

The two order reducing form symmetries defining the new variables t_n and s_n are apparent from the cofactor equations. In (3.33) it is easy to identify the factor chain

$$x_{n+1} = x_n + \frac{a(x_n - x_{n-1})^2}{x_{n-1} - x_{n-2}} \rightarrow t_{n+1} = \frac{at_n^2}{t_{n-1}} \rightarrow s_{n+1} = as_n.$$

In the remainder of this book we encounter both factor and cofactor chains. In particular, every linear nonhomogeneous difference equation of order $k+1$ with constant complex coefficients has the remarkable property that it can be factored repeatedly down to a system of $k+1$ linear nonhomogeneous equations of order one through either a *factor chain* (Chapter 5) or a *cofactor chain* (Chapter 6). In Chapter 7 the reader will find a comprehensive analysis of the SC factorizations and reductions in orders of linear nonhomogeneous equations, including those with variable coefficients.

3.4 SC factorizations as triangular systems

The pair of equations (3.18) and (3.19) that constitute a SC factorization of Eq. (3.1) have a special property in addition to each being of lower order. The factor equation (3.18) is entirely independent of the cofactor (3.19) and the variables appear in a layered or "triangular" form; i.e., the variables in the factor equation are a proper subset of the variables in the cofactor equation which can be restated as

$$x_{n+1} = g_n(t_n, \ldots, t_{n-m+1}) * h(x_n, \ldots, x_{n-k+m})^{-1}.$$

This feature uncouples the system of factor and cofactor equations in the following sense: A solution of the factor equation (3.18) defines the sequence $\{t_{n+1}\}$ in the cofactor equation which then generates a solution $\{x_n\}$ of Eq. (3.1). In principle, this layering of variables permits solving (3.1) by solving two lower order difference equations in turn, in the manner prescribed by the triangular nature of the SC factorization.

Triangular systems may be defined generally and independently of SC factorizations. We give an intuitive definition here.

DEFINITION 3.7 *A system of difference equations is **triangular** or a **triangular system** if, after a rearrangement of equations if necessary, the variables in each equation are a subset of the variables in the equation immediately succeeding it. If the system has the equation with the least number of variables on top then the system is in **descending** form and an equation that succeeds another is below it. If the equation with the least number of variables is at the bottom then the system is in **ascending** form.*

The SC factorization system of equations (3.18) and (3.19) is written in the descending form. If we switch the two equations then an ascending form is obtained. We generally write the SC factorizations in the descending form unless preferable otherwise.

In the SC factorization of a difference equation of order greater than two the factor or the cofactor equation may possess order-reducing form symmetries. In this case, the new system of three equations is again triangular. For instance, consider Eq. (3.33) again. The three first order equations that mark the ends of factor/cofactor chains in (3.33) can be arranged as follows:

$$s_{n+1} = as_n$$
$$t_{n+1} = as_n t_n \qquad\qquad (3.34)$$
$$x_{n+1} = as_n t_n + x_n.$$

This is a triangular system in descending form. We emphasize that the triangular system (3.34) is different from the system of three equations

$$x_{n+1} = x_n + \frac{a(x_n - y_n)^2}{y_n - z_n}$$
$$y_{n+1} = x_n \qquad\qquad (3.35)$$
$$z_{n+1} = y_n$$

that gives the unfolding of Eq. (3.33). While both systems consist of first-order equations, none of the equations in (3.35) are independent of each other. We may say that system (3.35) is more intertwined – or has a greater degree of coupling – than (3.34). Obtaining an explicit formula for the solutions of (3.33) directly from (3.35) is more difficult.

For an equation of order $k + 1$ the SC factorization process may continue until it eventually stops in at most k steps, given that a first-order difference equation cannot have order-reducing form symmetries. The result is a system of lower order equations marking the ends of factor/cofactor chains.

If a SC factorization can be found for each lower-order equation in turn then the original equation can be decomposed into a collection of first order equations. In this case we call the final system of first-order equations a *complete, or full SC factorization* of the original higher-order equation. System (3.34) represents a complete SC factorization of Eq. (3.33).

Certain difference equations always have a complete SC factorization; the class of linear nonhomogeneous equations with constant coefficients in an algebraically closed field is a familiar example that we will discuss in substantial detail later in this book. Whether a complete SC factorization as a triangular system exists for every higher-order difference equation is an open question that is equivalent to the Semiconjugacy Problem.

3.5   Order-preserving form symmetries

Often a difference equation can be transformed into a more tractable equation of the *same order* by means of a change of variables. For instance, the following equations

$$x_{n+1} = x_n^p x_{n-1}^q \quad \text{and} \quad x_{n+1} = \frac{x_n x_{n-1}}{a x_n + b x_{n-1}}$$

of order 2 on $(0, \infty)$ can be transformed into *linear* equations of order 2 by the substitutions $x_n = e^{y_n}$ and $x_n = 1/y_n$, respectively; see the Problems for this chapter. Our aim in this section is to show that these types of substitutions or changes of variables can be derived from semiconjugate relations.

If $m = k + 1$ in the preceding sections then the form symmetry H can no longer be "order-reducing" since the functions F_n and Φ_n both unfold difference equations of the *same* order. In this case, a class of link functions H that does *not* reduce the order is defined as follows

$$H(u_0, \ldots, u_k) = [h_0(u_0), \ldots, h_k(u_k)] \tag{3.36}$$

Since the coordinate functions h_j depend only on the time step of the same index in this case, the most distant time step k is preserved and the order of the difference equation is unchanged. To make H a SC link map and hence a form symmetry, a necessary restriction is supplied by the next result.

PROPOSITION 3.3
Suppose that H as defined by (3.36) is the SC link map of unfoldings F_n and Φ_n of difference equations, i.e., for each n there are functions f_n and ϕ_n such that

$$F_n(u_0, \ldots, u_k) = [f_n(u_0, \ldots, u_k), u_0, \ldots, u_{k-1}]$$
$$\Phi_n(u_0, \ldots, u_k) = [\phi_n(u_0, \ldots, u_k), u_0, \ldots, u_{k-1}].$$

Then all coordinate functions h_j are equal, i.e., $h_j = h_i$ for all $i, j = 0, 1, \ldots k$.

PROOF   If H is defined by (3.36) then

$$H(F_n(u_0, \ldots, u_k)) = [h_0(f_n(u_0, \ldots, u_k)), h_1(u_0), \ldots, h_k(u_{k-1})]$$
$$\Phi_n(H(u_0, \ldots, u_k)) = [\phi_n(h_0(u_0), \ldots, h_k(u_k)), h_0(u_0), \ldots, h_{k-1}(u_{k-1})].$$

If the left-hand sides are equal then from the right-hand sides we obtain

$$h_0 = h_1 = h_2 = \cdots = h_{k-1} = h_k.$$

∎

According to Proposition 3.3 the form symmetry H in (3.36) is determined by a single one-dimensional map. On the other hand, *it is unnecessary that the range of h be again G*, a fact that is important if we wish to retain the surjectivity of h (and hence of H) in the most common applications that involve changes of variables. This observation and additional information provided by the proof of Proposition 3.3 suggest the following definition.

DEFINITION 3.8   *Consider two difference equations of type (3.1) that are defined by functions $f_n : G^{k+1} \to G$ and $\phi_n : \Gamma^{k+1} \to \Gamma$ where Γ is a given nontrivial group (usually a subgroup of G). A surjective map $h : G \to \Gamma$ is an* **order-preserving** *form symmetry if*

$$h(f_n(u_0, \ldots, u_k)) = \phi_n(h(u_0), \ldots, h(u_k)). \tag{3.37}$$

In many cases of interest, order-preserving form symmetries yield *conjugate equivalents* (i.e., H is a bijection); e.g., if G is *finite* and $\Gamma = G$. This equivalence is not unexpected for two difference equations of the same order. The map h in Definition 3.8 defines a change of variables

$$t_j = h(u_j), \quad j = 0, 1, \ldots, k.$$

If h is a bijection then using (3.37) we obtain the transformed functions as

$$\phi_n(t_0, \ldots, t_k) = h(f_n(h^{-1}(t_0), \ldots, h^{-1}(t_k))). \tag{3.38}$$

Two common invertible choices for bijective h are the inverse or reciprocal transformation and the real exponential (or logarithmic) functions. Using the inverse function in (3.38) we obtain

$$h(u) = u^{-1} \Rightarrow \phi_n(t_0, \ldots, t_k) = f_n(t_0^{-1}, \ldots, t_k^{-1}).$$

For $G = \mathbb{R}$ under addition and $\Gamma = (0, \infty)$ under multiplication, using the exponential function in (3.38) yields

$$h(u) = e^u \Rightarrow \phi_n(t_0, \ldots, t_k) = e^{f_n(\ln t_0, \ldots, \ln t_k)}.$$

The next two examples illustrate the above choices. On the other hand, the complex exponential function $h(u) = e^{iu}$ is a form symmetry from the additive group $\mathbb{R}$ to the multiplicative circle group $\mathbb{T}$ that is surjective but not injective; recall Example 2.4. For another noninvertible (noninjective) example see the Problems for this chapter.

Example 3.9
Order-preserving form symmetries can be combined with order-reducing ones to simplify a given difference equation. Consider the rational difference equa-

tion

$$x_{n+1} = x_{n-k+1} + \frac{x_n - x_{n-k}}{a_n(x_n - x_{n-k}) + b_n} \tag{3.39}$$

where

$$a_n \geq 0, \ b_n > 0, \ x_0 > x_{-k}. \tag{3.40}$$

The substitution $y_n = x_n - x_{n-k}$ defines an order-reducing form symmetry $H(u_0, \ldots, u_k) = u_0 - u_k$ which transforms (3.39) into the first-order equation

$$y_{n+1} = \frac{y_n}{a_n y_n + b_n}, \quad y_0 = x_0 - x_{-k} > 0. \tag{3.41}$$

Note that $y_n > 0$ for all $n \geq 0$ so Eq. (3.41) is defined over the multiplicative group $G = (0, \infty)$. The reciprocal function $t_n = 1/y_n$ transforms the rational equation into a linear one as follows:

$$\frac{1}{t_{n+1}} = \frac{1/t_n}{a_n/t_n + b_n} \Rightarrow t_{n+1} = a_n + b_n t_n. \tag{3.42}$$

Now it is possible to obtain an explicit formula for the solutions of (3.39) subject to (3.40) using the formula for the solutions of the linear equation (3.42); see the Problems for this chapter. ☐

In the preceding example note that the map $f_n(u) = u/(a_n u + b_n)$ and the linear map $\phi_n(t) = a_n + b_n t$ are conjugates on $(0, \infty)$ for each $n \geq 1$ through $h(u) = 1/u$.

Example 3.10
Consider the following difference equation of order $k+1$ on the multiplicative group $G = (0, \infty)$

$$x_{n+1} = a_n x_n^p x_{n-k}^q \tag{3.43}$$

subject to parameter values

$$k \geq 0, \ p, q \in \mathbb{R}, \ a_n > 0 \text{ for all } n \geq 0,$$
$$x_i > 0 \text{ for } i = -k, \ldots, 0.$$

The substitution $y_n = \ln x_n$ which is represented by the form symmetry $h(u) = \ln u$, transforms Eq. (3.43) into the linear nonhomogeneous equation

$$y_{n+1} = p y_n + q y_{n-k} + \ln a_n \tag{3.44}$$

which can be solved using classical methods; see the Problems for this chapter. ☐

3.6   Notes

The concept of semiconjugacy that is not one-to-one (i.e., not a conjugacy) has appeared in different contexts in the literature; see, e.g., Milnor and Thurston (1977), Byers (1983), Rothe (1992), and Akin (1993). Semiconjugacy is used in an entirely new way in this book. There are several publications in which one sees elements of this approach, and many of these studies are referred to in later parts of this book. An earlier book that contains a more limited exposure to semiconjugacy is Sedaghat (2003). That book discusses situations in which behaviors of maps of higher-dimensional Euclidean spaces can be studied in terms of the properties of one-dimensional maps that are semiconjugates to the original maps. The current book's focus is not on maps, but depending on the underlying group, its ideas and methods are broader in nature, more powerful and potentially more applicable in a greater variety of contexts.

Invariants are discussed in various publications; see, e.g., Kocic and Ladas (1993), Kulenovic and Ladas (2002), and Sedaghat (2003) and the references cited in these books. In Sedaghat (2003) the similarities and differences among invariants, Liapunov functions, and semiconjugate relations are discussed in some detail and the relationships between these important concepts are highlighted.

Triangular maps and systems are discussed in, e.g., Alseda and Llibre (1993), Smital (2008), and the references in these articles. In particular, the results in Alseda and Llibre (1993) give information about periods of solutions of difference equations that possess semiconjugate factorizations since the factor-cofactor system is always triangular.

3.7   Problems

3.1 Let F and Φ be self-maps of sets S and M, respectively, such that $F \trianglerighteq \Phi$. Prove each of the following statements:

  (a) If F is surjective then Φ is surjective.

  (b) If F is a constant function then Φ is a constant function.

3.2 (a) Let S and M be finite sets and let F, Φ be self-maps of S and M, respectively. Prove that if $F \trianglerighteq \Phi$ and $\Phi \trianglerighteq F$ then $F \simeq \Phi$.

  (b) Let S be a finite set and $F : S \to S$ be a bijection. Prove that if $F \trianglerighteq F^{-1}$ or $F^{-1} \trianglerighteq F$ then $F \simeq F^{-1}$.

3.3 Let $S = \{1, 2, 3\}$ and $M = \{a, b\}$ with $a \neq b$.

  (a) List all self-maps of S and of M and all surjective link maps $H : S \to M$.

(b) By directly computing each of the compositions $H \circ F$ and $\Phi \circ H$ or by other means, determine those self-maps of S that are semiconjugate to some self-map of M.

3.4 Verify that the function H_L in Example 3.3 is an invariant.

3.5 Let $\mathbb{Z}_2$ be as in Example 3.2 and consider the Fibonacci equation modulo 2, i.e.,

$$x_{n+1} = x_n + x_{n-1}, \quad x_0, x_{-1} \in \{0, 1\}. \tag{3.45}$$

(a) Verify that the functions H_7 and H_8 in Example 3.2 or Example 3.4 are invariants for (the unfolding of) Eq. (3.45).

(b) Determine whether the unfolding of (3.45) has any other semiconjugate factors on $\mathbb{Z}_2$.

3.6 Let k be a positive integer and for $H : \mathbb{Z}_2^k \to \mathbb{Z}_2$ define its *dual* $H^* = H + 1$. Note that H and H^* have opposite values: $H = 0$ if and only if $H^* = 1$.

(a) Identify the duals for the four maps ι, ξ_0, ξ_1 and ψ when $k = 1$.

(b) Write each of the functions H_1-H_{14} in Example 3.2 in algebraic form using addition and multiplication in $\mathbb{Z}_2$ and identify the duals. For instance, $H_1(u, v) = uv$ for all $(u, v) \in \mathbb{Z}_2^2$ and $H_1^* = H_{14}$.

(c) How many surjective maps $H : \mathbb{Z}_2^k \to \mathbb{Z}_2$ are there for an arbitrary $k \geq 1$, excluding duals?

(d) Let $F : \mathbb{Z}_2^k \to \mathbb{Z}_2^k$. Prove that $H \circ F = \phi \circ H$ for some $\phi : \mathbb{Z}_2 \to \mathbb{Z}_2$ if and only if $H^* \circ F = \phi' \circ H^*$ for some $\phi' : \mathbb{Z}_2 \to \mathbb{Z}_2$. Thus duals of SC link maps are also SC link maps.

3.7 Let k, m be integers and define the *dual* of $F : \mathbb{Z}_2^k \to \mathbb{Z}_2^m$ as

$$F^* = [f_1 + 1, \ldots, f_m + 1]$$

where $f_i : \mathbb{Z}_2^k \to \mathbb{Z}_2$ for $i = 1, \ldots, m$ are the coordinate functions of F. Note that the values of F^* in $\mathbb{Z}_2^m$ are opposite to those of F with 0's and 1's switched.

(a) Verify that $(F^*)^* = F$.

Define a function $H : \mathbb{Z}_2^k \to \mathbb{Z}_2$ to be *dual preserving* if

$$H(u_1 + 1, \ldots, u_k + 1) = H(u_1, \ldots, u_k) + 1.$$

(b) Show that if H is dual preserving then so is its dual $H + 1$.

(c) Which of the four self-maps ι, ξ_0, ξ_1 and ψ of $\mathbb{Z}_2$ are dual preserving?

(d) Which of the functions H_1-H_{14} in Example 3.2 are dual preserving?

(e) Prove that if H is dual preserving then $H \circ F^* = H^* \circ F$ for each function $F : \mathbb{Z}_2^k \to \mathbb{Z}_2^m$.

(f) Let F be a self-map of $\mathbb{Z}_2^k$ and ϕ a self-map of $\mathbb{Z}_2$. If $F \trianglerighteq \phi$ with respect to a dual preserving map H then prove that $F^* \trianglerighteq \phi^*$ with the same link map H.

3.8 (a) Show that the only link maps in Example 3.2 that can be written in the recursive form $u + h(v)$ are H_3, H_6, H_9 and H_{12}. Determine the function h in each case.

(b) Determine all distinct autonomous second-order difference equations on $\mathbb{Z}_2$ for which one of the functions in (a) is a form symmetry.

(c) Call two difference equations of order $k+1$ on $\mathbb{Z}_2$ *dual equations* if their unfoldings are dual maps of $\mathbb{Z}_2^{k+1}$. Which of the difference equations in (b) are duals?

3.9 Let $\{a_n\}$ be an arbitrary binary sequence, i.e., $a_n \in \{0,1\}$ for all $n \geq 0$. Consider the nonautonomous difference equation

$$x_{n+1} = a_n + x_{n-1}, \quad x_0, x_{-1} \in \{0,1\} \tag{3.46}$$

which generalizes Eq. (3.23) when $G = \mathbb{Z}_2$.

(a) Show that the same form symmetry as in Example 3.5 still works and use Theorem 3.1 to obtain the SC factorization of (3.46). It is necessary to define suitable time dependent functions f_n on $\mathbb{Z}_2^2$ and ϕ_n on $\mathbb{Z}_2$.

(b) Let $a_n = 1$ for all $n \geq k$ where k is a positive integer. Show that every solution of (3.46) eventually has period 4 using the SC factorization in (a).

(c) Let $a_n = 0$ for all $n \geq k$ where k is a positive integer. Determine the asymptotic behaviors of solutions of (3.46) using the SC factorization in (a).

3.10 (a) Prove that each solution of a first-order, autonomous difference equation $t_{n+1} = g(t_n)$ on $\mathbb{Z}_2$ must either be eventually constant or have period 2.

(b) Let F be the unfolding of a second-order, autonomous difference equation $x_{n+1} = f(x_n, x_{n-1})$ on $\mathbb{Z}_2$ and suppose that $F \trianglerighteq \Phi$ for some self-map Φ of $\mathbb{Z}_2$ (we do **not** assume that Theorem 3.1 applies). Prove that the solutions of the second-order equation are eventually periodic with period $p = 1, 2, 3$, or 4 ($p = 1$ is interpreted as the constant solution).

3.11 Show that the third-order difference equation

$$x_{n+1} = x_n + \frac{a(x_n - x_{n-1})^2}{x_n - x_{n-2}}$$

has a complete SC factorization into a triangular system of first-order difference equations. Determine the reduction tree and the factor or cofactor chains for this equation.

3.12 In Example 3.9 assume that $a_n = a \geq 0$ and $b_n = b > 0$ are constants.

(a) Find the explicit formula for the solutions of Eq. (3.42).

(b) Use the formula in (a) to find explicit formulas for the solutions of equations (3.41) and (3.39).

3.13 In Example 3.10 assume that $k = 1$. Find the explicit formula for the solutions of Eq. (3.44) on $\mathbb{R}$ and use this formula to obtain an explicit formula for the solutions of Eq. (3.43) on G.

3.14 (a) Using the reciprocal transformation convert the difference equation

$$x_{n+1} = \frac{x_n x_{n-k}}{a x_n + b x_{n-k}}, \quad a, b \geq 0, \ a + b > 0. \tag{3.47}$$

to a linear equation of the same order.

(b) For $k = 1$ obtain an explicit formula for the positive solutions of Eq. (3.47).

3.15 Let $\mathbb{C}_0$ be the set of all nonzero complex numbers and let $h : \mathbb{C}_0 \to (0, \infty)$ be the modulus function $h(z) = |z|$. Recall that under ordinary multiplication of complex numbers $\mathbb{C}_0$ is a group having $(0, \infty)$ as a subgroup.

(a) Show that h is an order preserving form symmetry for

$$z_{n+1} = \frac{a z_n}{z_{n-k}}, \quad k \geq 1, \ a, z_{-j} \in \mathbb{C}_0 \text{ for } j = 0, 1, \ldots, k. \tag{3.48}$$

What are the functions f_n and ϕ_n in Eq. (3.37)?

(b) If $\{t_n\}$ is a solution of $t_{n+1} = \phi_n(t_n, \ldots, t_{n-k})$ with $t_n = h(z_n)$ describe the sets $h^{-1}(t_n)$.

(c) For $k = 1$, $a = 1 + i$, $z_0 = -2$ and $z_{-1} = i$ explicitly determine (i) the solution $\{z_n\}$ of (3.48), (ii) the solution of $t_{n+1} = \phi_n(t_n, t_{n-1})$, and (iii) the sets $h^{-1}(t_n)$ that contain the numbers z_n.

4

Homogeneous Equations of Degree One

In this chapter we discuss a special yet sizable class of recursive difference equations of type

$$x_{n+1} = f_n(x_n, x_{n-1}, \ldots, x_{n-k}) \tag{4.1}$$

that possess semiconjugate factorizations on groups. This class of equations is distinguished by the fact that the defining functions f_n can be identified by a simple algebraic rule.

All difference equations encountered in this chapter belong to a much broader class of difference equations that admit type-$(k, 1)$ reductions and are all characterized by the same form symmetry. We discuss this more general class in Chapter 5 but devote this chapter to the discussion of the special case, which is easier to understand and may serve as a preview of things to come.

4.1 Homogeneous equations on groups

In any group G we may define *integer* powers of elements as follows:

$$
u^p = \begin{cases}
\underbrace{u * \cdots * u}_{p \text{ times}}, & \text{if } p > 0 \\
\underbrace{u^{-1} * \cdots * u^{-1}}_{|p| \text{ times}}, & \text{if } p < 0 \\
1, & \text{if } p = 0
\end{cases}
$$

The following concept, borrowed from the theory of functional equations, also makes sense in the group context.

DEFINITION 4.1 *Let G be a nontrivial group and let p, m be integers with $m \geq 1$. If a function $f : G^m \to G$ satisfies the following*

$$f(u_1 * t, \ldots, u_m * t) = f(u_1, \ldots, u_m) * t^p, \quad \text{for all } u_i, t \in G$$

*then f is said to be (right) **homogeneous of degree** p or an **HDp** function. If G is noncommutative then a "left version" can be defined analogously.*

The developments below are stated for the right version; the parallel left versions can then be inferred easily and need not be stated explicitly. Therefore, we drop mention of parity when referring to homogeneous functions or equations.

We are interested only in the case $p = 1$ of the above definition although we may make occasional references to other values of p. Therefore, the following definition is stated only for case $p = 1$.

DEFINITION 4.2   *Equation (4.1) is said to be* **homogeneous of degree 1**, *or* **HD1**, *on a group G if for every $n = 0, 1, 2, \ldots$ the functions f_n are all homogeneous of degree 1 relative to G. Therefore, (4.1) is an HD1 difference equation if*

$$f_n(u_0 * t, \ldots, u_k * t) = f_n(u_0, \ldots, u_k) * t$$
$$\text{for all } t, u_i \in G, \ i = 0, \ldots, k, \text{ and all } n \geq 0.$$

Our first example indicates that HD1 equations can be quite general. Although commutativity of the group operation is not required, we assume it for familiarity of notation.

Example 4.1

Let G be a commutative group and let $\Delta s_n = s_n - s_{n-1}$ denote the backward difference operator (see Remark 2.1 in Chapter 2). Then for any sequence of functions $g_n : G^{k+1} \to G$ the difference equation

$$\Delta x_{n+1} = g_n(\Delta x_n, \Delta x_{n-1}, \ldots, \Delta x_{n-k+1}) \tag{4.2}$$

is HD1 on G for every integer $k \geq 1$. To see this, write Eq. (4.2) in the equivalent form

$$x_{n+1} = x_n + g_n(x_n - x_{n-1}, \ldots, x_{n-k+1} - x_{n-k}).$$

Now define

$$f_n(u_0, \ldots, u_k) = u_0 + g_n(u_0 - u_1, \ldots, u_{k-1} - u_k)$$

to see that for every $t \in G$,

$$f_n(u_0 + t, \ldots, u_k + t) = u_0 + t + g_n(u_0 - u_1, \ldots, u_{k-1} - u_k)$$
$$= f_n(u_0, \ldots, u_k) + t.$$

It follows that (4.2) is HD1.   ☐

Example 4.2

Let $\mathcal{F}$ be a nontrivial field and let $a_{j,n}, b_{j,n} \in \mathcal{F}$ for $j = 0, 1, \ldots, k$ and all $n \geq 0$. For nontriviality assume further that $b_{k,n} \neq 0$ for all n. It is readily verified using definitions that every rational difference equation of type

$$x_{n+1} = x_{n-j} \frac{a_{0,n}x_n + a_{1,n}x_{n-1} + \cdots + a_{k,n}x_{n-k}}{b_{0,n}x_n + b_{1,n}x_{n-1} + \cdots + b_{k,n}x_{n-k}}$$

is HD1 relative to the multiplicative group $\mathcal{F}\backslash\{0\}$ of units of $\mathcal{F}$ for each $j = 0, 1, \ldots, k$. ▯

Example 4.3
Let $\{a_n\}$ and $\{b_n\}$ be sequences of real numbers and let j, k be positive integers. Then it is easy to see that the difference equation

$$x_{n+1} = x_{n-j} + e^{a_n(x_n - x_{n-k}) + b_n}$$

is HD1 relative to the additive group of all real numbers. ▯

The next example concerns a generalization of the linear Eq. (3.26) in Example 3.7.

Example 4.4
Consider the difference equation

$$x_{n+1} = a_{0,n}x_n + a_{1,n}x_{n-1} + \cdots + a_{k,n}x_{n-k} + b_n \qquad (4.3)$$

where the sequences $\{a_{j,n}\}$ for $j = 0, 1, \ldots, k$ are in a given field $\mathcal{F}$ with $a_{k,n} \neq 0$ for all n. If $b_n = 0$ for all n, i.e., if Eq. (4.3) is linear homogeneous then it is readily seen that it is an HD1 equation relative to the multiplicative group $\mathcal{F}\backslash\{0\}$ of units of $\mathcal{F}$.

Linear equations such as the above also have a HDp property for arbitrary b_n if the coefficients $a_{j,n}$ are suitably restricted. For each integer p and each $t \in \mathcal{F}$ define the element pt (or t^p in multiplicative notation as in the beginning of this section) with $\mathcal{F}$ viewed as an additive group, i.e.,

$$pt = \begin{cases} \underbrace{t + t + \cdots + t}_{p \text{ times}} & \text{if } p > 0 \\ 0, & \text{if } p = 0 \\ \underbrace{(-t) + \cdots + (-t)}_{|p| \text{ times}} & \text{if } p < 0. \end{cases}$$

According to Definition 4.1, (4.3) is HDp with respect to $\mathcal{F}$ (as an additive group) for every p if and only if the coefficient sequences $\{a_{j,n}\}$ for $j = 0, 1, \ldots, k$ satisfy the following equality for every n:

$$a_{0,n} + a_{1,n} + \cdots + a_{k,n} = p1.$$

This claim is true because for all $(u_0, \ldots, u_k) \in \mathcal{F}^{k+1}$ and $t \in \mathcal{F}$,

$$a_{0,n}(u_0 + t) + a_{1,n}(u_1 + t) + \cdots + a_{k,n}(u_k + t) + b_n =$$
$$a_{0,n}u_0 + a_{1,n}u_1 + \cdots + a_{k,n}u_k + b_n + (a_{0,n} + a_{1,n} + \cdots + a_{k,n})t =$$
$$a_{0,n}u_0 + a_{1,n}u_1 + \cdots + a_{k,n}u_k + b_n + pt.$$

$\square$

An abundance of HD1 functions on groups is indicated in the next result. We leave the straightforward proof to the reader.

PROPOSITION 4.1

Let G be a nontrivial group and k a positive integer.
(a) If $g : G^k \to G$ is any given function, then $\bar{g} : G^{k+1} \to G$ defined by

$$\bar{g}(u_0, \ldots, u_k) = g(u_0 * u_1^{-1}, u_1 * u_2^{-1}, \ldots, u_{k-1} * u_k^{-1})$$

is a HD0 function and each of the following is HD1:

$$\bar{g}(u_0, \ldots, u_k) * u_j, \quad j = 0, 1, \ldots, m.$$

(b) If $g_1, g_2 : G^{k+1} \to G$ and $f : G^2 \to G$ are HD1 then so is the composition

$$f(g_1(u_0, \ldots, u_k), g_2(u_0, \ldots, u_k)).$$

(c) If f is HD1 and g is HD0 on G^{k+1} then the product

$$fg(u_0, \ldots, u_k) = f(u_0, \ldots, u_k) * g(u_0, \ldots, u_k)$$

is HD1.

4.2   Characteristic form symmetry of HD1 equations

As seen in Section 3.3.3, a type-$(k, 1)$ reduction has a form symmetry given by (3.27) with SC factorization given by the system (3.28) and (3.29). We now define a special form symmetry of this type.

DEFINITION 4.3   *If $h(u) = u^{-1}$ (i.e., the group inversion) in Eq. (3.27) in Section 3.3.3 then H is called the **inversion form symmetry**; i.e.,*

$$H(u_0, \ldots, u_k) = [u_0 * u_1^{-1}, u_1 * u_2^{-1} \ldots, u_{k-1} * u_k^{-1}]. \tag{4.4}$$

An interesting feature of the inversion form symmetry (4.4) is that we can precisely identify the class of all difference equations that have it. By the following theorem, the HD1 property characterizes the inversion form symmetry and its SC factorization completely. To avoid unnecessary repetition, we postpone the proof until Chapter 5 where it follows from a more general result.

THEOREM 4.1

Eq. (4.1) has the inversion form symmetry (4.4) if and only if f_n is HD1 relative to G for all n. In this case, (4.1) has a type-$(k,1)$ order-reduction with the SC factorization

$$t_{n+1} = f_n(1, t_n^{-1}, (t_n * t_{n-1})^{-1}, \ldots, (t_n * t_{n-1} * \cdots * t_{n-k+1})^{-1}) \qquad (4.5)$$

$$x_{n+1} = t_{n+1} * x_n. \qquad (4.6)$$

PROOF See Corollary 5.3 in Section 5.3. ∎

REMARK 4.1 1. From (4.6) we obtain the following formula for x_n in terms of a solution $\{t_n\}$ of (4.5) as follows:

$$x_n = \prod_{i=0}^{n-1} t_{n-i} * x_0 \quad n = 1, 2, 3, \ldots \qquad (4.7)$$

where the multiplicative notation is used for iterations of the group operation $*$. In additive (and commutative) notation, (4.7) takes the form

$$x_n = x_0 + \sum_{i=1}^{n} t_i. \qquad (4.8)$$

2. We can quickly construct Eq. (4.5) directly from (4.1) in the HD1 case by making the substitutions

$$1 \to x_n, \quad (t_n * t_{n-1} * \cdots * t_{n-i+1})^{-1} \to x_{n-i} \text{ for } i = 1, 2, \ldots, k. \qquad (4.9)$$

Recall that 1 represents the group identity in multiplicative notation. In additive notation (4.9) takes the form

$$0 \to x_n, \quad -t_n - t_{n-1} \cdots - t_{n-i+1} \to x_{n-i} \text{ for } i = 1, 2, \ldots, k. \qquad (4.10)$$

∎

4.3 Reductions of order in HD1 equations

In this section we discuss several types of HD1 equations and their SC factorizations using the inversion form symmetry. Our aim in part is to illustrate the variety of equations to which the methods of the previous section apply. In the next chapter we broaden our selection of equations after discovering other types of form symmetries that permit generalizing the methods of this chapter to non-HD1 equations.

We begin with the following corollary of Theorem 4.1 and (4.9).

COROLLARY 4.1

Let i, j, k be nonnegative integers such that $0 \leq i, j < k$ and let $g_n : G \to G$ be given functions on a commutative group G for $n \geq 0$. Then:

(a) The following difference equation is HD1

$$x_{n+1} = x_{n-i} * g_n(x_{n-j} * x_{n-k}^{-1}).$$ (4.11)

(b) If $i \geq 1$ then Eq. (4.11) has a factor equation

$$t_{n+1} = (t_n * t_{n-1} * \ldots * t_{n-i+1})^{-1} * g_n(t_{n-j} * t_{n-j-1} * \ldots * t_{n-k+1}).$$

For $i = 0$ the factor equation is

$$t_{n+1} = g_n(t_{n-j} * t_{n-j-1} * \ldots * t_{n-k+1}).$$

(c) If $k \geq i+1$ and $j = k - i - 1$ then Eq. (4.11) has a type-$(k-i, i+1)$ order reduction as well with the following SC factorization

$$t_{n+1} = g_n(t_{n-k+i+1}),$$
$$x_{n+1} = t_{n+1} * x_{n-i}.$$

PROOF   Parts (a) and (b) follow immediately from Theorem 4.1 and (4.9). Part (c) is easily seen to be true by first writing (4.11) as

$$x_{n+1} * x_{n-i}^{-1} = g_n(x_{n-k+i+1} * x_{n-k}^{-1})$$ (4.12)

and then substituting

$$t_n = x_n * x_{n-i-1}^{-1}$$

in (4.12).   ∎

Example 4.5
Consider the rational difference equation

$$x_{n+1} = x_n \left(a_n \frac{x_{n-k+1}}{x_{n-k}} + b_n \right),$$ (4.13)

where $\{a_n\}, \{b_n\}$ are sequences of positive real numbers. This equation is a special case of (4.11) relative to the group $(0, \infty)$ under ordinary multiplication. With $i = 0$, $j = k - 1$ and $g_n(u) = a_n u + b_n$ in Corollary 4.1 we obtain the reduction of order of (4.13) to factor as

$$t_{n+1} = a_n t_{n-k+1} + b_n.$$

This linear nonhomogeneous equation with a time delay of $k - 1$ can be solved to obtain an explicit solution of (4.13) through (4.7), if desired. ▯

Example 4.6
Consider the third-order equation

$$x_{n+1} = a_n + x_{n-1} + b_n e^{c_n(x_n - x_{n-2})}. \tag{4.14}$$

This equation is HD1 on the group of all real numbers under addition and a special case of (4.11) with $g_n(u) = a_n + b_n e^{c_n u}$. The factor equation of (4.14) is

$$t_{n+1} = a_n - t_n + b_n e^{c_n(t_n + t_{n-1})}. \tag{4.15}$$

Eq. (4.15) is not HD1, but it has the *identity form symmetry* that is discussed in the next chapter. Thus a complete SC factorization of (4.14) into a triangular system is possible. ▯

Example 4.7
Let $G = \mathbb{C} \backslash \{0\}$ be the group of nonzero complex numbers under ordinary multiplication and let p, q be positive integers. If $\{a_n\}$ is a sequence in G then the difference equation

$$x_{n+1} = \frac{a_n x_{n-p+1} x_{n-p-q}}{x_{n-q}} \tag{4.16}$$

is of the type defined in Corollary 4.1(c) with $i = p - 1$, $k = p + q$ and

$$g_n(u) = \frac{a_n}{u}.$$

Thus (4.16) has the type-$(q + 1, p)$ order reduction with the SC factorization

$$t_{n+1} = \frac{a_n}{t_{n-q}}, \quad x_{n+1} = t_{n+1} x_{n-p+1}. \tag{4.17}$$

▯

The basic type-$(k, 1)$ factorization given by Corollary 4.1(b) is also possible in Example 4.7 since Eq. (4.16) is HD1. However, the basic factorization is not simpler or more advantageous. Also G can be any commutative group.

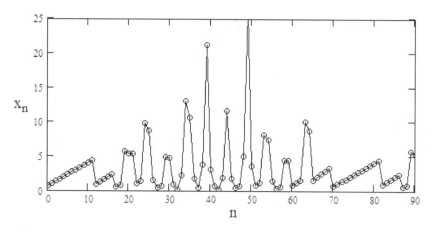

FIGURE 4.1

A solution of Eq. (4.16) with $p = 5$, $q = 6$, $a_n = 1$ for all n. This solution with initial values $x_j = (13 + j)/3$ for $j = -11, -10, \ldots, 0$ and all other solutions of (4.16) have period $2(q + 1)p = 70$; see the text.

There is another possible SC factorization for Eq. (4.16) based on the *identity form symmetry* that will be discussed in the next chapter.

REMARK 4.2   Factorization (4.17) yields some quick information about the solutions of Eq. (4.16). In particular, if $a_n = 1$ for all n then all non-constant solutions of the factor equation in (4.17) are periodic with period $2(q + 1)$. This indicates the existence of periodic solutions for (4.16) when p is odd; see the Problems for this chapter. This statement is not true for the slightly different rational recursive equation

$$x_{n+1} = \frac{a_n x_{n-p} x_{n-p-q}}{x_{n-q}} \tag{4.18}$$

which does not have a SC factorization of the type given in Example (4.7). However, (4.18) is clearly HD1 so Theorem 4.1 or Corollary 4.1(b) may be used to obtain a SC factorization.

Figures 4.1 and 4.2 show the different solutions of the two equations (4.16) and (4.18), respectively, with the same parameter values.

∎

HD1 difference equations of higher-order can be obtained from any given difference equation as follows: Let $g_n : G^k \to G$ denote a sequence of functions on a group $(G, *)$ and let

$$t_{n+1} = g_n(t_n, \ldots, t_{n-k+1})$$

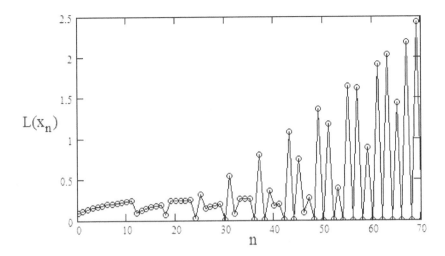

FIGURE 4.2
A solution of Eq. (4.18) with $p = 5$, $q = 6$, $a_n = 1$ for all n and initial values
$x_j = (13 + j)/3$ for $j = -11, -10, \ldots, 0$. To aid in visualizing both large and
small values, the double-log numbers $L(x_n) = log(log(x_n + 1) + 1)$ are
plotted rather than the actual values x_n.

be a given difference equation of order k. If we substitute $x_n * x_{n-1}^{-1}$ for t_n
in this equation then the following difference equation is HD1 by Proposition
4.1(a)

$$x_{n+1} = g_n(x_n * x_{n-1}^{-1}, \ldots, x_{n-k+1} * x_{n-k}^{-1}) * x_n.$$

Clearly this construction can be repeated to obtain higher-order difference
equations in new variables, e.g., $y_n * y_{n-1}^{-1} = x_n$. In particular, if G is commutative then the following equation is *doubly HD1* with respect to the same
group operation

$$y_{n+1} = g_n(y_n * y_{n-1}^{-2} * y_{n-2}, \ldots, y_{n-k+1} * y_{n-k}^{-2} * y_{n-k-1}) * y_n^2 * y_{n-1}^{-1}.$$

The same construction works using *different group operations*. For instance,
if $\mathcal{F}$ is a field and $g_n : \mathcal{F}^{k-1} \to \mathcal{F}$ is any sequence of functions then the
difference equation

$$t_{n+1} = t_{n-j}g_n\left(\frac{t_n}{t_{n-1}}, \ldots, \frac{t_{n-k+2}}{t_{n-k+1}}\right)$$

is HD1 with respect to multiplication for any $0 \le j \le k - 1$, by Proposition
4.1(a). Using the operation $+$ in the preceding construction with $x_n - x_{n-1}$
for t_n yields the following equation

$$x_{n+1} = x_n + (x_{n-j} - x_{n-j-1})g_n\left(\frac{x_n - x_{n-1}}{x_{n-1} - x_{n-2}}, \ldots, \frac{x_{n-k+2} - x_{n-k+1}}{x_{n-k+1} - x_{n-k}}\right)$$

that is HD1 with respect to addition. Reversing the roles of addition and multiplication in the preceding discussion results in additional equations. Clearly these constructions can be repeated to obtain higher-order difference equations. Conversely, if a given equation is multiply HD1 with respect to one or more group operations then its SC factors are also HD1, resulting in a factor chain of length two or greater.

Example 4.8

Consider the following difference equation of order three on the field of all real numbers

$$x_{n+1} = x_n + \frac{a(x_n - x_{n-1})^2}{x_{n-1} - x_{n-2}}, \quad a \neq 0. \tag{4.19}$$

As in the preceding discussion, substituting $t_n = x_n - x_{n-1}$ (or using Theorem 4.1) we obtain the SC factorization

$$t_{n+1} = \frac{a t_n^2}{t_{n-1}}, \quad t_0 = x_0 - x_{-1}, \ t_{-1} = x_{-1} - x_{-2} \tag{4.20}$$

$$x_{n+1} = t_{n+1} + x_n.$$

Note that $t_n \neq 0$ for $n \geq -1$ if initial values satisfy

$$x_0, x_{-2} \neq x_{-1}. \tag{4.21}$$

Relative to the group of all nonzero real numbers under multiplication, the second-order equation (4.20) is evidently HD1. Using Theorem 4.1 (or dividing both sides of this equation by t_n and substituting $r_n = t_n/t_{n-1}$) we obtain the SC factorization

$$r_{n+1} = a r_n, \quad r_0 = \frac{t_0}{t_{-1}}$$

$$t_{n+1} = r_{n+1} t_n.$$

We have therefore obtained a complete factorization of (4.19) into a triangular system of first-order equations:

$$r_{n+1} = a r_n$$

$$t_{n+1} = r_{n+1} t_n$$

$$x_{n+1} = t_{n+1} + x_n.$$

This factorization readily gives the following formula for solutions of (4.19) subject to (4.21):

$$x_n = x_0 + t_0 \sum_{j=1}^{n} r_0^j a^{j(j+1)/2}, \quad r_0 = \frac{t_0}{t_{-1}} = \frac{x_0 - x_{-1}}{x_{-1} - x_{-2}}.$$

We leave the details of calculations to the reader.   ▯

Example 4.9

The following difference equation of order three is easily seen to be HD1 relative to the multiplicative group of nonzero real numbers:

$$x_{n+1} = \frac{x_n^2 x_{n-1}}{a_n x_n x_{n-2} + b_n x_{n-1}^2}, \quad a_n, b_n \in \mathbb{R}, \ a_n \neq 0 \text{ for all } n.$$

From Theorem 4.1 we obtain the SC factorization

$$t_{n+1} = \frac{t_n t_{n-1}}{a_n t_n + b_n t_{n-1}}$$
$$x_{n+1} = t_{n+1} + x_n.$$

The second-order factor equation above is again HD1 relative to the multiplicative group of nonzero real numbers (hence, doubly HD1 relative to the same group) so we get the factorization

$$r_{n+1} = \frac{1}{a_n r_n + b_n}$$
$$t_{n+1} = r_{n+1} t_n.$$

▯

Example 4.10

A linear difference equation can be defined on any nontrivial field $\mathcal{F}$ as

$$x_{n+1} = a_{0,n} x_n + \cdots + a_{k,n} x_{n-k}, \quad a_{j,n} \in \mathcal{F}, \ a_{k,n} \neq 0 \tag{4.22}$$

for all $n \geq 1$, $0 \leq j \leq k$. The SC factorizations of general linear equations with variable coefficients is discussed in Chapter 7. These equations have the HD1 property relative to the group of units $\mathcal{F}\backslash\{0\}$ under multiplication. In general $\mathcal{F}\backslash\{0\}$ is not invariant and solutions of (4.22) may contain zero terms. However, the solutions that do not contain any zero terms can be generated through the factor equation

$$t_{n+1} = a_{0,n} + \frac{a_{1,n}}{t_n} + \frac{a_{2,n}}{t_n t_{n-1}} + \cdots \frac{a_{k,n}}{t_n t_{n-1} \cdots t_{n-k+1}}. \tag{4.23}$$

With $t_n = x_n/x_{n-1}$, each solution of the linear equation (4.22) that does not contain the origin generates a solution of (4.23) and conversely. The singularity set of (4.23) consists of initial values that generate solutions of (4.22) with zero terms.   ▯

Eq. (4.23) may be called the *Riccati difference equation* (or the *discrete Riccati equation*) of order k because of its relationship to the higher-order linear equation. In Chapter 7 we show that Riccati equations are closely related to the SC factorizations of linear equations. Section 5.4 illustrates how the known features of solutions of linear equations on the field $\mathbb{R}$ of real numbers can be used to derive the various interesting properties of the Riccati equation of order two on $\mathbb{R}$.

Example 4.11

The following *linear-like* equations that are defined on $\mathbb{R}$:

$$x_{n+1} = \max\{a_{0,n}x_n, \ldots, a_{k,n}x_{n-k}\}, \tag{4.24}$$

$$x_{n+1} = \min\{a_{0,n}x_n, \ldots, a_{k,n}x_{n-k}\}, \tag{4.25}$$

$$x_{n+1} = |a_{0,n}x_n + \cdots + a_{k,n}x_{n-k}|. \tag{4.26}$$

are readily verified to be HD1 on the multiplicative group $(0, \infty)$ of positive real numbers and their SC factorizations are obtained similarly to the linear equation (4.22).

The following linear-like equations are HD1 on the additive group of all real numbers:

$$x_{n+1} = x_{n-k_1} + \max\{a_n(x_{n-k_2} - x_{n-k_3}), b_n(x_{n-k_4} - x_{n-k_5})\} \tag{4.27}$$

$$x_{n+1} = x_{n-k_1} + \min\{a_n(x_{n-k_2} - x_{n-k_3}), b_n(x_{n-k_4} - x_{n-k_5})\} \tag{4.28}$$

and their SC factorizations are easily determined. In particular, if $k_1 = 0$, $k_3 = k_2 + 1 = j$ and $k_5 = k_4 + 1 = k$ then the factor equation of (4.27) is

$$t_{n+1} = \max\{a_n t_{n-j+1}, b_n t_{n-k+1}\}$$

which is of type (4.24) above. Therefore, Eq. (4.27) is doubly HD1 with respect to the two operations of $\mathbb{R}$. The same comment applies to Eqs. (4.28) and (4.25).

As a final example of linear-like equations, consider the difference equation

$$x_{n+1} = x_n + \min\{a(x_n - x_{n-1}), b(x_{n-k+1} - x_{n-k})\}, \quad a, b > 0, \ k \geq 2 \tag{4.29}$$

on $\mathbb{R}$. Being HD1 under addition the factor equation of (4.29) is found to be

$$t_{n+1} = \min\{a t_n, b t_{n-k+1}\}. \tag{4.30}$$

Eq. (4.30) is in turn HD1 on $(0, \infty)$ under multiplication and its factor equation is found to be

$$r_{n+1} = \min\left\{a, \frac{b}{r_n r_{n-1} \cdots r_{n-k+2}}\right\}.$$

☐

The next two examples discuss the SC factorizations of HD1 equations in noncommutative settings.

Example 4.12

Let G be the group of all invertible $l \times l$ matrices with real entries under matrix multiplication. Consider the matrix equation

$$x_{n+1} = Ax_{n-k+1}x_{n-k}^{-1}x_n, \quad A, x_0, x_{-1}, \ldots, x_{-k} \in G. \tag{4.31}$$

This equation is HD1 (from the right) since with the defining function

$$f(u_0, \ldots, u_k) = Au_{k-1}u_k^{-1}u_0$$

it follows that for every matrix $t \in G$,

$$f(u_0 t, \ldots, u_k t) = Au_{k-1}t(u_k t)^{-1}u_0 t = Au_{k-1}u_k^{-1}u_0 t = f(u_0, \ldots, u_k)t.$$

Now using (4.9) or by rewriting (4.31) as $x_{n+1}x_n^{-1} = Ax_{n-k+1}x_{n-k}^{-1}$, the SC factorization is

$$t_{n+1} = At_{n-k+1}, \quad t_{-j} = x_{-j}x_{-j-1}^{-1}, \ j = 0, 1, \ldots, k-1 \tag{4.32}$$
$$x_{n+1} = t_{n+1}x_n. \tag{4.33}$$

The linear factor equation (4.32) can be solved to obtain

$$t_n = A^m t_i \quad \text{where } m, i \text{ are integers such that} \tag{4.34}$$
$$n = km + i, \ 0 \le i < k.$$

We leave the details of calculation to the reader. Using this solution in the cofactor (4.33) yields the corresponding solution of (4.31). In the special case $k = 1$, $t_n = A^n t_0$ so that we obtain the formula

$$x_n = A^n t_0 A^{n-1} t_0 \cdots At_0 x_0.$$

This solution can be simplified further if the initial value matrix $t_0 = x_0 x_{-1}^{-1}$ commutes with the coefficient matrix A. In this case the solution reduces to

$$x_n = A^{n(n-1)/2}t_0^n x_0.$$

☐

It is worth noting that generating a solution of (4.31) directly by iteration may not be very inspiring unless we can identify and substitute for the form symmetry as we go forward. We also note that rearranging the terms in the right-hand side of (4.31) may result in a non-HD1 equation. For example,

$$x_{n+1} = Ax_n x_{n-k+1}x_{n-k}^{-1}$$

is not HD1 either from the right or from the left (but also see the next example).

Example 4.13
Consider the following variation of (4.31) on the group G of all invertible $l \times l$ matrices with real entries under matrix multiplication:

$$x_{n+1} = A x_n x_{n-k}^{-1} x_{n-k+1}, \quad A, x_0, x_{-1}, \ldots, x_{-k} \in G. \tag{4.35}$$

As in the preceding example, it can be shown that this equation is HD1. We show that (4.35) has two different order reductions. In addition to the type-$(k, 1)$ reduction due to the HD1 property, there is also a type-$(1, k)$ reduction. For the latter, rewriting (4.35) as

$$x_{n+1} x_{n-k+1}^{-1} = A x_n x_{n-k}^{-1}$$

reveals the form symmetry $H(u_0, \ldots, u_k) = u_0 u_k^{-1}$ which is a scalar function on G^{k+1}. The substitution $r_n = x_n x_{n-k}^{-1}$ gives the SC factorization

$$r_{n+1} = A r_n,$$
$$x_{n+1} = r_{n+1} x_{n-k+1}.$$

This type-$(1, k)$ reduction belongs to the variety of order reductions that is discussed in Chapter 6. By way of comparison, the SC factorization for the type-$(k, 1)$ reduction is calculated using (4.9) as

$$t_{n+1} = A t_{n-k+1}^{-1},$$
$$x_{n+1} = t_{n+1} x_n$$

and it has the form symmetry specified in Theorem 4.1.   ⬚

The next example illustrates the importance of the specific group structure in reducing the order of a difference equation.

Example 4.14
Let $\mathbb{Z}_m$ be the finite ring of integers modulo m where both addition and multiplication are modulo m and consider the difference equation

$$x_{n+1} = x_n + (m - 2) p x_n x_{n-1} + p x_n^2 + p x_{n-1}^2, \quad p \in \{0, 1, \ldots, m - 1\}. \tag{4.36}$$

This equation is HD1 relative to the additive group $(\mathbb{Z}_m, +)$. To see this, let

$$f(u_0, u_1) = u_0 + (m - 2) p u_0 u_1 + p u_0^2 + p u_0^2, \tag{4.37}$$

and note that because $m - 2$ is in fact the additive inverse -2 in $\mathbb{Z}_m$, we have

$$\begin{aligned}
f(u_0 + t, u_1 + t) &= u_0 + t + p[(u_0 + t) - (u_1 + t)]^2 \\
&= u_0 + t - 2 u_0 u_1 + p u_0^2 + p u_1^2 \\
&= u_0 + (m - 2) u_0 u_1 + p u_0^2 + p u_1^2 + t \\
&= f(u_0, u_1) + t.
\end{aligned}$$

Thus, using (4.10) or rewriting (4.36) as

$$x_{n+1} - x_n = p(x_n^2 - 2x_n x_{n-1} + x_{n-1}^2) = p(x_n - x_{n-1})^2$$

we obtain the SC factorization

$$t_{n+1} = pt_n^2, \quad t_0 = x_0 - x_{-1}, \qquad (4.38)$$
$$x_{n+1} = t_{n+1} + x_n.$$

Note that this SC factorization is **not** valid in $\mathbb{Z}_j$ if $j \neq m$ and also not valid if (4.36) is taken over rings such as $\mathbb{Z}$ or $\mathbb{R}$. ☐

4.4 *Absolute value equation

In order to illustrate that simple HD1 difference equations are capable of generating a rich variety of solutions on familiar, one-dimensional groups, this section presents a detailed study of the set of solutions of a second-order difference equation that is HD1 on the multiplicative group of positive real numbers. The solutions of this *absolute value equation* can be fruitfully studied using its SC factorization and the resulting pair of first-order equations (the factor and the cofactor). This section contains a substantial amount of detailed analysis and a number of intriguing results but because this material is not essential to understanding SC factorization and reduction of order, this section may be omitted without loss of continuity.

Consider the autonomous, second-order absolute value difference equation

$$x_{n+1} = |ax_n - bx_{n-1}|, \qquad a, b \geq 0, \ a + b > 0, \ n = 0, 1, 2, \dots \qquad (4.39)$$

We may assume without loss of generality that the initial values x_{-1}, x_0 in Eq. (4.39) are nonnegative and for nontriviality, at least one is not zero. This equation is clearly HD1 on the multiplicative group $(0, \infty)$ and its factor equation is easily calculated as

$$r_{n+1} = \left| a - \frac{b}{r_n} \right|, \qquad n = 0, 1, 2, \dots \qquad (4.40)$$

with a cofactor equation

$$x_{n+1} = r_{n+1} x_n. \qquad (4.41)$$

We may think of (4.40) as the one-dimensional recursion $r_{n+1} = \phi(r_n)$ where ϕ is the piecewise smooth mapping

$$\phi(r) = \left| a - \frac{b}{r} \right|, \qquad r > 0.$$

Although (4.39) is a rather simple equation, we see in this section that it exhibits interesting dynamics for certain values of the parameters a, b. By way of comparison, consider other simple HD1 equations such as

$$u_{n+1} = au_n + bu_{n-1}$$
$$v_{n+1} = \max\{av_n, bv_{n-1}\} \qquad (4.42)$$
$$w_{n+1} = \min\{aw_n, bw_{n-1}\}$$

that are related to (4.39) through

$$|\alpha - \beta| = 2\max\{\alpha, \beta\} - (\alpha + \beta) = \max\{\alpha, \beta\} - \min\{\alpha, \beta\}.$$

In contrast to (4.39), each of the equations in (4.42) produces solutions with simple behaviors for all parameter values a, b. Some indication of the major difference between these equations and (4.39) may be had by looking at their semiconjugate factor equations with $r_n = x_n/x_{n-1}$:

$$\phi_u(r) = a + \frac{b}{r}, \quad \phi_v(r) = \max\left\{a, \frac{b}{r}\right\}, \quad \phi_w(r) = \min\left\{a, \frac{b}{r}\right\}. \qquad (4.43)$$

It is evident from a quick examination of these mappings that only the factor ϕ, which was defined for (4.39), is capable of generating complex behavior on the half-line. Indeed, an important feature of ϕ is that it is nonsmooth at its critical point, the minimum value of zero occurring at b/a, whose pre-image is the origin, i.e., the point of discontinuity of ϕ. This source of complexity is lacking in the factors ϕ_u, ϕ_v and ϕ_w all of which are monotonically nonincreasing maps.

The region $b \leq a^2/4$ and the fixed points of ϕ

The mapping ϕ always has a fixed point $\bar{r}$ through its *left half*

$$\phi_L(r) = \frac{b}{r} - a, \quad 0 < r \leq \frac{b}{a}$$

(see Figure 4.3). Solving $\phi_L(r) = r$ gives the value of $\bar{r}$ as

$$\bar{r} = \frac{1}{2}\left[\sqrt{a^2 + 4b} - a\right].$$

It may be noted that $\bar{r}$ is an unstable fixed point for all positive values of a, b, since

$$|\phi_L'(\bar{r})| = \frac{b}{\bar{r}^2} = \frac{b}{b - a\bar{r}} > 1.$$

The *right half* of ϕ, namely,

$$\phi_R(r) = a - \frac{b}{r}, \quad r \geq \frac{b}{a}$$

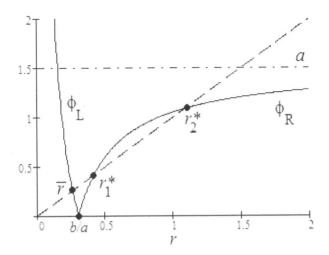

FIGURE 4.3
A graph of the factor function ϕ, its left and right halves and its fixed points when $b \leq a^2/4$.

can have up to two fixed points if the equation $\phi_R(r) = r$ has positive solutions; see Figure 4.3. Solving the latter equation yields two values

$$r_1^* = \frac{1}{2}\left[a - \sqrt{a^2 - 4b}\right], \quad r_2^* = \frac{1}{2}\left[a + \sqrt{a^2 - 4b}\right]$$

both of which are real if and only if $b \leq a^2/4$. Stated differently, ϕ has up to two additional fixed points for all points (a, b) on or below the parabola $b = a^2/4$ in the parameter plane. On this parabola,

$$r_1^* = r_2^* = \frac{a}{2}$$

and the single fixed point of ϕ is semi-stable. In the sub-parabolic region $b < a^2/4$ we find that

$$\frac{b}{a} < r_1^* < r_2^*$$

with r_1^* unstable and r_2^* asymptotically stable as may be readily verified, e.g., by calculating the derivative values $\phi_R'(r_1^*), \phi_R'(r_2^*)$. The tangent bifurcation of the factor map ϕ that produces r_1^*, r_2^* when a parameter point (a, b) crosses the parabola $b = a^2/4$ corresponds, via the cofactor equation (4.41), to significant behavioral changes in the solutions of (4.39) as well.

Figure 4.4 shows the parameter plane. We begin our examination of solutions when (a, b) is in the sub-parabolic region $b \leq a^2/4$ of the parameter plane with the following lemma whose straightforward proof is omitted.

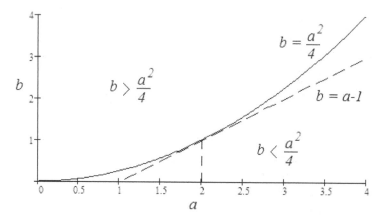

FIGURE 4.4
The (a, b) parameter plane and regions defined by the parabola $b = \frac{a^2}{4}$ and the line $b = a - 1$; see the text for details.

LEMMA 4.1
Let $b \leq a^2/4$.
(a) $r_2^* \leq a$ with equality holding if and only if $b = 0$.
(b) Let $a < 2$. Then $r_1^* < 1$; also $r_2^* < 1$ if and only if $b > a - 1$. Further, $r_2^* = 1$ if $b = a - 1$.
(c) Let $a \geq 2$. Then $r_2^* \geq 1$; also $r_1^* \leq 1$ if and only if $b \leq a - 1$. Further, $r_1^* = 1$ if $b = a - 1$.

In Lemma 4.1 notice that the line $b = a - 1$ in the parameter plane is tangent to the parabola $b = a^2/4$ at $a = 2$; see Figure 4.4.

LEMMA 4.2
Let $b \leq a^2/4$ and let $\{x_n\}$ be a solution of (4.39) such that $x_k \geq r_1^* x_{k-1}$ for some integer $k \geq 0$.
(a) If $b \neq a^2/4$ then there are real constants c_1, c_2 such that

$$x_n = c_1(r_1^*)^n + c_2(r_2^*)^n, \quad n \geq k. \tag{4.44}$$

(b) If $b = a^2/4$ then there are real constants c_1, c_2 such that

$$x_n = (c_1 + c_2 n)(a/2)^n, \quad n \geq k \tag{4.45}$$

with $r_1^* = r_2^* = a/2$.

PROOF   By hypothesis, $r_k = x_k/x_{k-1} \geq r_1^*$. If $r_n = \phi(r_{n-1})$ for $n > k$, then since $r_n \geq r_1^*$ for $n \geq k$, we may write

$$\frac{x_{n+1}}{x_n} = r_{n+1} = \phi_R(r_n) = a - \frac{bx_{n-1}}{x_n}$$

i.e., $x_{n+1} = ax_n - bx_{n-1}$ for $n \geq k$. The eigenvalues of the preceding linear equation are none other than r_1^* and r_2^* so (4.44) or (4.45) holds as appropriate for $n \geq k$ and suitable constants c_1, c_2. ∎

THEOREM 4.2
Let $a < 2$.

(a) If $a - 1 < b \leq a^2/4$ then every nontrivial solution of Eq. (4.39) converges to zero eventually monotonically.

(b) If $b = a - 1$ then every solution of (4.39) converges to a nonnegative constant eventually monotonically.

(c) If $b < a - 1$ with $a > 1$ then any solution $\{x_n\}$ of Eq. (4.39) for which $x_k > r_1^* x_{k-1}$ for some integer $k \geq 0$ is unbounded and strictly increasing eventually. On the other hand, if a solution of (4.39) is bounded, then it is strictly decreasing to zero; such solutions do exist.

PROOF (a) First, let $\{x_n\}$ be a positive solution of (4.39), i.e., $x_n > 0$ for $n \geq -1$. Then the sequence $\{r_n\}$ where $r_n = \phi(r_{n-1})$ and $r_0 = x_0/x_{-1}$ is well defined for all n. If $x_n < r_1^* x_{n-1}$ for all $n \geq 0$ then $x_n \to 0$ as $n \to \infty$ because $r_1^* < 1$ by Lemma 4.1. Otherwise, $x_k \geq r_1^* x_{k-1}$ for some $k \geq 0$ so by Lemma 4.2, (4.44) or (4.45) holds as appropriate and since $r_2^* < 1$ the proof is completed. Next, suppose that $x_m = 0$ for some $m \geq -1$. Then $x_{m+2} = ax_{m+1}$ and therefore,

$$r_{m+2} = \frac{x_{m+2}}{x_{m+1}} = a.$$

By Lemma 4.1, $r_2^* \leq a$ so $r_{m+2} \geq r_2^* > r_1^*$. Letting $k = m + 2$ and applying the above argument once more we obtain (4.44) or (4.45) and complete the proof.

(b) Here $r_2^* = 1$ and $r_1^* < 1$; the proof is similar to that for Part (a).

(c) With $r_2^* > 1 > r_1^*$, if $\{x_n\}$ is a solution with $r_k = x_k/x_{k-1} > r_1^*$ for some k, then apply Lemma 4.2 with (4.44) to show that $\{x_n\}$ is unbounded and eventually increasing. This argument also shows that if $\{x_n\}$ is bounded, then it has to be strictly decreasing with $x_n \leq r_1^* x_{n-1}$ for all $n \geq 0$. It is clear in this case that $x_n \to 0$ as $n \to \infty$. Finally, to verify that the latter type of solution does exist, let $x_0/x_{-1} = r_0$ belong to either of the sets

$$B = \bigcup_{i=0}^{\infty} \phi^{-i}(\bar{r}), \qquad B_1 = \bigcup_{i=0}^{\infty} \phi^{-i}(r_1^*)$$

of backward iterates of $\bar{r}$ or r_1^* respectively under the inverse map ϕ^{-1}. Then $r_n = \bar{r} < 1$ or $r_n = r_1^* < 1$ for all large n and x_n converges to zero exponentially. ∎

LEMMA 4.3
(a) $\bar{r} \geq 1$ if and only if $b \geq a+1$ with $\bar{r} = 1$ if $b = a+1$.
(b) If $\bar{r} \geq 1$ and $b \leq a^2/4$ then $a \geq 2 + \sqrt{8} \approx 4.828$.

Note that the line $b = a+1$ in the parameter plane intersects the boundary parabola at $a = 2 + \sqrt{8}$. Further, on both of the lines $b = a \pm 1$, (4.39) is degenerate in the sense that if $x_{-1} = x_0$ then $x_n = x_0$ for all $n \geq 0$. In particular, the origin is not the only fixed point of (4.39).

THEOREM 4.3
Let $a \geq 2$ and $b \leq a^2/4$.
(a) Any solution $\{x_n\}$ of(4.39) for which $x_k > r_1^* x_{k-1}$ for some integer $k \geq 0$ is unbounded and strictly increasing eventually.
(b) If $b < a + 1$, then there are solutions of (4.39) that converge to zero eventually monotonically, and if $b = a \pm 1$, then there are also solutions that are eventually constant and positive.

PROOF   (a) This is proved similarly to Theorem 4.2(c), the only difference being that here it is possible that $r_1^* \geq 1$.

(b) If $b < a + 1$, then by Lemma 4.3, $\bar{r} < 1$ so if $x_0/x_{-1} = r_0 \in B$ where B is the set of backward iterates defined in the proof of Theorem 4.2(c), then $x_n \to 0$ eventually monotonically. If $b = a + 1$, then $\bar{r} = 1$ so $r_0 \in B$ implies that $r_n = 1$ for all large n and thus x_n is eventually constant. For $b = a - 1$ it is the case that $r_1^* = 1$ so letting $r_0 \in B_1$ we obtain an eventually constant solution. ∎

The region $b > a^2/4$ and the periodic points of ϕ

In this region, fixed points such as r_1^* and r_2^* do not exist; see Figure 4.5.

The absence of such fixed points mandates the use of different tools and methods. In particular, periodic solutions present themselves as interesting substitutes for the missing fixed points. We observe that a positive, p-periodic solution $\{x_n\}$ of (4.39) induces a periodic ratio sequence $\{x_n/x_{n-1}\}$ of the same period that satisfies the identity

$$\prod_{i=1}^{p} \frac{x_i}{x_{i-1}} = 1.$$

It follows that positive, period-p solutions of (4.39) correspond in a one-to-one fashion to the p-cycles $\{r_n\}$ of (4.40) with the property that $\prod_{i=1}^{p} r_i = 1$.

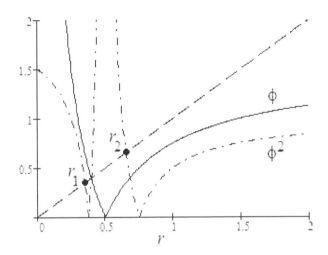

FIGURE 4.5
A graph of the factor function ϕ when $b > a^2/4$. The fixed points r_1^* and r_2^* are absent now but periodic points exist; e.g., r_1, r_2 that are the fixed points of the iterate ϕ^2.

We begin with the next lemma, which shows in particular that unlike the fixed points r_1^* and r_2^*, we can always expect to find 2-cycles for (4.40) for all points in the parameter plane.

LEMMA 4.4
(a) For fixed $a, b > 0, (4.40)$ has a unique 2-cycle $\{r_1, r_2\}$ where

$$r_1 = \frac{b}{a} - \frac{\sqrt{a^4 + 4b^2} - a^2}{2a}, \quad r_2 = \frac{b}{a} + \frac{\sqrt{a^4 + 4b^2} - a^2}{2a}. \tag{4.46}$$

(b) For fixed $a^2 > b > 0, (4.40)$ has a unique 3-cycle $\{r_1, r_2, r_3\}$ where

$$r_2 = \phi_L(r_1) < \frac{b}{a}, \ r_3 = \phi_L(r_2) > \frac{b}{a}, \ r_1 = \phi_R(r_3) < \frac{b}{a}$$

and r_1 is given by

$$r_1 = \frac{a(a^2 + 3b) - \sqrt{a^2(a^2 + b)^2 - 4b^3}}{2(a^2 + b)}. \tag{4.47}$$

PROOF   (a) If $\{r_1, r_2\}$ is a 2-cycle of (4.40) with $r_1 < r_2$, then it is the case that $r_1 < b/a < r_2$. This is true because ϕ_R is increasing and also $\phi_L^2(r) = r$ implies that $r = \bar{r}$. Hence both r_1 and r_2 cannot be on one side of b/a. Now setting $r_2 = \phi_L(r_1)$ and $r_1 = \phi_R(r_2)$ gives

$$r_1 = a - \frac{b}{\phi_L(r_1)} = a - \frac{br_1}{b - ar_1} = \frac{ab - (a^2 + b)r_1}{b - ar_1}.$$

Solving the above equation for r_1 we obtain the value in (4.46) as the only solution in the interval $(0, b/a)$. Then $r_2 = \phi_L(r_1)$ is the other value (4.46).

(b) We calculate the 3-cycle indicated in the statement of the lemma similarly to Part (a), although the algebraic manipulations are more extensive. The unique value for $r_1 \in (0, b/a)$ in the form (4.47) is defined only when the expression under the square root is nonnegative; i.e., r_1 exists if and only if

$$a(a^2 + b) \geq 2b^{3/2}.$$

This inequality can be written as a cubic polynomial inequality in a

$$a^3 + ab - 2b\sqrt{b} \geq 0.$$

Noting that $a = \sqrt{b}$ is a root of the polynomial on the left-hand side, we have a factorization

$$(a - \sqrt{b})(a^2 + a\sqrt{b} + 2b) \geq 0.$$

Since the quadratic factor is positive for $a, b > 0$ we conclude that r_1 (hence also the 3-cycle) exist when $a \geq \sqrt{b}$ or equivalently, $a^2 \geq b$. If the equality holds, then we calculate from (4.47)

$$r_1 = a = \frac{b}{a}$$

where the last equality is equivalent to $b = a^2$. However, b/a maps to zero and cannot be a periodic point of ϕ; therefore, the strict inequality $a^2 > b$ must hold.   ∎

In Lemma 4.4 we point out that $b > 0$ because it is necessary that $r_1 > 0$ in both (4.46) and (4.47). We need one more lemma before presenting our main theorem on periodicity. The next lemma concerns solutions that contain zeros, i.e., they "pass through the origin" repeatedly. Note that such solutions cannot be represented by the mapping ϕ.

LEMMA 4.5
Assume that $b > 0$ and let $\{x_n\}$ be a solution of (4.39) such that $x_k = 0$ for some $k \geq 0$ and $x_{k-1} > 0$.
(a) If $x_{k+2} = 0$, then $a = 0$ and $x_{k+2n-1} = b^n x_{k-1}$ for $n = 1, 2, 3, \ldots$
(b) If $x_{k+3} = 0$, then $a^2 = b$ and $x_{k+n} = a^{n+1} x_{k-1}$ for $n \neq 3j$, $j = 1, 2, 3, \ldots$

PROOF   (a) We have $x_{k+1} = bx_{k-1} > 0$ and $0 = x_{k+2} = ax_{k+1}$. Thus $a = 0$ and the stated formula is easily established by induction on n.

(b) Since $x_{k+3} = |ax_{k+2} - bx_{k+1}| = 0$, it follows that $ax_{k+2} = bx_{k+1} \neq 0$. In particular, $a \neq 0$ and $x_{k+2} \neq 0$. In fact,

$$x_{k+2} = ax_{k+1} = abx_{k-1}$$

so

$$0 = x_{k+3} = b|a^2 - b|x_{k-1}$$

and it follows that $b = a^2$. We may now write $x_{k+1} = a^2 x_{k-1}$ and $x_{k+2} = a^3 x_{k-1}$. Further,

$$x_{k+4} = bx_{k+2} = a^5 x_{k-1},$$
$$x_{k+5} = ax_{k+4} = a^6 x_{k-1}.$$

Induction on n now completes the proof. ∎

THEOREM 4.4

(a) Eq.(4.39) has a period-2 solution if and only if

$$b^2 - a^2 = 1 \text{ or equivalently } b = \sqrt{a^2 + 1} \text{ for } a \geq 0. \tag{4.48}$$

Further, if $a > 0$, then the period-2 solutions are positive and confined to the pair of lines $y = r_1 x$ and $y = r_2 x$ in the state-space (or the phase plane), where r_1, r_2 are given by

$$r_1 = \frac{b-1}{a}, \quad r_2 = \frac{b+1}{a} \tag{4.49}$$

(see Figure 4.7). On the other hand, the only period-2 solutions of (4.39) that pass through the origin occur at $a = 0$ where $b = 1$.

(b) Eq.(4.39) has a period-3 solution if and only if

$$a^3 + ab - b^3 = 1, \quad a \geq 1. \tag{4.50}$$

Further, if $a > 1$, then the period-3 solutions are positive and confined to the three lines $y = r_i x$ in the state-space (or the phase plane) where for $i = 1, 2, 3$, r_i are given by

$$r_1 = \frac{ab+1}{a^2+b}, \quad r_2 = \frac{b^2-a}{ab+1}, \quad r_3 = \frac{b+a^2}{b^2-a} \tag{4.51}$$

(see Figure 4.8). On the other hand, the only period-3 solutions of (4.39) that pass through the origin occur at $a = 1$ where $b = 1$ also.

PROOF (a) First, let us assume that $a > 0$ and use (4.46) to compute

$$r_1 r_2 = \frac{1}{2} \left[\sqrt{a^4 + 4b^2} - a^2 \right]. \tag{4.52}$$

Setting $r_1 r_2 = 1$ and simplifying, we obtain the hyperbola (4.48) in the parameter plane. Next, using (4.48) we may simplify the formulas in (4.46) to obtain the values in (4.49). For parameter values on the quadratic curve (4.48), if x_0/x_{-1} equals either r_1 or r_2 then $\phi(x_0/x_{-1}) = r_2$ and $\phi(x_1/x_0) = r_1$ or conversely, so the corresponding solution $\{x_n\}$ of (4.39) with period 2 is confined to the lines $y = r_1 x$ and $y = r_2 x$ in phase space.

Now consider the case $a = 0$ in (4.48). Although r_1, r_2 are not defined, it is clear that for $(a, b) = (0, 1)$ every solution of (4.39) has period 2 and by Lemma 4.5(a) these include all the period-2 solutions that pass through the origin. Since every positive period-2 solution of (4.39) induces a 2-cycle in the ratios, it follows from Lemmas 4.4 and 4.5 that there are no other period-2 solutions of (4.39) than the ones already mentioned.

(b) With r_1 given by (4.47) and

$$r_2 = \phi_L(r_1) = \frac{b - a r_1}{r_1}, \quad r_3 = \phi_L(r_2) = \frac{(a^2 + b)r_1 - ab}{b - a r_1}$$

we see that

$$r_1 r_2 r_3 = (a^2 + b)r_1 - ab = \frac{1}{2}\left[a(a^2 + b) - \sqrt{a^2(a^2 + b)^2 - 4b^3}\right]. \quad (4.53)$$

Setting $r_1 r_2 r_3 = 1$ and re-arranging terms, gives the equation

$$\sqrt{a^2(a^2 + b)^2 - 4b^3} = a(a^2 + b) - 2 \quad (4.54)$$

which can hold only when (i) $a(a^2 + b) - 2 \geq 0$ or equivalently, $b \geq 2/a - a^2$ and (ii) $b < a^2$ so that the square root is real and $r_1 > 0$. The curve $2/a - a^2$ is strictly decreasing and intersects the parabola $b = a^2$ at $a = 1$. It follows that (4.54) holds for $a > 1$. Now, if we square both sides of (4.54) and simplify, we get the cubic equation in (4.50). This equation in turn may be used to simplify the values r_1, r_2, r_3 in Lemma 4.4(b) to get the values in (4.51). It follows from Lemma 4.4 that positive period-3 solutions of (4.39) exist if and only if the parameters a, b satisfy (4.50) with $a > 1$ and in this case the solution in the phase space is confined to the three straight lines mentioned in the statement of the theorem.

Next, consider $a = 1$ in (4.50). Here $b = 1$ also so Lemma 4.5(b) shows that all period-3 solutions that pass through the origin are accounted for. ∎

In Figure 4.6 we see that the quadratic curve (4.48) intersects the cubic curve (4.50) in the (a, b) parameter plane at a point where the given value of, say, a generates both a 2-cycle and a 3-cycle of ϕ at the *same* value of b. At other points on each of the two curves, the value of b generating a 2-cycle is different from that generating a 3-cycle for the same value of a. The coordinates of the intersection point seen in Figure 4.6 are easily determined by inserting $b = \sqrt{a^2 + 1}$ in (4.50) and solving for a as follows:

$$a^3 + a\sqrt{a^2 + 1} - (a^2 + 1)\sqrt{a^2 + 1} = 1$$

$$(a^2 - a + 1)\sqrt{a^2 + 1} = a^3 - 1.$$

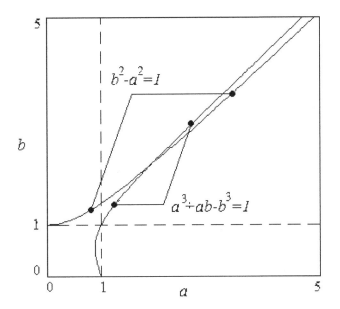

FIGURE 4.6

Curves of periodic points of ϕ in the (a, b) parameter plane; see Theorem 4.4.

Taking note of the fact that $a > 1$ and $a^2 - a + 1 > 0$ and squaring both sides yields

$$(a^4 - 2a^3 + 3a^2 - 2a + 1)(a^2 + 1) = a^6 - 2a^3 + 1. \tag{4.55}$$

Now simplifying the above expression we get

$$a^4 - 2a^3 + a^2 - 2a + 1 = 0. \tag{4.56}$$

Comparing the left-hand side of this equality with the similar expression on the left of (4.55) shows that (4.56) is equivalent to

$$(a^2 - a + 1)^2 - 2a^2 = 0.$$

From this we readily obtain

$$a = \frac{1}{2}\left(\sqrt{2} + 1 + \sqrt{2\sqrt{2} - 1}\,\right) \approx 1.8832$$

and

$$b = \sqrt{a^2 + 1} \approx 2.1322.$$

Figures 4.7 and 4.8 illustrate the 2-cycles and 3-cycles in Theorem 4.4 using the preceding values of a, b at the intersection point.

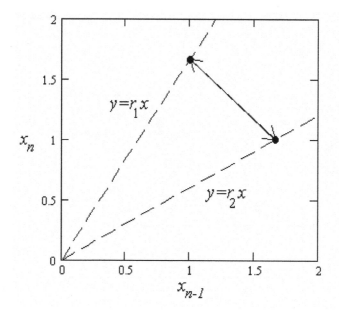

FIGURE 4.7

A 2-cycle of the absolute-value equation with its locus of two lines in
state-space; see the text for parameter values.

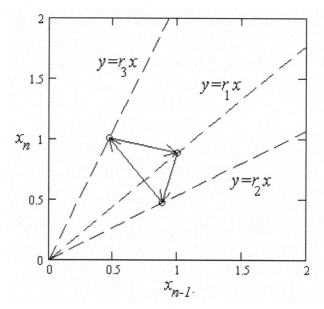

FIGURE 4.8

A 3-cycle of the absolute-value equation with its locus of three lines in
state-space; see the text for parameter values.

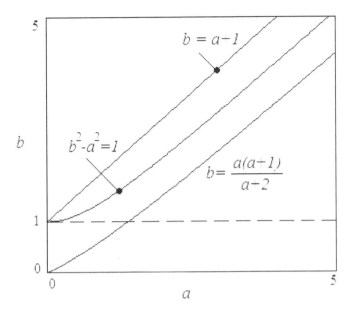

FIGURE 4.9
The various curves in the parameter plane discussed in Theorem 4.5.

As the preceding results show, periodic ratios are useful in answering various questions about the dynamics of (4.39). The next result is another example of this usage.

THEOREM 4.5

Let $b > a^2/4$.

(a) If $b \leq (a^2 + a)/(a + 2)$ then there are solutions of (4.39) that converge to zero eventually monotonically.

(b) If $(a^2 + a)/(a + 2) < b < \sqrt{a^2 + 1}$ then for all $a \geq 0$ there are solutions of (4.39) that converge to zero in an oscillatory fashion.

(c) If $b > \sqrt{a^2 + 1}$ then for all $a \geq 0$ there are solutions of (4.39) that are oscillatory and unbounded. However, if $b < a + 1$, then there are also solutions that converge to zero monotonically (eventually monotonically if $b < 2a^2$).

(d) If $b = a$, then for all $a > 0$ there are solutions of (4.39) that converge to zero in an oscillatory fashion as well as solutions that converge to zero eventually monotonically.

PROOF   (a) and (b): Setting $r_2 \leq 1$ in (4.46) and simplifying gives

$$b \leq \frac{a^2 + a}{a + 2} \qquad (4.57)$$

with equality holding if and only if $r_2 = 1$. Using (4.48) we find that $r_1 r_2 < 1$ if and only if $b < \sqrt{a^2 + 1}$. Further, for $a > 0$, we have $\sqrt{a^2 + 1} > a$ whereas $(a^2 + a)/(a + 2) < a$. It follows that the region in Part (b) is nonempty and in there, $r_1 r_2 < 1$ but $r_2 > 1$. Hence, if e.g., $x_0/x_{-1} = r_1$ then the ratio sequence is 2-periodic. In this case, $x_1 = r_2 x_0 > x_0$ while $x_2 = r_1 r_2 x_0 < x_0$. Inductively, for all $n \geq 1$, it follows that

$$x_{2n-1} > r_2 r_1 x_{2n-1} = r_2 x_{2n} = x_{2n+1} > x_{2n} > r_1 r_2 x_{2n} = x_{2n+2}.$$

It is clear that $\{x_n\}$ in this case converges to zero in an oscillatory fashion. This proves statement (b); we also note that the region of the parameter plane in this case contains a part of the unit square. On the other hand, if (4.57) holds, then both r_2 and $r_1 r_2$ are less than unity and the solution $\{x_n\}$ with $x_0/x_{-1} = r_1$ converges to zero monotonically. This concludes the proof of Part (a).

(c) In this case $r_1 r_2 > 1$ so solutions of (4.39) with $x_0/x_{-1} = r_1$ are unbounded and they oscillate because $r_1 < 1$. In fact, this last inequality holds provided that $b > (a^2 - a)/(2a - 1)$ which is true because

$$(a^2 - a)/(2a - 1) = a[(a - 1)/(2a - 1)] < a < \sqrt{a^2 + 1} < b.$$

If $b < a + 1$, then by Lemma 4.3, $\bar{r} < 1$ so if $x_0 = \bar{r} x_{-1}$ then $x_n = \bar{r} x_{n-1}$ for all n and $\{x_n\}$ converges to zero monotonically. If also $b < 2a^2$ (possible for $a > [(1 + \sqrt{17})/8]^{1/2}$) then we find that $\bar{r} < a$ so that $\phi^{-1}(\bar{r})$ is not empty. In this case, we can find solutions $\{x_n\}$ of (4.39) where there is $k \geq 0$ such that $x_k = \bar{r} x_{k-1}$ and thus $\{x_n\}$ converges to zero eventually monotonically. Note that these conclusions extend similar ones made in Theorem 4.3(b) for the region $b \leq a^2/4$.

(d) The line $b = a$ in the parameter space lies within the region in Part (b) so there are solutions of (4.39) in this case that converge to zero in an oscillatory fashion. Further, $b < a + 1$ so $\bar{r} < 1$ and if $\{x_n\}$ is a solution of (4.39) with x_0/x_{-1} in the set B of backward iterates of $\bar{r}$ under ϕ, then $x_n = \bar{r} x_{n-1}$ for all sufficiently large n. Clearly, such $\{x_n\}$ converges to zero eventually monotonically. ∎

If we raise the lower limit in Theorem 4.5(c) to $b > a + 1$ the following stronger conclusion is obtained.

THEOREM 4.6
Let $b > a^2/4$. If $b > a + 1$, then every solution of (4.39) is unbounded. If a solution $\{x_n\}$ has the property that $x_k \leq x_{k-1}$ for some $k \geq 0$ then $\{x_n\}$ is

unbounded and oscillatory. There are also unbounded solutions of (4.39) that are monotonically increasing.

PROOF Since the case $a = 0$ is discussed in Lemma 4.5, we assume from here on that $a > 0$. Define $c = b - a$ and note that $c > 1$. Let $\{x_n\}$ be a nonzero solution of (4.39) where there is $k \geq 0$ such that $x_k \leq x_{k-1}$. Then

$$x_{k+1} = |bx_{k-1} - ax_k| = |cx_{k-1} - a(x_k - x_{k-1})| \geq cx_{k-1} \geq cx_k > x_k. \quad (4.58)$$

Notice that a decrease is always followed by an increase in such a way that the post-decrease high value $x_{k+1} \geq cx_{k-1}$ with x_{k-1} representing the pre-decrease high.

Next, $x_{k+2} = |cx_k - a(x_{k+1} - x_k)|$ and here are two possible cases: If

$$x_{k+2} = cx_k - a(x_{k+1} - x_k) \quad (4.59)$$

then clearly $x_{k+2} < cx_k \leq cx_{k-1} \leq x_{k+1}$ which implies that

$$x_{k+3} = |cx_{k+1} - a(x_{k+2} - x_{k+1})| > cx_{k+1} > x_{k+1} > x_{k+2}.$$

Therefore, the situation in (4.58) is repeated with the new low and high values x_{k+2} and x_{k+3}. If (4.59) does not hold, then its negative holds for possibly more than one index value:

$$x_{k+j} = a(x_{k+j-1} - x_{k+j-2}) - cx_{k+j-2} = ax_{k+j-1} - bx_{k+j-2}, \quad j = 2, 3, 4 \ldots \quad (4.60)$$

Eq. (4.60) is linear with complex eigenvalues since $b > a^2/4$. If $\lambda^{\pm}$ are these eigenvalues, then $|\lambda^{\pm}| = \sqrt{b} > 1$ if $b > a + 1$. It follows that there is an integer $j_{\max}(a, b) \geq 2$ such that

$$x_{k+j_{\max}+1} \leq x_{k+j_{\max}}$$

and once again a situation analogous to that in (4.59) is obtained. We have shown that after at most a finite number of terms (determined only by the values of a, b) there will be a drop in the value of x_n for all n, i.e., $\{x_n\}$ is not eventually monotonic. It follows that $\{x_n\}$ oscillatory. Further, If $\{n_m\}$ where $m = 0, 1, 2, \ldots$ is the sequence of indicies at which drops in x_n occur, then $x_{-1+n_m} \geq x_{1+n_{m-1}}$, so from the preceding arguments and (4.58) it may be concluded that

$$x_{1+n_m} \geq cx_{-1+n_m} \geq cx_{1+n_{m-1}}.$$

Therefore, inductively

$$x_{1+n_m} \geq cx_{1+n_{m-1}} > c^2 x_{1+n_{m-2}} > \cdots > c^m x_{1+n_0}$$

from which we may conclude that $\{x_n\}$ is unbounded.

Finally, if $b > a + 1$ then any strictly increasing solution of (4.39) must approach infinity, since the origin is the only possible fixed point of (4.39).

Such strictly increasing solutions do exist; for example, since $\bar{r} > 1$ by Lemma 4.3, a solution $\{x_n\}$ with $x_0/x_{-1} = \bar{r}$ converges to infinity monotonically.

We note here that if $a + 1 < b \leq a^2/4$, then by Theorem 4.3 the solutions of (4.39) are generally unbounded, approaching infinity eventually monotonically.

The special role of period 3 in the well-known Li-Yorke Theorem gives the following result on the complex nature of the mapping ϕ. The significant consequence of the following corollary and its extensions is that regardless of whether a solution $\{x_n\}$ of (4.39) converges or not, it may oscillate in a complicated way.

THEOREM 4.7

Let $b \leq a^2$. Then the mapping ϕ is chaotic in the sense of Li and Yorke; i.e., it has an uncountable set S called a scrambled set with the following properties:
(i) S contains no periodic points of ϕ and $\phi(S) \subset S$;
(ii) For every $x, y \in S$ and $x \neq y$,

$$\limsup_{k \to \infty} \left\| \phi^k(x) - \phi^k(y) \right\| > 0, \quad \liminf_{k \to \infty} \left\| \phi^k(x) - \phi^k(y) \right\| = 0.$$

(iii) For every $x \in S$ and periodic y,

$$\limsup_{k \to \infty} \left\| \phi^k(x) - \phi^k(y) \right\| > 0.$$

PROOF   Let $[0, \infty]$ be the one-point compactification of the closed half-line $[0, \infty)$ and extend ϕ continuously to $[0, \infty]$ as follows:

$$\phi^*(0) = \infty, \quad \phi^*(\infty) = a, \quad \phi^*(r) = \phi(r) \text{ if } r \in (0, \infty).$$

Then ϕ^* defines a continuous dynamical system on the compact interval $[0, \infty]$. Now, if $b < a^2$ then by Lemma 4.4(b) ϕ has a 3-cycle, which is also clearly a 3-cycle for ϕ^*. Also if $b = a^2$ then since

$$a = \frac{b}{a} \xrightarrow{\phi} 0 \xrightarrow{\phi^*} \infty \xrightarrow{\phi^*} a$$

it follows that ϕ^* again has a 3-cycle, namely, $\{a, 0, \infty\}$. In either case, since $[0, \infty]$ is homeomorphic to $[0,1]$, by the Li-Yorke theorem ϕ^* has a scrambled set $S^* \subset [0, \infty]$. Define

$$S = S^* - \left[\{\infty\} \cup \bigcup_{n=0}^{\infty} \phi^{-n}(0) \right].$$

Since $\cup_{n=0}^{\infty} \phi^{-n}(0)$ is countable for each n, it follows that S is an uncountable subset of $(0, \infty) \cap S^*$ which satisfies conditions (i)–(iii) above because

$$\phi|_S = \phi^*|_S.$$

Hence, S is a scrambled set for ϕ as required. ∎

The unit square $a, b \leq 1$ and stability of the origin

We now consider the parameter values (a, b) in the unit square of the parameter plane where some additional features exist.

LEMMA 4.6
If $a, b < 1$ then the origin is a globally asymptotically stable fixed point of (4.39).

PROOF Define $f(x, y) = |ax - by|$ and note that for $x, y \in [0, \infty)^2$ we have

$$f(x, y) \leq \max\{ax, by\} \leq \max\{a, b\} \max\{x, y\}.$$

Hence f is a weak contraction on $[0, \infty)^2$. The proof is completed upon recalling that for each solution $\{x_n\}$ of (4.39) $x_n \geq 0$ for all $n \geq 1$. ∎

LEMMA 4.7
Let $0 < a, b \leq 1$ and either $a = 1$ or $b = 1$. For each solution $\{x_n\}$ of (4.39) if $x_k > x_{k-1} > 0$ for some $k \geq 0$ then $x_n < x_k$ for all $n > k$.

PROOF First, let $b = 1$ and $0 < a \leq 1$. If $ax_k > x_{k-1}$ then

$$x_{k+1} = ax_k - x_{k-1} < ax_k \leq x_k. \tag{4.61}$$

From this we infer that $ax_{k+1} < ax_k \leq x_k$ so

$$x_{k+2} = x_k - ax_{k+1} < x_k. \tag{4.62}$$

Now from (4.61) and (4.62) it follows that

$$x_{k+3} \leq \max\{ax_{k+2}, x_{k+1}\} < x_k.$$

The last step is easily extended by induction to all $n > k$. Next, assume that $ax_k \leq x_{k-1}$. Then

$$x_{k+1} = x_{k-1} - ax_k < x_{k-1} < x_k$$

and it follows that $ax_{k+1} < ax_k \leq x_k$. This again implies (4.62) and we argue as before.

Now, assume that $a = 1$ and $0 < b \leq 1$. Since $x_k > x_{k-1} \geq bx_{k-1} > 0$ by hypothesis, it follows that

$$0 < x_{k+1} = x_k - bx_{k-1} < x_k.$$

If $x_{k+1} \geq bx_k$, then $x_{k+2} = x_{k+1} - bx_k < x_{k+1}$ so that

$$x_{k+3} \leq \max\{x_{k+2}, bx_{k+1}\} < x_k.$$

Inductively, $x_n > x_k$ for $n > k$. If $x_{k+1} < bx_k$ then

$$x_{k+2} = bx_k - x_{k+1} < bx_k \leq x_k$$

which implies $x_{k+3} < x_k$ and by induction the proof is completed.   ∎

THEOREM 4.8
For all $(a, b) \in [0, 1]^2$ except at the three boundary points (1,1), (1,0) and (0,1), the origin is a globally asymptotically stable fixed point of Eq. (4.39).

PROOF   In light of Lemma 4.6 it remains only to consider the boundaries where either $a = 1$ or $b = 1$. For such points (a, b), first assume that $0 < ab < 1$ and let $\{x_n\}$ be a solution of (4.39). Either $x_n < x_{n-1}$ for all n in which case x_n converges to zero monotonically, or there is $k_1 \geq 0$ such that $x_{k_1} > x_{k_1-1} > 0$. In the latter case, Lemma 4.7 implies that $x_n < x_{k_1}$ for all $n > k_1$. If the sequence $\{x_n\}$ is not eventually decreasing, then there is an increasing sequence k_i of positive integers such that

$$x_{k_1} > x_{k_2} > \cdots > x_{k_i} > \cdots$$

and for $i = 1, 2, 3, \ldots$

$$x_n < x_{k_i} \quad \text{if} \quad k_i < n \leq k_{i+1}.$$

These facts imply that $x_n \to 0$ as $n \to \infty$. Stability of the origin follows from Lemma 4.7 also since $x_n \leq \max\{x_0, x_{-1}\}$ for all $n \geq 1$.   ∎

Let us now examine the three boundary points mentioned in the statement of Theorem 4.8. At $(a, b) = (1, 0)$, Eq. (4.39) reduces to the trivial equation $x_{n+1} = x_n$ whose solutions are constants given by the values of x_0. At $(a, b) = (0, 1)$ each solution of $x_{n+1} = x_{n-1}$ trivially has period 2: $\{x_{-1}, x_0\}$. What about the parameter point $(a, b) = (1, 1)$?

Eq. (4.39) exhibits rather interesting properties when $a = b = 1$ that we now discuss. We start with the following result concerning the bifurcations of solutions of (4.39) at the parameter point $(a, b) = (1, 1)$.

COROLLARY 4.2
Every neighborhood of (1,1) in the parameter plane contains a pair (a,b) for which one of the following is true:
(a) Every solution of (4.39) converges to zero;
(b) There are unbounded solutions of (4.39) that are equal to zero infinitely often;
(c) There are positive solutions of (4.39) that have period three (hence are bounded but not convergent).

PROOF (a) For each (a, b) in the open unit square every solution of (4.39) converges to zero by Theorem 4.8.

(b) On the parabola $b = a^2$ which contains $(1,1)$, Lemma 4.5 shows that the desired solutions exist for every value $a > 1$. In fact, Lemma 4.5 shows that as (a, b) passes $(1,1)$ on the curve $b = a^2$, solutions of (4.39) change qualitatively from bounded and convergent to unbounded and nonconvergent, with an "uneasy compromise" taking place at $(1,1)$.

(c) For each (a, b) on the cubic curve $a^3 + ab - b^3 = 1$ which contains $(1,1)$, Theorem 4.4 establishes the existence of positive period-3 solutions for every value $a > 1$. ∎

From Corollary 4.2 it is clear that $(1, 1)$ is an interesting point in the parameter plane of (4.39). We now consider the equation

$$x_{n+1} = |x_n - x_{n-1}|, \qquad n = 0, 1, 2, \ldots \qquad (4.63)$$

The factor equation is, of course,

$$r_{n+1} = \left| \frac{1}{r_n} - 1 \right|, \qquad n = 0, 1, 2, \ldots \qquad (4.64)$$

if we define $r_n = x_n/x_{n-1}$ for every $n \geq 0$. The corresponding mapping of the half-line is

$$\phi(r) = \left| \frac{1}{r} - 1 \right|, \qquad r > 0.$$

Define the set of backward iterates of zero as

$$C = \cup_{i=0}^{\infty} \phi^{-i}(0) = \{r > 0 : r_i = \phi^i(r) = 0 \text{ for some positive integer } i\} \cup \{0\}.$$

In the above definition, we interpret ϕ^0 as the identity mapping. The next result establishes a basic property of the set C with respect to the solutions of (4.63).

LEMMA 4.8
If $\{x_n\}$ is a solution of (4.63) with $x_0/x_{-1} \in C$, then $\{x_n\}$ eventually has period 3.

PROOF   By assumption $r_0 = x_0/x_{-1} \in C$; therefore, $x_{-1} \neq 0$ and there is $k \geq 0$ such that $r_k = \phi^k(r_0) = 0$ for some least integer k. Hence, $x_k = 0$ and it readily follows that

$$\{x_n\} = \{x_{-1}, x_0, \ldots x_{k-1}, 0, x_{k-1}, x_{k-1}, 0, x_{k-1}, x_{k-1}, 0, \ldots\}.$$

∎

The left and right halves

$$\phi_L(r) = \frac{1}{r} - 1, \ 0 < r \leq 1, \quad \phi_R(r) = 1 - \frac{1}{r}, \ r \geq 1$$

are both one-to-one maps and their inverses are easily computed as

$$\phi_L^{-1}(r) = \frac{1}{1+r}, \ r \geq 0, \quad \phi_R^{-1}(r) = \frac{1}{1-r}, \ 0 \leq r < 1.$$

The mapping ϕ has a unique fixed point

$$\bar{r} = \frac{\sqrt{5} - 1}{2}$$

which is the same as the unique fixed point for ϕ_L; ϕ_R has no fixed points in this case. Next, we define the set of backward iterates of $\bar{r}$ as

$$D = \cup_{i=0}^{\infty} \phi^{-i}(\bar{r}) = \{r > 0 : r_i = \phi^i(r) = \bar{r} \text{ for some positive integer } i\}.$$

Note that $\bar{r} = \phi^0(\bar{r}) \in D$ and that $D \cap C$ is empty. The next result establishes a basic property of the set D based on the fact that $\bar{r} < 1$.

LEMMA 4.9
If $\{x_n\}$ is a solution of (4.63) with $x_0/x_{-1} \in D$, then $\{x_n\}$ is eventually decreasing monotonically to zero.

PROOF   If $x_0/x_{-1} \in D$ then $x_j/x_{j-1} = r_j = \bar{r}$ for some $j \geq 1$. Thus $x_n/x_{n-1} = r_n = \bar{r}$ for all $n \geq j$ and since $\bar{r} < 1$ it follows that $x_n < x_{n-1}$ for all $n \geq j$. Therefore, $\{x_n\}$ is monotonically decreasing. If the limit $\lim_{n \to \infty} x_n = L \geq 0$ then by continuity we must have $L = 0$. ∎

Since ϕ is a rational form, we see that $C \subset \mathbb{Q}^+$ but $D \cap \mathbb{Q}^+$ is empty, where $\mathbb{Q}^+$ is the set of all nonnegative rational numbers. Therefore, by Lemmas 4.8 and 4.9, period-3 solutions of (4.63) may exist when the initial values x_0, x_{-1} are rational, whereas solutions that converge to zero may occur when the initial values have an irrational ratio. Theorem 4.9 below shows that this dichotomy is descriptive of *all* solutions of (4.63). We need one more lemma before stating the theorem.

LEMMA 4.10

Let $\{x_n\}$ be a solution of (4.63) such that $x_n > 0$ for all n. If $x_k > x_{k-1}$ for some $k \geq 0$, then $x_n < x_k$ for all $n > k$; i.e., $\{x_n\}$ displays a downward trend.

PROOF Under the given hypotheses we have that $x_{k+1} = x_k - x_{k-1} < x_k$. Therefore, $x_{k+2} = x_k - x_{k+1} < x_k$, and thus, $x_{k+3} \leq \max\{x_{k+1}, x_{k+2}\} < x_k$. The last step by induction extends to $n > k + 3$ and completes the proof. ∎

THEOREM 4.9

(a) If $x_0/x_{-1} \notin \mathbb{Q}^+$ then the corresponding solution $\{x_n\}$ of (4.63) converges to zero.

(b) If $x_0/x_{-1} \notin \mathbb{Q}^+ \cup D$ then the solution $\{x_n\}$ converges to zero but it is not eventually monotonic.

(c) $C = \mathbb{Q}^+$; thus if $x_0/x_{-1} \in \mathbb{Q}^+$ then the corresponding solution $\{x_n\}$ of (4.63) has period 3 eventually.

PROOF (a) Since $r_0 = x_0/x_{-1} \notin C$, it follows that $r_n \neq 0, 1$ for all n. This implies that $x_n \neq 0, x_{n-1}$ for all n. Therefore, $x_n > 0$ and either $x_n < x_{n-1}$ for all n in which case x_n converges to zero monotonically, or there is $k_1 \geq 0$ such that $x_{k_1} > x_{k_1-1} > 0$. In the latter case, Lemma 4.10 implies that $x_n < x_{k_1}$ for all $n > k_1$. Thus, if the sequence $\{x_n\}$ is not eventually decreasing then there is an increasing sequence k_i of positive integers such that

$$x_{k_1} > x_{k_2} > \cdots > x_{k_i} > \cdots$$

and $\{x_n\}$ is decreasing for $k_i \leq n < k_{i+1}$, $i = 1, 2, 3, \ldots$, i.e.,

$$x_{k_i} > x_{k_i+1} > \cdots > x_{k_{i+1}-1}.$$

These facts imply that $x_n \to 0$ as $n \to \infty$.

(b) Convergence follows from Part (a). If $\{x_n\}$ is eventually monotonic, then there is $k \geq 0$ such that $x_n < x_{n-1}$ or equivalently, $r_n < 1$ for all $n \geq k$. We show that this leads to a contradiction. Since $r_0 \notin D$, it follows that $r_n \neq \bar{r}$ for all n. Note that for $r \in (1/2, \bar{r})$,

$$\phi^2(r) = \phi_1^2(r) = \phi_1(\phi_1(r)) = \frac{2r - 1}{1 - r} < r.$$

Thus, if $r_k \in (1/2, \bar{r})$ then there is j with $\phi^j(r_k) = \phi_1^j(r_k) \leq 1/2$; i.e., $r_{k+j} \leq 1/2$ and therefore, $r_{k+j+1} \geq \phi_1(1/2) = 1$ which is a contradiction. We conclude that $\{x_n\}$ is not eventually monotonic.

(c) Since ϕ is a rational form, $C \subset \mathbb{Q}^+$. Now we show that $\mathbb{Q}^+ \subset C$. To this end, let $r_0 \in \mathbb{Q}^+$ where $r_0 = x_0/x_{-1}$. First, let us assume that both x_0 and x_{-1} are integers. Then the corresponding solution of (4.63) also has integer

terms x_n. For each n, either $x_n \leq x_{n-1}$ or $x_n > x_{n-1}$. In the latter case,
Lemma 4.10 implies that $x_{n+i} < x_n$ for $i \geq 1$ and in the former case, either
$x_n < x_{n-1}$ or $x_n = x_{n-1}$. That is, either $r_n = 1 \in C$ or x_n must decrease.
Since there are only finitely many integers involved, it follows that $r_n = 1$ or
$x_n = 0$ for some n; i.e., $r_n = 1$ or 0 for a sufficiently large integer n which
means that $r_0 \in C$.

Next, let x_0 and x_{-1} be any pair of positive real numbers such that $r_0 = x_0/x_{-1}$ is rational. Then $r_0 = q_0/q_{-1}$ where q_0, q_{-1} are positive integers so
by the preceding argument, $r_0 \in C$ and the proof is complete. ∎

COROLLARY 4.3
Let $\{x_n\}$ be a solution of (4.63). Then:
(a) $\{x_n\}$ has period 3 eventually if and only if $x_0/x_{-1} \in \mathbb{Q}^+$ or $x_{-1} = 0$.
(b) $x_n = x_k(\bar{r})^{n-k}$ for some $k \geq 0$ with $x_k \leq x_0$ if and only if $x_0/x_{-1} \in D$.
(c) Let $x_{-1} \neq 0$. Then $x_n \to 0$ as $n \to \infty$ if and only if $x_0/x_{-1} \notin \mathbb{Q}^+$.
(d) $\{x_n\}$ is unstable in all cases; i.e., (4.63) has no stable solutions.

The next corollary is the ratios version of Corollary 4.3.

COROLLARY 4.4
Let $\{r_n\}$ be a solution of (4.40). Then:
(a) $r_k = 0$ for some $k \geq 0$ (so r_n is undefined for $n > k$) if and only if $r_0 \in \mathbb{Q}^+$.
(b) For $r_0 \notin \mathbb{Q}^+$, $\{r_n\}$ is unstable.

Let us now take a closer look at the solutions of (4.64) when r_0 is irrational.
We begin by showing that equation (4.64) has periodic solutions of all possible
periods except 3. With minor modifications, the next theorem applies to
eventually periodic solutions as well.

THEOREM 4.10
(a) Equation (4.64) has a p-periodic solution $\{r_1,\ldots,r_p\}$ for every $p \neq 3$
given by

$$r_1 = \frac{1+\sqrt{5}}{2}, \; r_2 = \frac{\sqrt{5}-1}{\sqrt{5}+1} \quad (p=2)$$

$$r_1 = \frac{1}{2}\left[y_{p-4} + \sqrt{y_{p-4}^2 + 4y_{p-4}y_{p-1}}\right], \; r_k = \frac{y_{k-4}r_1 - y_{k-2}}{y_{k-3} - y_{k-5}r_1}, \; 2 \leq k \leq p, \; (p \geq 4)$$

where y_n is the n-th term of the Fibonacci sequence; i.e., $y_{n+1} = y_n + y_{n-1}$
for $n \geq 0$ where we define

$$y_{-3} = -1, \; y_{-2} = 1 \; y_{-1} = 1 \; y_0 = 1.$$

(b) If $\{r_1, \ldots, r_p\}$ is a periodic solution of (4.64) then for the corresponding solution $\{x_n\}$ of (4.63) it is true that

$$x_n = x_0 \rho^{n/p}, \quad \text{if } n/p \text{ is an integer}$$

$$x_n \leq x_0 \alpha \rho^{n/p}, \quad \text{otherwise}$$

where

$$\rho = \prod_{i=1}^{p} r_i < 1, \quad \alpha = \max\{r_1, \ldots, r_p\} \rho^{-(1-1/p)} > 1.$$

PROOF   (a) Let $r_1 > 1$. Then $r_2 = \phi(r_1) = \phi_2(r_1) = 1 - 1/r_1 < 1$ and

$$r_3 = \phi(r_2) = \phi_L(r_2) = \frac{1}{r_2} - 1 = \frac{1}{r_1 - 1}.$$

Though it is possible that $r_3 = r_1$, to examine potential 3-cycles, let us assume that $r_3 < 1$. Then

$$r_4 = \frac{1}{r_3} - 1 = r_1 - 2.$$

Clearly $r_4 \neq r_1$, so a period-3 solution cannot occur with two points less than 1. Since ϕ_2 maps the interval $(1, \infty)$ into $(0, 1)$, a period-3 solution cannot have two or more points greater than 1. We can also rule out a period-3 solution having all three points less than 1, since ϕ_1 is strictly decreasing on the interval $(0,1)$. Therefore, (4.64) cannot have a period-3 solution. Next, we seek cycles of the form

$$r_1 > 1, \ 0 < r_k < 1, \ k = 2, 3, \ldots, p. \tag{4.65}$$

To explicitly determine a 2-cycle, set

$$r_1 > 1, r_2 = \phi_R(r_1) = \frac{r_1 - 1}{r_1}, \ r_3 = \phi_L(r_2) = \frac{1}{r_1 - 1} \tag{4.66}$$

and solve the equation $r_3 = r_1$ to obtain

$$r_1 = \frac{1 + \sqrt{5}}{2} = \gamma, \ r_2 = \frac{\sqrt{5} - 1}{\sqrt{5} + 1} = \frac{1}{\gamma^2}.$$

The number γ is of course the "golden ratio or mean." For explicitly listing cycles of length $p \geq 4$ that satisfy conditions (4.65) we need the famous sequence of Fibonacci numbers

$$y_1 = 1, \ y_2 = 2, \ y_3 = 3, \ y_4 = 5, \ y_5 = 8, \ y_6 = 13, \ldots$$

that are generated by the linear initial value problem

$$y_{n+1} = y_n + y_{n-1}, \quad y_0 = 1, \ y_{-1} = 0. \tag{4.67}$$

Following the pattern that was started above, namely,

$$r_4 = \frac{r_1 - 2}{1 - 0}, \qquad r_5 = \frac{r_1 - 3}{2 - r_1}, \quad \cdots$$

we claim that

$$r_k = \frac{y_{k-4}r_1 - y_{k-2}}{y_{k-3} - y_{k-5}r_1}, \qquad k = 4, 5, \ldots, p. \tag{4.68}$$

with r_k given by (4.66) for $k = 1, 2, 3$. If we assume that (4.68) holds for some k, then

$$\begin{aligned}
r_{k+1} &= \frac{1}{r_k} - 1 \\
&= \frac{y_{k-3} - y_{k-5}r_1 - y_{k-4}r_1 + y_{k-2}}{y_{k-4}r_1 - y_{k-2}} \\
&= \frac{y_{k-1} - y_{k-3}r_1}{y_{k-4}r_1 - y_{k-2}}
\end{aligned}$$

where we used (4.67) for the last equality. This establishes (4.68) by induction. Next, using (4.68) we can solve the equation $r_{p+1} = r_1$ or

$$\frac{y_{p-3}r_1 - y_{p-1}}{y_{p-2} - y_{p-4}r_1} = r_1$$

to obtain the value

$$r_1 = \frac{1}{2}\left[y_{p-4} + \sqrt{y_{p-4}^2 + 4y_{p-4}y_{p-1}} \right], \qquad p \geq 4$$

which together with (4.66), (4.67), and (4.68) completely determines the p-cycle that satisfies conditions (4.65) for $p \neq 3$.

(b) Without loss of generality, let $r_1 = x_1/x_0$. If $\{r_1, \ldots, r_p\}$ is a solution with period p, and

$$\rho = r_1 r_2 \cdots r_p$$

then for each positive integer k,

$$x_{kp} = r_1 r_2 \cdots r_p x_{(k-1)p} = x_{(k-1)p}\rho = \cdots = x_0 \rho^k.$$

More generally, writing $n = kp + l$ where $0 \leq l \leq p - 1$, we get

$$\begin{aligned}
x_n &= r_n r_{n-1} \cdots r_{n-l+1} x_{kp} \\
&\leq \max\{r_1, \ldots, r_p\} x_0 \rho^{n/p - l/p} \\
&\leq x_0 \max\{r_1, \ldots, r_p\} \rho^{-(p-1)/p} \rho^{n/p}
\end{aligned}$$

which establishes the assertion about x_n. Clearly, if $\rho < 1$ then $\alpha > 1$ since at least one of the p points of the cycle must exceed 1. Finally, $\rho < 1$ for otherwise the subsequence $\{x_0\rho^k\}$ of $\{x_n\}$ with $n = pk$ would be unbounded if $\rho > 1$, or $\{x_n\}$ would be periodic with period p if $\rho = 1$. But neither of these cases is possible. ∎

We conclude the discussion of the absolute value equation with two remarks.

REMARK 4.3 *(Period 3 solutions)* As we have just seen, for $(a, b) = (1, 1)$ the period 3 solutions of (4.39) pass through the origin and are not strictly positive. Hence, they cannot correspond to 3-cycles of (4.64), a fact that is consistent with Theorem 4.8(c). On the other hand, note that as (a, b) approaches $(1, 1)$ along the curve (4.50) the quantities r_1, r_2, r_3 in (4.51) have the following limits

$$\lim_{(a,b)\to(1,1)} r_1 = 1, \qquad \lim_{(a,b)\to(1,1)} r_2 = 0, \qquad \lim_{(a,b)\to(1,1)} r_3 = \infty.$$

These correspond to, respectively, the line $y = x$, the x-axis and the y-axis in the phase plane of (4.39) which are indeed lines that are the loci of the period-3 solutions of Theorem 4.4(b) in the state-space (or the phase plane). As (a, b) approaches $(1, 1)$, we may imagine the three lines in Figure 4.8 moving towards the coordinate axes and the 45-degree line and coinciding with them in the limit. Therefore, the nonpositive period-3 solutions of (4.39) may be interpreted as limiting values of the positive period-3 solutions as (a, b) approaches $(1, 1)$ along the curve (4.42). ∎

REMARK 4.4 (Stability) No period-2 or period-3 solution of (4.39) is stable, whether asymptotically or structurally. Theorem 4.4 shows the structural instability; as for asymptotic instability, recall that since a p-cycle $\{r_1, \ldots, r_p\}$ of (4.40) does not contain the minimum point of ϕ, it is *unstable* if

$$1 < \prod_{i=1}^{p} |\phi'(r_i)| = \prod_{i=1}^{p} \frac{b}{r_i^2} = \frac{b^p}{\prod_{i=1}^{p} r_i^2}$$

that is, if

$$\prod_{i=1}^{p} r_i < b^{p/2}.$$

If $p = 2$ and $a > 0$, then $r_1 r_2 < (a^2 + 2b - a^2)/2 = b$ by (4.52) so the positive 2-periodic solutions are not asymptotically stable. The same conclusion clearly holds when $a = 0$ and $b = 1$. For $p = 3$ and $a > 1$ we have from (4.53) that $r_1 r_2 r_3 < b^{3/2}$ if and only if

$$a(a^2 + b) - \sqrt{a^2(a^2 + b)^2 - 4b^3} < 2b^{3/2}.$$

The preceding inequality reduces to $2b^{3/2} < a(a^2 + b)$ which is true if r_1 is real. Hence, positive period-3 solutions of (4.39) are not asymptotically stable. Also, Theorem 4.9 shows that the period-3 solutions passing through the origin are unstable, though ironically these are the only solutions that appear in computer simulations because of their rationality!   ∎

4.5   Notes

The results in this chapter have been used in prior literature, e.g., in Dehghan, et al. (2008b), and Sedaghat (2009a, 2009b). The appeal of HD1 equations is in their easy identifiability and subsequent factorization. The factor equations are not always easy to deal with, just because they have lower order. However, the reduction in order through SC factorization can simplify the study of equations in some cases. The absolute value equation is a case in point. The results presented here are largely taken from Sedaghat (2004a) and Kent and Sedaghat (2004a). Related results, background material, and additional comments appear in the next chapter.

4.6   Problems

4.1 Show that Example 4.1 can be extended to all nontrivial groups.

4.2 Prove Proposition 4.1.

4.3 Let $\{a_n\}$, $\{b_n\}$ be sequences of real numbers with $a_n \neq 0$ for all n. Show that the rational difference equation

$$x_{n+1} = x_{n-1}\left(\frac{a_n x_{n-k}}{x_n} + b_n\right)$$

is HD1 on the group of all nonzero real numbers under ordinary multiplication and determine its SC factorization.

4.4 Let $\{a_n\}$ be a sequence in the group $G = (0, \infty)$ under ordinary multiplication and let $\{b_n\}$ be any sequence of nonnegative real numbers.
  (a) Show that the following generalization of Eq. (4.16) in Example 4.7 is HD1 on G and find its SC factorization:

$$x_{n+1} = \frac{a_n x_{n-p+1} x_{n-p-q}}{x_{n-q} + b_n x_{n-p-q}}.$$

  (b) Show that the following variation of the equation in (a) is also HD1 on G and find its SC factorization:

$$x_{n+1} = \frac{a_n x_{n-p+1} x_{n-q}}{b_n x_{n-q} + x_{n-p-q}}.$$

(c) Show that the following HD1 difference equation

$$x_{n+1} = \frac{a_n x_{n-p} x_{n-q}}{b_n x_{n-q} + x_{n-p-q}}$$

has a SC factorization given by Corollary 4.1, Part (b) but not (c).

4.5 Let $a_n = 1$ for all $n \geq 0$ in Eq. (4.16) of Example 4.7.

(a) If $p = 1$ and q is any nonnegative integer in (4.16) use the SC factorization (4.17) to show that all solutions of (4.16) are eventually periodic with period $2(q+1)$.

(b) Is the assertion in Part (a) true if $a_n = a \neq 1$ is some nonzero real constant for all $n \geq 0$? Explain the changes in the behaviors of solutions in this case.

(c) Generalize the result in Part (a) to all odd positive integers p.

(d) Use the SC factorization (4.17) to discuss the behaviors of solutions of (4.16) for an even positive integer p.

4.6 Recall the backward difference operator Δ on a commutative group G

$$\Delta x_n = x_n - x_{n-1}.$$

For $m \geq 1$ define $\Delta^{m+1} x_n = \Delta^m(\Delta x_n)$. If $g_n : G \to G$ for each $n \geq 1$ and $0 \leq j \leq k$ then show that the difference equation

$$x_{n+1} = x_{n-j} + g_n(\Delta^k x_n)$$

is HD1 on G and find its SC factorization.

4.7 Derive the solution of Eq. (4.19) from the triangular system in Example 4.8.

4.8 Consider the following variation of Eq. (4.19):

$$x_{n+1} = x_n + \frac{a(x_n - x_{n-1})^2}{x_n - x_{n-2}}, \qquad a \neq 0. \tag{4.69}$$

(a) Show that Eq. (4.69) is HD1 relative to the additive group of real numbers and find its SC factorization.

(b) Prove that the second-order factor equation obtained in (a) is HD1 relative to the relevant group of nonzero real numbers and find its SC factorization.

(c) Using the order-preserving substitution $s_n = r_n^{-1}$ to transform the first-order factor equation in (b) into a linear equation, determine the complete factorization of (4.69) into a triangular system of three first-order equations.

4.9 (a) Verify that the difference equation

$$x_{n+1} = x_n + \frac{a(x_n - x_{n-1})}{x_{n-1} - x_{n-2}}, \qquad a \neq 0$$

is HD1 relative to the additive group of real numbers and determine its SC factorization.

(b) Use the factorization in (a) to show that the solutions of the difference equation in (a) are given by the formula

$$x_n = x_0 + \sigma \delta_n + \sum_{i=6\delta_n+1}^{n} t_i$$

for a suitable real number σ and integers $\delta_n \geq 0$ such that $n = 6\delta_n + \rho_n$ for integers $0 \leq \rho_n \leq 5$. Example 2.9 in Chapter 2 is relevant to this problem.

4.10 Verify that the linear-like equations (4.24)-(4.26) are HD1 on $(0, \infty)$ under multiplication and find their SC factorizations.

4.11 Assume that $k = 2$ in Eq. (4.29).

(a) Determine the complete SC factorization into a triangular system and use it to determine the behaviors of solutions of (4.29) with initial values $x_0 > x_{-1} > x_{-2}$.

(b) If the three initial values and a, b are arbitrary real numbers with $b \neq 0$ then use the SC factor equation to determine the behaviors of solutions of the third-order Eq. (4.29).

4.12 Supply the details of calculation for the solution (4.34) in Example 4.12.

4.13 Explain the dynamics of the three simple maps in (4.43) on $(0, \infty)$ with $a, b > 0$.

5

Type-(k, 1) Reductions

Recall from Section 3.3.3 that type-$(k, 1)$ order reductions of the recursive equation

$$x_{n+1} = f_n(x_n, x_{n-1}, \ldots, x_{n-k}) \tag{5.1}$$

are generally based on the form symmetry

$$H(u_0, u_1, \ldots, u_k) = [u_0 * h(u_1), u_1 * h(u_2), \ldots, u_{k-1} * h(u_k)] \tag{5.2}$$

where $h : G \to G$ is a given self-map of G. In Chapter 4 we considered a special case where h is the group inversion map $h(u) = u^{-1}$. In this chapter we discuss a much more general case in which it is assumed only that *the mapping h is invertible.*

Before proceeding with details, we make some basic observations. From the semiconjugate relation it follows that a type-$(k, 1)$ reduction with a form symmetry of type (5.2) exists if and only if there are functions ϕ_n such that

$$\phi_n(u_0 * h(u_1), u_1 * h(u_2), \ldots, u_{k-1} * h(u_k)) = f_n(u_0, u_1, \ldots, u_k) * h(u_0). \tag{5.3}$$

If such ϕ_n exists then a change of variables to

$$t_n = x_n * h(x_{n-1}) \tag{5.4}$$

yields a SC factorization of (5.1) with the factor and cofactor equations

$$t_{n+1} = \phi_n(t_n, \ldots, t_{n-k+1}), \tag{5.5}$$

$$x_{n+1} = t_{n+1} * h(x_n)^{-1}. \tag{5.6}$$

It is evident from this SC factorization that type-$(k, 1)$ reductions are characterized by a one-dimensional cofactor equation whereas the factor equation is k-dimensional. In this chapter we discover that a large variety of difference equations, including all HD1 equations, fall into this category.

5.1   Invertible-map criterion

We found earlier that if $h(u) = u^{-1}$ then identity (5.3) holds if and only if the functions f_n are homogeneous of degree 1 for all n. A question that naturally arises now is if h is a given function then for which types of functions f_n does the identity (5.3) holds.

The next result answers this question when h is an invertible map. Though not as succinct as the HD1 property, the necessary and sufficient condition that it gives applies to a much broader class of functions. To avoid notational conflict, we write h^{-1} for the inverse of the function h and $h(u)^{-1}$ for the inverse of the group element $h(u)$.

THEOREM 5.1

(Invertible map criterion) Assume that $h : G \to G$ is a bijection. For $u_0, v_1, \ldots, v_k \in G$ let $\zeta_0 = u_0$ and define

$$\zeta_j = h^{-1}(\zeta_{j-1}^{-1} * v_j), \quad j = 1, \ldots, k. \tag{5.7}$$

Then Eq. (5.1) has the form symmetry (5.2) with SC factors ϕ_n satisfying (5.3) if and only if the quantity

$$f_n(u_0, \zeta_1, \ldots, \zeta_k) * h(u_0) \tag{5.8}$$

is independent of u_0 for every n. In this case the factor functions are defined as

$$\phi_n(v_1, \ldots, v_k) = f_n(u_0, \zeta_1, \ldots, \zeta_k) * h(u_0).$$

PROOF   First assume that (5.8) is independent of u_0 for all $v_1, \ldots, v_k$ so that the function

$$\phi_n(v_1, \ldots, v_k) = f_n(u_0, \zeta_1, \ldots, \zeta_k) * h(u_0) \tag{5.9}$$

is well defined. Next, if H is given by (5.2) then for all $u_0, u_1, \ldots, u_k$

$$\phi_n(H(u_0, u_1, \ldots, u_k)) = \phi_n(u_0 * h(u_1), u_1 * h(u_2), \ldots, u_{k-1} * h(u_k)).$$

Define

$$v_j = u_{j-1} * h(u_j), \quad j = 1, \ldots, k. \tag{5.10}$$

Then by (5.7)

$$\zeta_1 = h^{-1}(u_0^{-1} * v_1) = h^{-1}(u_0^{-1} * u_0 * h(u_1)) = u_1.$$

We show that in fact $\zeta_j = u_j$ for every j if (5.10) holds. Suppose by way of induction that $\zeta_l = u_l$ for $1 \le l < j$. Then

$$\zeta_j = h^{-1}(\zeta_{j-1}^{-1} * v_j) = h^{-1}(u_{j-1}^{-1} * u_{j-1} * h(u_j)) = u_j$$

as claimed. Thus by (5.9)

$$\phi_n(H(u_0, u_1, \ldots, u_k)) = f_n(u_0, \ldots, u_k) * h(u_0)$$

Now if F_n and Φ_n are the unfoldings of f_n and ϕ_n respectively, then

$$H(F_n(u_0, \ldots, u_k)) = [f_n(u_0, \ldots, u_k) * h(u_0), u_0 * h(u_1), \ldots, u_{k-2} * h(u_{k-1})]$$
$$= [\phi_n(H(u_0, u_1, \ldots, u_k)), u_0 * h(u_1), \ldots, u_{k-2} * h(u_{k-1})]$$
$$= \Phi_n(H(u_0, \ldots, u_k))$$

and it follows that H is a semiconjugate form symmetry for Eq. (5.1).

Conversely, if H as given by (5.2) is a form symmetry then the semiconjugate relation implies that there are functions ϕ_n such that

$$f_n(u_0, \ldots, u_k) * h(u_0) = \phi_n(u_0 * h(u_1), \ldots, u_{k-1} * h(u_k)). \tag{5.11}$$

For every $v_1, \ldots, v_k$ in G and with ζ_j as defined in (5.7),

$$f_n(u_0, \zeta_1, \ldots, \zeta_k) * h(u_0) = \phi_n(u_0 * h(\zeta_1), \zeta_1 * h(\zeta_2), \ldots, \zeta_{k-1} * h(\zeta_k))$$
$$= \phi_n(v_1, \ldots, v_k)$$

which is clearly independent of u_0. ∎

It is worth nothing that (5.7) is a backwards version of the cofactor equation (5.6) that is obtained by solving (5.4) for x_{n-1} instead of x_n. To do this we required h to be invertible. Of course, in (5.7) it is necessary to iterate only k times.

Theorem 5.1 is a significant extension of Theorem 4.1 to all cases with invertible h. In the rest of this chapter we discuss a few special cases of (5.2) where h is invertible.

5.2 Identity form symmetry

The identity map or function h on G, i.e., $h(u) = u$ defines the *identity form symmetry* i.e., the following special case of (5.2):

$$H(u_0, u_1, \ldots, u_k) = [u_0 * u_1, u_1 * u_2, \ldots, u_{k-1} * u_k]. \tag{5.12}$$

This type of form symmetry is characterized by a change of variables to

$$t_n = x_n * x_{n-1}.$$

The factor and cofactor equations for the SC factorization of (5.1) with the identity form symmetry are given by (5.5) and (5.6) as

$$t_{n+1} = \phi_n(t_n, \ldots, t_{n-k+1}), \tag{5.13}$$
$$x_{n+1} = t_{n+1} * x_n^{-1}. \tag{5.14}$$

From the cofactor equation we obtain the following relations for odd and even terms of a solution $\{x_n\}$ of (5.1) in terms of a solution $\{t_n\}$ of the factor equation:

$$x_{2n} = t_{2n} * \cdots * t_4 * t_2 * x_0 * (t_{2n-1} * \cdots * t_3 * t_1)^{-1}$$
$$x_{2n+1} = t_{2n+1} * \cdots * t_3 * t_1 * (t_{2n} * \cdots * t_4 * t_2 * x_0)^{-1}.$$

Example 5.1

Each of the following equations has the identity form symmetry

$$x_{n+1} = \frac{a_n}{x_n(x_n x_{n-1} + b_n)}, \qquad x_{n+1} = e^{a_n + b_n(x_n + x_{n-1})} - x_n.$$

In the case of the exponential equation, adding x_n to both sides and substituting $t_n = x_n + x_{n-1}$ we readily obtain the following SC factorization on $\mathbb{R}$ under ordinary addition:

$$t_{n+1} = e^{a_n + b_n t_n}$$
$$x_{n+1} = t_{n+1} - x_n.$$

For the rational equation, if a_n, b_n are sequences of positive real numbers then multiplying the equation on both sides by x_n and setting $t_n = x_n x_{n-1}$ it is easy to see that the rational equation has the following SC factorization over $(0, \infty)$ under ordinary multiplication:

$$t_{n+1} = \frac{a_n}{t_n + b_n},$$
$$x_{n+1} = \frac{t_{n+1}}{x_n}.$$

☐

The next result uses Theorem 5.1 to give a condition for verifying whether a difference equation has the identity form symmetry.

COROLLARY 5.1

For every $u_0, v_1, \ldots, v_k \in G$ let $\zeta_0 = u_0$ and define $\zeta_j = \zeta_{j-1}^{-1} * v_j$ for $j = 1, \ldots, k$. Then Eq. (5.1) has the identity form symmetry (5.12) if and only if the quantity

$$f_n(u_0, \zeta_1, \ldots, \zeta_k) * u_0 \tag{5.15}$$

is independent of u_0 for every n. In this case the factor functions are defined as

$$\phi_n(v_1, \ldots, v_k) = f_n(u_0, \zeta_1, \ldots, \zeta_k) * u_0.$$

The following examples illustrate Corollary 5.1.

Example 5.2

Consider the third-order difference equation on $(0, \infty)$ under ordinary multiplication:

$$x_{n+1} = \frac{x_n x_{n-1}}{a_n x_n + b_n x_{n-2}}, \qquad a_n \geq 0, b_n > 0. \tag{5.16}$$

Here $f_n(u_0, u_1, u_2) = u_0 u_1 / (a_n u_0 + b_n u_2)$ and if

$$\zeta_1 = \frac{v_1}{u_0}, \qquad \zeta_2 = \frac{v_2}{\zeta_1} = \frac{u_0 v_2}{v_1}$$

are substituted in f_n we obtain

$$f_n\left(u_0, \frac{v_1}{u_0}, \frac{u_0 v_2}{v_1}\right) u_0 = \frac{u_0(v_1/u_0)}{a_n u_0 + b_n(u_0 v_2/v_1)} u_0 = \frac{v_1^2}{a_n v_1 + b_n v_2}.$$

Since the last expression is independent of u_0 and a function of the new variables v_1 and v_2 only, by Corollary 5.1 Eq. (5.16) has the identity form symmetry. Its factor equation is

$$t_{n+1} = \frac{t_n^2}{a_n t_n + b_n t_{n-1}} \tag{5.17}$$

with cofactor $x_{n+1} = t_{n+1}/x_n$. Now let us examine whether the second-order equation (5.17) also has the identity form symmetry. Substituting $\zeta_1 = v_1/u_0$ in the functions

$$\phi_n(u_0, u_1) = \frac{u_0^2}{a_n u_0 + b_n u_1}$$

gives

$$\phi_n\left(u_0, \frac{v_1}{u_0}\right) u_0 = \frac{(u_0)^2 u_0}{a_n u_0 + b_n(v_1/u_0)} = \frac{u_0^4}{a_n u_0^2 + b_n v_1}$$

which is clearly not independent of u_0. Thus Corollary 5.1 implies that (5.17) does not have the identity form symmetry. On the other hand, (5.17) is HD1 so it has the inversion form symmetry. Theorem 4.1 yields the following SC factorization for (5.17):

$$r_{n+1} = \frac{1}{a_n + b_n/r_n} \tag{5.18}$$

$$t_{n+1} = r_{n+1} t_n.$$

The substitution $s_n = 1/r_n$ (an order-preserving form symmetry) in (5.18) produces a linear equation $s_{n+1} = a_n + b_n s_n$. Thus we have a full factorization of Eq. (5.16) into the following triangular system of first-order equations

$$s_{n+1} = a_n + b_n s_n$$

$$t_{n+1} = \frac{t_n}{s_{n+1}}$$

$$x_{n+1} = \frac{t_{n+1}}{x_n}.$$

⧠

Example 5.3

We show that the difference equation

$$x_{n+1} = \frac{a_n}{x_n} e^{b_n x_n / x_{n-2}}, \quad a_n, b_n \in \mathbb{R}, \ a_n \neq 0 \text{ for all } n \tag{5.19}$$

has the identity form symmetry over the group of nonzero real numbers under multiplication. Applying Corollary 5.1 to the function on the right-hand side of (5.19) gives

$$f_n\left(u_0, \frac{v_1}{u_0}, \frac{u_0 v_2}{v_1}\right) u_0 = a_n \exp\left(b_n \frac{u_0}{u_0 v_2 / v_1}\right) = a_n \exp\left(b_n \frac{v_1}{v_2}\right).$$

Since the above expression is independent of u_0 Eq. (5.19) has the identity form symmetry and its factor equation is

$$t_{n+1} = a_n e^{b_n t_n / t_{n-1}}.$$

We note in passing that Eq. (5.19) is HDp with $p = -1$, showing that homogeneous equations with degrees other than one may possess non-inversion form symmetries and SC factorizations.   ⧠

The next example discusses a higher-order version of the identity form symmetry that leads to an order reduction that is not of type-$(k, 1)$.

Example 5.4

Consider the equation (4.16), i.e.,

$$x_{n+1} = \frac{a_n x_{n-p+1} x_{n-p-q}}{x_{n-q}}$$

from Example 4.7 again. This equation can be rewritten as

$$x_{n+1} x_{n-q} = a_n x_{n-p+1} x_{n-p-q}$$

so the substitution

$$t_n = x_n x_{n-q-1} \tag{5.20}$$

gives the pair of equations

$$t_{n+1} = a_n t_{n-p}, \quad x_{n+1} = \frac{t_{n+1}}{x_{n-q}}.$$

⧠

5.3   Inversion form symmetry

As previously defined in Section 4.2, group inversion on G, i.e., $h(u) = u^{-1}$ defines the *inversion form symmetry* i.e., the following special case of (5.2):

$$H(u_0, u_1, \ldots, u_k) = [u_0 * u_1^{-1}, u_1 * u_2^{-1}, \ldots, u_{k-1} * u_k^{-1}].$$

Note that h is an invertible map and in fact a *self-inverse* map, i.e., $h = h^{-1}$. The inversion form symmetry is characterized by a change of variables to

$$t_n = x_n * x_{n-1}^{-1}. \tag{5.21}$$

The next result is an immediate consequence of Theorem 5.1.

COROLLARY 5.2

For every $u_0, v_1, \ldots, v_k \in G$ let $\zeta_0 = u_0$ and define $\zeta_j = v_j^{-1} * \zeta_{j-1}$ for $j = 1, \ldots, k$. Then Eq. (5.1) has the inversion form symmetry if and only if the quantity

$$f_n(u_0, \zeta_1, \ldots, \zeta_k) * u_0^{-1} \tag{5.22}$$

is independent of u_0 for every n. In this case the factor functions are defined as

$$\phi_n(v_1, \ldots, v_k) = f_n(u_0, \zeta_1, \ldots, \zeta_k) * u_0^{-1}.$$

As stated in Chapter 4 there is a simple characterization of functions f_n which satisfy (5.22). Recall that Eq. (5.1) is *homogeneous of degree one (HD1)* if for every $n = 0, 1, 2, \ldots$ the functions f_n are all homogeneous of degree one relative to the group G, i.e.,

$$f_n(u_0 * t, \ldots, u_k * t) = f_n(u_0, \ldots, u_k) * t$$
$$\text{for all } t, u_i \in G, \ i = 0, \ldots, k, \text{ and all } n \geq 0.$$

We now prove Theorem 4.1 in Section 4.2 as a consequence of Corollary 5.2 and Theorem 3.1.

COROLLARY 5.3

Eq. (5.1) has the inversion form symmetry if and only if f_n is HD1 relative to G for all n. In this case, the factor equation for (5.1) is given by

$$t_{n+1} = f_n(1, t_n^{-1}, (t_n * t_{n-1})^{-1}, \ldots, (t_n * t_{n-1} * \cdots * t_{n-k+1})^{-1}) \tag{5.23}$$

with the cofactor $x_{n+1} = t_{n+1} * x_n$.

PROOF First if $\zeta_j = v_j^{-1} * \zeta_{j-1}$ as in Corollary 5.2 then by straightforward iteration

$$\zeta_j = v_j^{-1} * \cdots * v_1^{-1} * u_0, \quad j = 1, \ldots, m. \tag{5.24}$$

Now if f_n is HD1 for every n then

$$f_n(u_0, \zeta_1, \ldots, \zeta_k) * u_0^{-1} = f_n(1, \zeta_1 * u_0^{-1}, \ldots, \zeta_k * u_0^{-1})$$

which by (5.24) is independent of u_0. Thus by Corollary 5.2 Eq. (5.1) has the inversion form symmetry.

Conversely, assume that (5.1) has the inversion form symmetry. Then by Corollary 5.2 for every $u_0, v_1, \ldots, v_k \in G$ the quantity in (5.22) is independent of u_0. There are functions ϕ_n where

$$f_n(u_0, \zeta_1, \ldots, \zeta_k) * u_0^{-1} = \phi_n(v_1, \ldots, v_k) \tag{5.25}$$
$$= \phi_n(\zeta_0 * \zeta_1^{-1}, \ldots, \zeta_{k-1} * \zeta_k^{-1}).$$

Note that (5.25) holds for arbitrary values of $u_0, \zeta_1, \ldots, \zeta_k$ since $v_1, \ldots, v_k$ are arbitrary. Thus for all $t, s_0, \ldots, s_k \in G$ and all n,

$$f_n(s_0 * t, \ldots, s_k * t) = \phi_n((s_0 * t) * (s_1 * t)^{-1}, \ldots, (s_{k-1} * t) * (s_k * t)^{-1}) * (s_0 * t)$$
$$= [\phi_n(s_0 * s_1^{-1}, \ldots, s_{k-1} * s_k^{-1}) * s_0] * t$$
$$= f_n(s_0, \ldots, s_k) * t.$$

It follows that f_n is HD1 relative to G for all n. The factor equation (5.23) is obtained using the change of variables (5.21) and the HD1 property as follows: If $\{x_n\}$ is a solution of (5.1) then

$$t_{n+1} = x_{n+1} * x_n^{-1}$$
$$= f_n(x_n, x_{n-1}, x_{n-2}, \ldots, x_{n-k}) * x_n^{-1}$$
$$= f_n(1, x_{n-1} * x_n^{-1}, x_{n-2} * x_n^{-1}, \ldots, x_{n-k} * x_n^{-1})$$
$$= f_n(1, x_{n-1} * x_n^{-1}, (x_{n-2} * x_{n-1}^{-1})(x_{n-1} * x_n^{-1}), \ldots$$
$$(x_{n-k} * x_{n-k+1}^{-1}) \cdots (x_{n-2} * x_{n-1}^{-1})(x_{n-1} * x_n^{-1}))$$
$$= f_n(1, t_n^{-1}, (t_n * t_{n-1})^{-1}, \ldots, (t_n * t_{n-1} * \cdots * t_{n-k+1})^{-1}).$$

Finally, the equivalence of the SC factorization here with Eq. (5.1) follows from Theorem 3.1. ∎

The next example presents an alternative way of finding the SC factorization and reducing the order of Eq. (5.16) in Example 5.2.

Example 5.5
Consider again the third-order difference equation (5.16), i.e.,

$$x_{n+1} = \frac{x_n x_{n-1}}{a_n x_n + b_n x_{n-2}}, \quad a_n, b_n > 0.$$

Here for all n, $f_n(u_0, u_1, u_2) = u_0 u_1/(a_n u_0 + b_n u_2)$ is HD1 on $(0, \infty)$ under ordinary multiplication. By Corollary 5.3 Eq. (5.16) has the following factor equation

$$t_{n+1} = \frac{t_{n-1}}{a_n t_n t_{n-1} + b_n} \tag{5.26}$$

with a cofactor $x_{n+1} = t_{n+1} x_n$. Now, (5.26) which has order two is not HD1 on $(0, \infty)$; however, substituting $\zeta_1 = v_1/u_0$ in the functions $\phi_n(u_0, u_1) = u_1/(a_n u_0 u_1 + b_n)$ gives

$$\phi_n\left(u_0, \frac{v_1}{u_0}\right) u_0 = \frac{v_1/u_0}{a_n u_0 v_1/u_0 + b_n} u_0 = \frac{v_1}{a_n v_1 + b_n}$$

which is independent of u_0. Thus Corollary 5.1 implies that Eq. (5.26) has the identity form symmetry. Its SC factorization is found by substituting $r_n = t_n t_{n-1}$ to get

$$r_{n+1} = \frac{r_n}{a_n r_n + b_n}$$
$$t_{n+1} = \frac{r_{n+1}}{t_n}.$$

Note that the factor equation above is the same as (5.18). $\square$

REMARK 5.1 A change of variables such as $s_n = x_{n-1} * x_n^{-1}$ is also essentially the inversion form symmetry since the mapping $s_n \to s_n^{-1} = x_n * x_{n-1}^{-1}$ is a conjugacy. $\blacksquare$

5.4 *Discrete Riccati equation of order two

As we saw in Example 4.10 in Chapter 4 every linear difference equation on a nontrivial field $\mathcal{F}$ is HD1 relative to the group of units of $\mathcal{F}$ and thus admits a type-$(k, 1)$ reduction in order. The resulting factor equation, i.e., the discrete Riccati equation (4.23), plays a more subtle role in the theory of linear difference equations on arbitrary fields than we can discuss at this stage. This basic role, which is *not* limited to the inversion form symmetry or the HD1 property, is discussed later in this chapter and more completely in Chapter 7. This section is starred because its content is not essential to understanding the material on factorization and reduction of order that appear in later sections and chapters. However, the reader is urged to at least read through and understand the statements of results in this section so as to not feel entirely unfamiliar with the very basic concept of the Riccati equation.

In this section we give a detailed analysis of the Riccati difference equation of order two with constant coefficients in the field $\mathbb{R}$ of real numbers (the Riccati equation of order one is simpler and similarly treated in existing literature; see the notes for this chapter). Our purpose is to gain a deeper understanding of Riccati equations of higher-order and their real solutions before encountering them again later in relation to reductions of order. The study in this section uses well-known results from the classical theory of difference equations; specifically, we use the linear difference equation of order three

$$y_{n+1} = ay_n + by_{n-1} + cy_{n-2}. \tag{5.27}$$

whose SC factor is the following discrete Riccati equation of order two

$$x_{n+1} = a + \frac{b}{x_n} + \frac{c}{x_n x_{n-1}}. \tag{5.28}$$

To simplify the discussion and reduce the number of possible cases that need to be considered, we assume that

$$a, b \geq 0, \ c > 0, \ x_0, x_{-1} \in (-\infty, \infty). \tag{5.29}$$

If we define the initial values for (5.27) as

$$y_0 = x_0 y_{-1}, \ y_{-1} = x_{-1} y_{-2} \text{ and set} \tag{5.30}$$
$$y_{-2} = 1 \text{ (or any fixed nonzero real number)}$$

then we obtain a one to one correspondence between the solutions of (5.28) and those solutions of (5.27) that do not contain zero; i.e., each solution of (5.28) uniquely defines a solution of (5.27) that does not pass through the origin and vice versa. If $\{y_n\}$ is a solution of (5.27) with $y_k = 0$ for some least k then $x_{k+1} = y_{k+1}/y_k$ is not defined. Under conditions (5.30) the correspondence between solutions of (5.27) that pass through the origin and those of (5.28) that become undefined is also one to one.

In the next section we use the solutions of the linear equation (5.27) to determine the singularity set $\mathcal{S}$ of (5.28), i.e., the set of all initial values in $\mathbb{R}^2$ that lead to a zero in the denominator of (5.28) after a finite number of iterations. After that, we prove that the unique positive fixed point of (5.28) under conditions (5.29) is globally asymptotically stable for initial points (x_0, x_{-1}) outside a set M of Lebesgue measure zero that contains $\mathcal{S}$. Finally, we discuss solutions of (5.28) that under conditions (5.29) originate in M but do not converge to the positive fixed point. These solutions may be called *exceptional* since M has measure zero; they are not typically observed in numerical simulations. Nevertheless, these solutions include periodic orbits of all possible periods as well as an uncountable number of oscillatory, nonperiodic ones.

The singularity set

We now determine the singularity set of Eq. (5.28) under conditions (5.29) using the properties of solutions of the linear equation (5.27). The characteristic polynomial of (5.27) is

$$P(\lambda) = \lambda^3 - a\lambda^2 - b\lambda - c. \tag{5.31}$$

Note that the real solutions of (5.31) also give the fixed points of the Riccati equation (5.28). The cubic polynomial P has at least one real root. The next two results give more precise information about all roots of P.

LEMMA 5.1
Assume that conditions (5.29) hold. Then the polynomial P has precisely one positive real root ρ that satisfies

$$\rho \geq \max\left\{ \sqrt[3]{c}, \frac{a + \sqrt{a^2 + 4b}}{2} \right\} \tag{5.32}$$

with equality holding if and only if $a = b = 0$.

PROOF   By the Descartes rule of signs P has only one positive root ρ. Further

$$P(\lambda) = \lambda(\lambda^2 - a\lambda - b) - c$$

and the roots of $\lambda^2 - a\lambda - b$ are $(a \pm \sqrt{a^2 + 4b})/2$. If λ_0 is the nonnegative one of these roots then since $P(\lambda_0) = -c < 0$ it follows that $\rho > \lambda_0$. Next, from $P(\rho) = 0$ we obtain

$$\rho^2 - a\rho - b = \frac{c}{\rho}$$

which implies that $c/\rho \leq \rho^2$, i.e., $\rho \geq \sqrt[3]{c}$. Finally, if equality holds in (5.32) then $\rho = \sqrt[3]{c} \neq \lambda_0$ since $P(\lambda_0) = -c \neq 0$. But then $P(\sqrt[3]{c}) = 0$ implies that $a\sqrt[3]{c} + b = 0$ which implies $a = b = 0$ because $a, b \geq 0$. Conversely if $a = b = 0$ then $\lambda^3 - c = 0$ so $\rho = \sqrt[3]{c}$ and equality holds in (5.32).   ∎

It is possible to find a formula for ρ in terms of radicals, but that information is neither necessary nor particularly useful here.

LEMMA 5.2
Assume that conditions (5.29) hold and let ρ be the positive root of (5.31).
    (a) Eq. (5.31) has two other roots that can be calculated in terms of ρ as

$$r^{\pm} = -\frac{\rho - a}{2} \pm \sqrt{\left(\frac{\rho + a}{2}\right)^2 - \rho^2 + b}. \tag{5.33}$$

(b) If $(\rho + a)^2 \geq 4(\rho^2 - b)$ then the real roots $r^{\pm}$ are negative and

$$-\rho < r^- \leq -\frac{\rho - a}{2} \leq r^+ < 0.$$

(c) If $(\rho + a)^2 < 4(\rho^2 - b)$ (e.g., if $b = 0$) then the complex roots $r^{\pm}$ satisfy

$$|r^{\pm}| = \sqrt{\rho^2 - a\rho - b} = \sqrt{\frac{c}{\rho}}. \tag{5.34}$$

PROOF   (a) Dividing $P(\lambda)$ by $\lambda - \rho$ gives the quadratic polynomial $Q(\lambda) = \lambda^2 + (\rho - a)\lambda + \rho^2 - a\rho - b$. The two roots $r^{\pm}$ of Q are given by (5.33).
(b) In this case,

$$r^- > -\rho \quad \text{iff} \quad \rho - \frac{\rho - a}{2} > \sqrt{\left(\frac{\rho + a}{2}\right)^2 - \rho^2 + b} \quad \text{iff} \quad \rho^2 > b.$$

The last inequality is true by (5.32). Similarly,

$$r^+ < 0 \quad \text{iff} \quad \sqrt{\left(\frac{\rho + a}{2}\right)^2 - \rho^2 + b} < \frac{\rho - a}{2} \quad \text{iff} \quad \rho^2 - a\rho - b > 0.$$

The last inequality is true again by (5.32).
(c) In this case the moduli of $r^{\pm}$ are easily found to be given by (5.34). If $b = 0$ then since by (5.32) $\rho > a$ it follows that

$$(\rho + a)^2 < (2\rho)^2 = 4\rho^2$$

and roots $r^{\pm}$ are complex.   ∎

Based on Lemma 5.2 the next result summarizes the standard facts about the solutions of the linear equation (5.27). Of particular interest to us is the fact that the coefficients of solutions all have the same general formula in terms of initial values.

LEMMA 5.3
Suppose that conditions (5.29) hold.
(a) If $(\rho + a)^2 > 4(\rho^2 - b)$ then

$$y_n = C_1 \rho^n + C_2 (r^+)^n + C_3 (r^-)^n$$

where the coefficients C_j, $j = 1, 2, 3$ are given by

$$C_j(x_0, x_{-1}) = \alpha_{1j} x_0 x_{-1} + \alpha_{2j} x_{-1} + \alpha_{3j} \tag{5.35}$$

for suitable constants α_{ij}, $i, j = 1, 2, 3$ that do not depend on the initial values.

(b) If $(\rho + a)^2 = 4(\rho^2 - b)$ then

$$y_n = C_1\rho^n + (C_2 + C_3 n)r^n \quad \text{where } r = r^+ = r^- = -\frac{\rho - a}{2}$$

where the coefficients C_j are given by (5.35) with constants α_{ij}, $i, j = 1, 2, 3$ appropriate to this case.

(c) If $(\rho + a)^2 < 4(\rho^2 - b)$ then

$$y_n = C_1\rho^n + (\rho^2 - a\rho - b)^{n/2}(C_2 \cos n\theta + C_3 \sin n\theta)$$

where $\theta \in (\pi/2, \pi)$ is a constant and the coefficients C_j are given by (5.35) with constants α_{ij}, $i, j = 1, 2, 3$ appropriate to this case.

PROOF The solutions $\{y_n\}$ in each case are obtained routinely from the classical theory of linear difference equations so we only explain about (5.35) and the range of θ in (c).

(a) The coefficients C_j satisfy the linear system

$$C_1 + C_2 + C_3 = x_0 x_{-1}, \quad C_1/\rho + C_2/(r^+) + C_3/(r^-) = x_{-1},$$
$$\text{and} \quad C_1/\rho^2 + C_2/(r^+)^2 + C_3/(r^-)^2 = 1.$$

This system which is linear in the C_j can be easily solved to obtain

$$C_1 = \frac{\rho^2[x_0 x_{-1} - (r^+ + r^-)x_{-1} + r^+ r^-]}{(\rho - r^+)(\rho - r^-)} \tag{5.36}$$

from which we can read off the values of the constants α_{1j}. Further,

$$C_2 = \frac{-(r^+)^2[x_0 x_{-1} - (r^+ + r^-)x_{-1} + r^+ r^-]}{(\rho - r^+)(r^+ - r^-)}$$

gives the constants α_{2j} and

$$C_3 = x_0 x_{-1} - C_1 - C_2$$
$$= (1 - \alpha_{11} - \alpha_{12})x_0 x_{-1} - (\alpha_{21} + \alpha_{22})x_{-1} - (\alpha_{31} + \alpha_{32})$$

which gives the numbers α_{3j}.

(b) In this case the coefficients C_j satisfy

$$C_1 + C_2 = x_0 x_{-1}, \quad C_1/\rho - 2(C_2 - C_3)/(\rho - a) = x_{-1},$$
$$\text{and} \quad C_1/\rho^2 + 4(C_2 - 2C_3)/(\rho - a)^2 = 1.$$

In this case we obtain

$$C_1 = \frac{4\rho^2(\rho - a)}{(3\rho - a)^2}x_0 x_{-1} + \frac{4\rho^2}{(\rho - a)(3\rho - a)}x_{-1} + \frac{\rho^2(\rho - a)^2}{(3\rho - a)^2} \tag{5.37}$$

from which we can read off the values of the constants α_{1j}. Further,

$$C_2 = x_0 x_{-1} - C_1 = (1 - \alpha_{11})x_0 x_{-1} - \alpha_{21} x_{-1} - \alpha_{31}$$

gives the constants α_{2j} for this case and

$$C_3 = \frac{\rho - a}{2} x_{-1} - \frac{\rho - a}{2\rho} C_1 + C_2$$

from which α_{3j} can be calculated.

(c) In this case the coefficients C_j satisfy

$$C_1 + C_2 = x_0 x_{-1}, \quad C_1/\rho + (C_2 \cos\theta - C_3 \sin\theta)/\sqrt{\rho^2 - a\rho - b} = x_{-1}, \quad (5.38)$$
$$\text{and} \quad C_1/\rho^2 + (C_2 \cos 2\theta - C_3 \sin 2\theta)/(\rho^2 - a\rho - b) = 1$$

where θ is defined by the equalities

$$\cos\theta = -\sqrt{\frac{\rho}{c}}\frac{\rho - a}{2}, \quad \sin\theta = \sqrt{\frac{\rho}{c}}\sqrt{\rho^2 - b - \left(\frac{\rho + a}{2}\right)^2} \quad (5.39)$$

which also show that $\theta \in (\pi/2, \pi)$. From (5.38) we obtain using $\rho^2 - a\rho - b = c/\rho$,

$$C_1 = \frac{\rho^2 c}{\rho^3 + c - 2\sqrt{\rho^3 c}\cos\theta}\left[\frac{\rho}{c} x_0 x_{-1} - 2\sqrt{\frac{\rho}{c}}(\cos\theta)x_{-1} + 1\right] \quad (5.40)$$

from which we can read off the values of the constants α_{1j}. Further,

$$C_2 = x_0 x_{-1} - C_1$$
$$C_3 = \frac{c \sin 2\theta}{\rho^3} C_1 + \frac{\cos 2\theta}{\sin 2\theta} C_2 - \frac{c}{\rho}\sin 2\theta$$

from which α_{ij}, $i = 2, 3$ can be calculated. ∎

From (5.35) it appears that the initial points (x_0, x_{-1}) that lead to singularities or the singular states of (5.28) must be located on a family of hyperbolas in the two-dimensional state-space (or the phase plane). Since each of the coefficients C_j depends on the two intial values, each solution of the linear equation (5.27) is a function $y_n(u, v)$ of two variables, all other parameters being fixed. Thus the singularity set S of Eq. (5.28) can be written as

$$S = \bigcup_{n=-1}^{\infty} \{(u, v) : y_n(u, v) = 0\}.$$

We note that $S \subset \mathbb{R}^2\backslash(0, \infty)^2$ because under conditions (5.29) each solution $\{x_n\}$ of (5.28) with $(x_0, x_{-1}) \in (0, \infty)^2$ satisfies $x_n > 0$ for all $n \geq -1$ and thus there are no undefined values. Now the next result is an immediate consequence of the preceding lemma.

THEOREM 5.2
Suppose that conditions (5.29) hold. Then the singularity set of Eq. (5.28) is the following sequence of hyperbolas

$$S = \bigcup_{n=-1}^{\infty} \{(u, v) : \beta_{1n} uv + \beta_{2n} v + \beta_{3n} = 0\} \subset \mathbb{R}^2 \backslash (0, \infty)^2$$

where the sequences β_{in} are defined as follows:
  (a) If $(\rho + a)^2 > 4(\rho^2 - b)$ then

$$\beta_{in} = \alpha_{i1} + \alpha_{i2}(r^+/\rho)^n + \alpha_{i3}(r^-/\rho)^n$$

where α_{ij} are the constants in Lemma 5.3(a).
  (b) If $(\rho + a)^2 = 4(\rho^2 - b)$ then

$$\beta_{in} = \alpha_{i1} + (-1/2)^n (1 - a/\rho)^n (\alpha_{i2} + \alpha_{i3}n)$$

where α_{ij} are the constants in Lemma 5.3(b).
  (c) If $(\rho + a)^2 < 4(\rho^2 - b)$ then

$$\beta_{in} = \alpha_{i1} + (c/\rho^3)^{n/2} (\alpha_{i2} \cos n\theta + \alpha_{i3} \sin n\theta)$$

where α_{ij} are the constants in Lemma 5.3(c).

Global asymptotic stability

In this section we use the preceding results to show that under conditions (5.29) the positive fixed point ρ is stable and almost all solutions of Eq. (5.28) converge to ρ if at least one of the parameters a or b is positive.

LEMMA 5.4
Under conditions (5.29) ρ is the unique positive fixed point of (5.28) and if $a + b > 0$ then ρ is locally asymptotically stable.

PROOF   Define
$$f(u, v) = a + \frac{b}{u} + \frac{c}{uv}.$$

Since the fixed points of (5.28) correspond to the roots of the polynomial P in (5.31), the uniqueness of ρ follows from Lemma 5.1. Next, the characteristic equation of the linearization of (5.28) at the fixed point (ρ, ρ) is

$$\lambda^2 - f_u(\rho, \rho)\lambda - f_v(\rho, \rho) = 0 \tag{5.41}$$

where

$$f_u = \frac{-1}{u^2}\left(b + \frac{c}{v}\right), \quad f_v = \frac{-c}{uv^2}.$$

These and the fact that $b\rho + c = \rho^3 - a\rho^2$ determine Eq. (5.41) as

$$\lambda^2 + \frac{\rho - a}{\rho}\lambda + \frac{c}{\rho^3} = 0.$$

The zeros of this quadratic equation are

$$\lambda^{\pm} = \frac{\rho - a}{2\rho}\left[-1 \pm \sqrt{1 - \frac{4c}{\rho(\rho - a)^2}}\right].$$

If

$$\rho(\rho - a)^2 \geq 4c \qquad (5.42)$$

then the numbers $\lambda^{\pm}$ are real and $\lambda^- \leq \lambda^+ < 0$. Further, a little algebra shows that $\lambda^- > -1$ if and only if

$$\sqrt{1 - \frac{4c}{\rho(\rho - a)^2}} < \frac{\rho + a}{\rho - a}$$

which is obviously true since the left side is less than 1 and the right side greater than 1. Thus if (5.42) holds then ρ is a stable node for (5.28). Next suppose that (5.42) is false. Then $\lambda^{\pm}$ are complex with $|\lambda^{\pm}| = \sqrt{c/\rho^3} < 1$ where the inequality holds by Lemma 5.1 when $a + b > 0$. Thus if (5.42) is false then ρ is a stable focus for (5.28). These cases exhaust all possibilities so ρ is locally asymptotically stable. ∎

In considering the global behavior of solutions of Eq. (5.28) the following set must be considered:

$$M = \mathcal{S} \cup \{(u,v) : C_1(u,v) = 0\} = \mathcal{S} \cup \{(u,v) : \alpha_{11}uv + \alpha_{21}v + \alpha_{31} = 0\} \quad (5.43)$$

where $\mathcal{S}$ is the singularity set of (5.28) as determined in Theorem 5.2 and α_{i1} are the constants defined in Lemma 5.3.

THEOREM 5.3
Assume that conditions (5.29) hold with $a + b > 0$. Then the positive fixed point ρ is globally asymptotically stable relative to $\mathbb{R}^2 \backslash M$ where the set $M \subset \mathbb{R}^2 \backslash (0,\infty)^2$ defined by (5.43) has Lebesgue measure zero.

PROOF By Lemma 5.4 ρ is stable so it only remains to prove global attractivity. If $\{x_n\}$ is a solution of Eq. (5.28) then we claim that $\lim_{n\to\infty} x_n = \rho$ if $(x_0, x_{-1}) \notin M$.

First, consider the case where $r^\pm$ are real and distinct. In this case, Lemma 5.3 implies that

$$x_n = \frac{y_n}{y_{n-1}} = \frac{C_1\rho^n + C_2(r^+)^n + C_3(r^-)^n}{C_1\rho^{n-1} + C_2(r^+)^{n-1} + C_3(r^-)^{n-1}}. \tag{5.44}$$

Since $(x_0, x_{-1}) \notin M$ we have $C_1 = C_1(x_0, x_{-1}) \neq 0$. Now dividing by $C_1\rho^{n-1}$ yields

$$x_n = \frac{\rho + (C_2\rho/C_1)(r^+/\rho)^n + (C_3\rho/C_1)(r^-/\rho)^n}{1 + (C_2/C_1)(r^+/\rho)^{n-1} + (C_3/C_1)(r^-/\rho)^{n-1}}$$

which implies, by Lemma 5.2(b), that $\lim_{n\to\infty} x_n = \rho$. Next, in the case of equal real roots a similar calculation gives

$$x_n = \frac{\rho + r(C_2/C_1 + C_3 n/C_1)(r/\rho)^{n-1}}{1 + [C_2/C_1 + C_3(n-1)/C_1](r/\rho)^{n-1}}.$$

Since by Lemma 5.2(b) $|r/\rho| < 1$ it follows that $\lim_{n\to\infty} x_n = \rho$. Next, in the case of complex roots

$$x_n = \frac{\rho + \sqrt{\rho^2 - a\rho - b}(1 - a/\rho - b/\rho^2)^{(n-1)/2}(C_2\cos n\theta + C_3\sin n\theta)/C_1}{1 + (1 - a/\rho - b/\rho^2)^{(n-1)/2}[C_2\cos(n-1)\theta + C_3\sin(n-1)\theta]/C_1}$$

so again we obtain $\lim_{n\to\infty} x_n = \rho$.

Finally, since M is a countable collection of hyperbolas it has Lebesgue measure zero in $\mathbb{R}^2$. To establish that $M \subset \mathbb{R}^2\backslash(0,\infty)^2$ it remains to show that the set

$$\{(u,v) : C_1(u,v) = 0\} = \{(u,v) : \alpha_{11}uv + \alpha_{21}v + \alpha_{31} = 0\} \tag{5.45}$$

does not intersect the positive quadrant $(0,\infty)^2$. From expressions (5.36), (5.37) and (5.40) above we see that $\alpha_{i1} > 0$ for $i = 1, 2, 3$ in each of the three possible cases. Thus the set (5.45) cannot contain points (u,v) with $u, v > 0$. ∎

REMARK 5.2 Theorem 5.3 also reveals an interesting property of solutions of the linear difference equation (5.27): All solutions $\{y_n\}$ of (5.27) except for those whose initial values are in the following set of measure zero

$$M_1 = \left\{(y_0, y_{-1}, y_{-2}) : y_{-2}y_{-1} = 0 \text{ or } \frac{y_0}{y_{-1}}, \frac{y_{-1}}{y_{-2}} \in M\right\}$$

have the ratios of their consecutive terms y_n/y_{n-1} converge to the real number ρ. Thus, for all large values of n we have $y_n \approx \rho y_{n-1}$. For initial values in M_1 more varied types of asymptotic or long-term behavior can occur which we discuss in the next section. ∎

In the boundary case $a = b = 0$ in (5.29) Theorem 5.3 is false; as the next proposition shows the solutions of (5.28) exhibit a completely different behavior in this case.

PROPOSITION 5.1

If neither of the initial values x_0, x_{-1} is zero then the corresponding solution of

$$x_{n+1} = \frac{c}{x_n x_{n-1}}, \quad c \neq 0 \tag{5.46}$$

is the period-3 sequence

$$\left\{ x_{-1}, x_0, \frac{c}{x_0 x_{-1}}, x_{-1}, x_0, \frac{c}{x_0 x_{-1}}, \dots \right\}.$$

In particular, every nonconstant solution (i.e., with $x_0 \neq \sqrt[3]{c}$ or $x_{-1} \neq \sqrt[3]{c}$) of (5.46) has period 3.

The next result applies Theorem 5.3 to an equation that is similar to (5.28).

COROLLARY 5.4

Assume that conditions (5.29) hold for the following equation

$$z_{n+1} = \frac{1}{a + b z_n + c z_n z_{n-1}}. \tag{5.47}$$

If $a > 0$ and $z_0, z_{-1} \geq 0$ or if $a + b > 0$ and $z_0 > 0$, $z_{-1} \geq 0$ then $\lim_{n \to \infty} z_n = 1/\rho$.

PROOF   Since the change of variables $x_n = 1/z_n$ transforms (5.47) into (5.28), if $z_0, z_{-1} > 0$ then Theorem 5.3 implies that

$$\lim_{n \to \infty} z_n = \lim_{n \to \infty} \frac{1}{x_n} = \frac{1}{\rho}.$$

If $a > 0$ and either $z_0 = 0$ or $z_{-1} = 0$ then from (5.47) we find that $z_1, z_2 > 0$ so again Theorem 5.3 applies. The last case is argued similarly.   ∎

A variety of exceptional solutions

The proof of Theorem 5.3 contains information about solutions that do not converge to ρ. These are exceptional solutions of (5.28) since they can only originate in a set with Lebesgue measure zero, i.e., $M \backslash S$. Yet, they can be either convergent or oscillatory, periodic or nonperiodic. Thus they exhibit a significant variety of qualitatively different types of behavior.

THEOREM 5.4

Assume that conditions (5.29) hold with $a + b > 0$.

(a) The hyperbola $H = \{(u,v) : uv - (r^+ + r^-)v + r^+ r^- = 0\}$ is an invariant subset of M.

(b) If $(\rho + a)^2 \geq 4(\rho^2 - b)$ and $(x_0, x_{-1}) \in H \backslash \{(r^+, r^+)\}$ then $\lim_{n \to \infty} x_n = r^-$.

(c) If $(\rho + a)^2 < 4(\rho^2 - b)$ and:

(i) there are positive integers p, q that are relatively prime such that $\theta = \pi q / p$ satisfies (5.39) then for each $(x_0, x_{-1}) \in H$ the corresponding solution $\{x_n\}$ of (5.28) is periodic with period p.

(ii) θ is an irrational multiple of π, then the corresponding solution of (5.28) is oscillatory but not periodic and the orbit $\{(x_n, x_{n-1})\}$ is dense in H.

PROOF (a) Notice from (5.36) that the expression for C_1 is real even if $r^\pm$ are complex and that $C_1(x_0, x_{-1}) = 0$ if and only if

$$x_0 x_{-1} - (r^+ + r^-)x_{-1} + r^+ r^- = 0. \tag{5.48}$$

Indeed, from (5.40) we obtain $C_1 = 0$ if and only if

$$x_0 x_{-1} - 2\sqrt{\frac{c}{\rho}}(\cos \theta)x_{-1} + \frac{c}{\rho} = 0$$

which is identical to (5.48) if $r^\pm$ are complex. Thus $C_1 = 0$ in all cases if it is shown that $C_1(x_{n+1}, x_n) = 0$ for all $n \geq 0$ whenever x_0 and x_{-1} satisfy (5.48).

Now, if (5.48) holds then

$$C_1(x_1, x_0) = x_0 \left(a + \frac{b}{x_0} + \frac{c}{x_0 x_{-1}} \right) - (r^+ + r^-)x_0 + r^+ r^-$$

$$= ax_0 + b + \frac{c}{x_{-1}} + (\rho - a)x_0 + \rho^2 - a\rho - b$$

$$= \frac{c}{x_{-1}} + \rho x_0 + \rho^2 - a\rho. \tag{5.49}$$

From (5.48) we obtain

$$x_0 = -(\rho - a) - \frac{\rho^2 - a\rho - b}{x_{-1}} = -(\rho - a) - \frac{c}{\rho x_{-1}}$$

which if inserted into (5.49) yields $C_1(x_1, x_0) = 0$. The proof of (a) can now be completed by induction.

(b) In this case the roots $r^\pm$ are real. First suppose that $(\rho + a)^2 > 4(\rho^2 - b)$. If $(x_0, x_{-1}) \in H$ then $C_1 = 0$ in (5.44) and thus

$$x_n = \frac{C_2(r^+)^n + C_3(r^-)^n}{C_2(r^+)^{n-1} + C_3(r^-)^{n-1}}.$$

If $C_3 = 0$ then $x_n = r^+$ for all n which can occur only if $x_1 = x_0 = r^+$. If $C_3 \neq 0$ then dividing by $C_3(r^-)^{n-1}$ and taking the limit gives

$$\lim_{n \to \infty} x_n = \lim_{n \to \infty} \frac{(C_2/C_3)(r^+/r^-)^n + r^-}{(C_2/C_3)(r^+/r^-)^{n-1} + 1} = r^-.$$

The argument for the case $(\rho + a)^2 = 4(\rho^2 - b)$ is similar but using Lemma 5.3(b); we omit the straightforward details.

(c) In this case the roots $r^{\pm}$ are complex and if $C_1 = 0$ then from Lemma 5.3 we obtain

$$x_n = \frac{(\rho^2 - a\rho - b)(C_2 \cos n\theta + C_3 \sin n\theta)}{C_2 \cos(n-1)\theta + C_3 \sin(n-1)\theta}$$

$$= \frac{c}{\rho} \cos \theta + \frac{c}{\rho} \sin \theta \frac{C_3 \cos(n-1)\theta - C_2 \sin(n-1)\theta}{C_2 \cos(n-1)\theta + C_3 \sin(n-1)\theta}.$$

Define $\cos \phi = C_2/\sqrt{C_2^2 + C_3^2}$ and $\sin \phi = C_3/\sqrt{C_2^2 + C_3^2}$. Then

$$x_n = \frac{c}{\rho} \cos \theta + \frac{c}{\rho} \sin \theta \frac{\sin \phi \cos(n-1)\theta - \cos \phi \sin(n-1)\theta}{\cos \phi \cos(n-1)\theta + \sin \phi \sin(n-1)\theta}$$

$$= \frac{c}{\rho} \cos \theta - \frac{c}{\rho} \sin \theta \frac{\sin[(n-1)\theta - \phi]}{\cos[(n-1)\theta - \phi]}$$

$$= \frac{c}{\rho} \cos \theta - \frac{c}{\rho} \sin \theta \tan[(n-1)\theta - \phi]. \qquad (5.50)$$

Now if $\theta = \pi q/p$ is a rational multiple of π then it follows from (5.50) that x_n is periodic with period p if q/p is in reduced form, i.e., if p, q are relatively prime. If θ is not a rational multiple of π then the angles $(n-1)\theta - \phi$ form a dense subset of the circle as $n \to \infty$. Given that $\tan x$ is a homeomorphism from $(-\pi/2, \pi/2)$ to $\mathbb{R}$ we conclude from (5.50) that the sequence $\{x_n\}$ is dense in $\mathbb{R}$. It follows that the orbit $\{(x_n, x_{n-1})\}$ is dense in H. ∎

Figure 5.1 shows the characteristic polynomials P_A and P_B of (5.28) for two different sets of parameters:

$$(A) \qquad a = b = c = 1$$

and

$$(B) \qquad a = c = 0.25, \ b = 2.$$

The roots of the characteristic polynomial P_A are estimated to be

$$\rho_A = 1.8393, \quad r_A^+ = -0.42 + 0.606i, \quad r_A^- = -0.42 - 0.606i$$

and for the polynomial P_B we obtain the estimates

$$\rho_B = 1.5988, \quad r_B^+ = -0.128, \quad r_B^- = -1.221.$$

In Figure 5.2 we see the invariant hyperbolas H_A and H_B of exceptional solutions corresponding to the sets (A) and (B) of parameter values above.

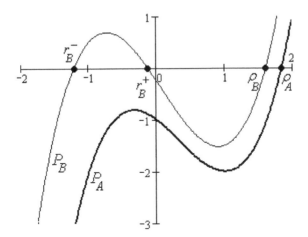

FIGURE 5.1

Two characteristic polynomials of the Riccati equation (5.28); see the text for parameter values.

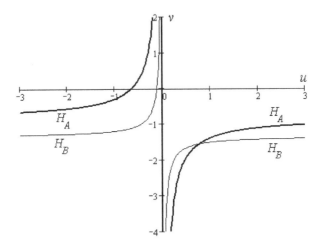

FIGURE 5.2

Two invariant hyperbolas of exceptional solutions of the Riccati equation (5.28); see the text for parameter values.

5.5   Linear form symmetry

The one-variable function h may be defined using structures that are more complex than groups. In this section we consider the function

$$h(u) = -\alpha u, \quad \alpha \neq 0$$

on a nontrivial field $\mathcal{F}$ where α is a fixed element of the field. For convenience, we represent the field operations by the symbols for ordinary addition and multiplication. The linear function h defines an order-reducing form symmetry of type (5.2)

$$H(u_0, u_1, \ldots, u_k) = [u_0 - \alpha u_1, u_1 - \alpha u_2, \ldots, u_{k-1} - \alpha u_k]. \qquad (5.51)$$

We call H the *linear form symmetry*. Relative to the (additive) group of a field, the linear form symmetry generalizes both the identity form symmetry ($\alpha = -1$) and the inversion form symmetry ($\alpha = 1$). As we see in this section and the next, large classes of difference equations possess the linear form symmetry (5.51). These classes include the linear equations as might be expected as well as several types of nonlinear equations.

5.5.1   Determination criterion

The linear form symmetry is characterized by a change of variables to

$$t_n = x_n - \alpha x_{n-1}.$$

The factor and cofactor equations for the SC factorization of (5.1) with the linear form symmetry are given by (5.5) and (5.6) as

$$t_{n+1} = \phi_n(t_n, \ldots, t_{n-k+1}), \qquad (5.52)$$
$$x_{n+1} = t_{n+1} + \alpha x_n. \qquad (5.53)$$

For each sequence $\{t_n\}$ in $\mathcal{F}$ the general solution of the cofactor equation (5.53) is readily calculated:

$$x_n = \alpha^n x_0 + \sum_{j=1}^{n} \alpha^{n-j} t_j, \quad n \geq 1. \qquad (5.54)$$

By Theorem 3.1 this equation gives a solution of Eq. (5.1) if $\{t_n\}$ is a solution of (5.52) with the form symmetry (5.51).

The following corollary of Theorem 5.1 gives a necessary and sufficient condition for verifying whether Eq. (5.1) has the linear form symmetry.

COROLLARY 5.5

For every $u_0, v_1, \ldots, v_k$ in a field $\mathcal{F}$ define $\zeta_0 = u_0$ and

$$\zeta_j = \frac{u_0}{\alpha^j} - \sum_{i=1}^{j} \frac{v_i}{\alpha^{j-i+1}}, \quad j = 1, \ldots, k. \tag{5.55}$$

Eq. (5.1) has the linear form symmetry (5.51) if and only if the quantity

$$f_n(u_0, \zeta_1, \ldots, \zeta_k) - \alpha u_0 \tag{5.56}$$

is independent of u_0. In this case, the factor functions are defined as

$$\phi_n(v_1, \ldots, v_k) = f_n(u_0, \zeta_1, \ldots, \zeta_k) - \alpha u_0.$$

Note that (5.55) defines ζ_j in Corollary 5.5 directly rather than recursively. It is obtained from the recursive definition (5.7) by a simple calculation; since

$$\zeta_j = h^{-1}(\zeta_{j-1}^{-1} * v_j) = -\frac{1}{\alpha}(-\zeta_{j-1} + v_j) = \frac{\zeta_{j-1} - v_j}{\alpha},$$

equality (5.55) can be established by straightforward iteration.

The next result gives necessary and sufficient conditions for the existence of a linear form symmetry for certain rational difference equations of order two in an arbitrary algebraic field.

COROLLARY 5.6

The rational recursive equation

$$x_{n+1} = \frac{a x_n^2 + b x_n x_{n-1} + c x_{n-1}^2 + d x_n + e x_{n-1} + \sigma_n}{A x_n + B x_{n-1} + C} \tag{5.57}$$

where the constant parameters a, b, c, d, e, A, B, C and the sequence $\{\sigma_n\}$ are in a nontrivial field $\mathcal{F}$, has the linear form symmetry $u - \alpha v$ with the constant $\alpha = -B/A$ if and only if $A, B \neq 0$ and the following equalities hold:

$$cA^2 - aB^2 - B^3 = 0, \tag{5.58}$$

$$2cA^2 - bAB - B^3 = 0, \tag{5.59}$$

$$eA^2 - dAB - B^2C = 0. \tag{5.60}$$

The corresponding factorization of (5.57) subject to (5.58)–(5.60) is

$$t_{n+1} = \frac{(CA^2/B^2)t_n^2 + (eA/B)t_n + \sigma_n}{A t_n + C}, \tag{5.61}$$

$$x_{n+1} = t_{n+1} - \frac{B}{A} x_n.$$

PROOF   Define the functions

$$f_n(u, v) = \frac{au^2 + buv + cv^2 + du + ev + \sigma_n}{Au + Bv + C}$$

and

$$\zeta_1 = \frac{u_0 - v_1}{\alpha}.$$

By Corollary 5.5 $u - \alpha v$ is a form symmetry of (5.57) if and only if the following quantity is independent of u_0:

$$f_n(u_0, \zeta_1) - \alpha u_0 = \left[\left(A + \frac{B}{\alpha} \right) u_0 - \frac{B}{\alpha} v_1 + C \right]^{-1} \left[\left(d - \alpha C + \frac{e}{\alpha} \right) u_0 + \right.$$

$$+ \left(a - B + \frac{b}{\alpha} + \frac{c}{\alpha^2} \right) u_0^2 + \left(B - \frac{b}{\alpha} - \frac{2c}{\alpha^2} \right) u_0 v_1 +$$

$$\left. + \frac{c}{\alpha^2} v_1^2 - \frac{e}{\alpha} v_1 + \sigma_n \right]. \tag{5.62}$$

Setting the coefficients of terms containing u_0 equal to zero gives

$$A + \frac{B}{\alpha} = 0, \tag{5.63}$$

$$d - \alpha C + \frac{e}{\alpha} = 0 \tag{5.64}$$

$$a - B - \alpha A + \frac{b}{\alpha} + \frac{c}{\alpha^2} = 0, \tag{5.65}$$

$$B - \frac{b}{\alpha} - \frac{2c}{\alpha^2} = 0. \tag{5.66}$$

From (5.63) we obtain the value

$$\alpha = -\frac{B}{A}. \tag{5.67}$$

Adding (5.65) and (5.66) gives

$$0 = a - \alpha A - \frac{c}{\alpha^2} = a + B - \frac{cA^2}{B^2}$$

from which (5.58) follows. Next, using the value of α in (5.67) in (5.66) and (5.64) gives the equalities (5.59) and (5.60), respectively. Finally, as in Corollary 5.5, the remaining terms in (5.62) that do not contain u_0 give the factor equation (5.61).   ∎

Conditions (5.58)–(5.60) are not as restrictive as they may seem when it comes to the existence of interesting solutions for (5.57). For $\mathcal{F} = \mathbb{R}$ the field of real numbers, we discover in Section 5.6.5 that Eq. (5.57) has a rich variety of solutions, including chaotic ones, even when (5.58)–(5.60) all hold.

In the next example we establish the existence of a linear form symmetry in a third-order nonlinear difference equation.

Example 5.6

Let $\mathcal{F}$ be a nontrivial algebraic field, $a_0, a_1, a_2, b, c \in \mathcal{F}$, and let $g_n : \mathcal{F}^2 \to \mathcal{F}$ be a given sequence of functions. Consider the third-order equation

$$x_{n+1} = a_0 x_n + a_1 x_{n-1} + a_2 x_{n-2} + g_n(x_n - b x_{n-1}, x_n - c x_{n-2}). \quad (5.68)$$

We use Corollary 5.5 to determine conditions on the parameters that imply the existence of the linear form symmetry for some nonzero real constant α. Define the functions

$$f_n(u_0, u_1, u_2) = a_0 u_0 + a_1 u_1 + a_2 u_2 + g_n(u_0 - b u_1, u_0 - c u_2)$$

which lead to the following expression in (5.56) via (5.55)

$$f_n(u_0, \zeta_1, \zeta_2) - \alpha u_0 = \left(a_0 - \alpha + \frac{a_1}{\alpha} + \frac{a_2}{\alpha^2}\right) u_0 - \frac{a_1}{\alpha} v_1 - \frac{a_2}{\alpha^2} v_1 - \frac{a_2}{\alpha} v_2$$
$$+ g_n \left(u_0 - \frac{b}{\alpha} u_0 + \frac{b}{\alpha} v_1, u_0 - \frac{c}{\alpha^2} u_0 + \frac{c}{\alpha^2} v_1 + \frac{c}{\alpha} v_2\right).$$

The preceding expression is independent of u_0 for all choices of v_1, v_2 if and only if

$$\alpha^3 - a_0 \alpha^2 - a_1 \alpha - a_2 = 0, \quad (5.69)$$
$$b = \alpha, \quad c = \alpha^2.$$

Using the value $\alpha = b \in \mathcal{F}$ in the polynomial (5.69) we conclude that (5.68) has the linear form symmetry $u - bv$ if and only if

$$b \neq 0, \quad c = b^2, \quad b^3 - a_0 b^2 - a_1 b - a_2 = 0. \quad (5.70)$$

If the above conditions are satisfied then by Corollary 5.5 the factor equation of (5.68) is the following second-order equation

$$t_{n+1} = -\frac{1}{\alpha^2}(a_1 \alpha + a_2) t_n - \frac{a_2}{\alpha} t_{n-1} + g_n \left(\frac{b}{\alpha} t_n, \frac{c}{\alpha^2} t_n + \frac{c}{\alpha} t_{n-1}\right)$$
$$= -\frac{1}{b^2}(a_1 b + a_2) t_n - \frac{a_2}{b} t_{n-1} + g_n(t_n, t_n + b t_{n-1}).$$

Since by (5.70), $a_1 b + a_2 = b^3 - a_0 b^2$ the above equation further reduces to

$$t_{n+1} = (a_0 - b) t_n - \frac{a_2}{b} t_{n-1} + g_n(t_n, t_n + b t_{n-1}).$$

□

5.5.2   Periodic solutions

In this section we assume that $\mathcal{F}$ is the field of complex numbers $\mathbb{C}$ and derive some general results about periodic solutions of equations having a linear form symmetry. These results are not generally true for the more special types of field symmetries such as the inversion or the identity. The next basic result concerns the cofactor equation (5.53).

LEMMA 5.5
Let p be a positive integer and let $\alpha \in \mathbb{C}$, $\alpha \neq 0$.
(a) If for a given sequence $\{t_n\}$ Eq. (5.53) has a solution $\{x_n\}$ of period p then $\{t_n\}$ is periodic with period p.
(b) Let $\{t_n\}$ be a periodic sequence in $\mathbb{C}$ with period p and assume that α is not a p-th root of unity; i.e., $\alpha^p \neq 1$. If $\{\tau_0, \ldots, \tau_{p-1}\}$ is one cycle of $\{t_n\}$ and

$$\xi_i = \frac{1}{1-\alpha^p} \sum_{j=0}^{p-1} \alpha^{p-j-1} \tau_{(i+j)\bmod p} \quad i = 0, 1, \ldots, p-1 \qquad (5.71)$$

then the solution $\{x_n\}$ of Eq. (5.53) with $x_0 = \xi_0$ and $t_1 = \tau_0$ has period p and $\{\xi_0, \ldots, \xi_{p-1}\}$ is a cycle of $\{x_n\}$. If p is a minimal (or prime) period of $\{t_n\}$ then $\{x_n\}$ has minimal period p.

PROOF   (a) Suppose that for a given sequence $\{t_n\}$ of complex numbers the corresponding solution of (5.53) is periodic with period p. Let $t_1 = x_1 - \alpha x_0$ and from (5.53) obtain inductively for $i = 1, \ldots, p$

$$t_{p+i} = x_{p+i} - \alpha x_{p+i-1} = x_i - \alpha x_{i-1} = t_i.$$

Therefore, $\{t_n\}$ is periodic with period p.
(b) With $x_0 = \xi_0$ and $t_1 = \tau_0$ we get $x_1 = \alpha x_0 + t_1 = \alpha \xi_0 + \tau_0$. Using (5.71) for ξ_0 gives

$$x_1 = \frac{\alpha}{1-\alpha^p}\left(\sum_{j=0}^{p-1}\alpha^{p-j-1}\tau_j\right) + \tau_0 = \frac{1}{1-\alpha^p}\left(\sum_{j=0}^{p-2}\alpha^{p-j-1}\tau_{j+1} + \tau_0\right) = \xi_1.$$

Proceeding in an inductive fashion, we see in this way that $x_i = \xi_i$ for $i = 0, \ldots, p-1$. Next, we show that $x_p = x_0$. Using (5.54) we have

$$x_p = \alpha^p \xi_0 + \sum_{j=0}^{p-1}\alpha^{p-j-1}\tau_j = \frac{\alpha^p}{1-\alpha^p}\sum_{j=0}^{p-1}\alpha^{p-j-1}\tau_j + \sum_{j=0}^{p-1}\alpha^{p-j-1}\tau_j = \xi_0 = x_0.$$

Now, again by induction, for $i = 1, \ldots, p$

$$x_{p+i} = t_{p+i} + \alpha x_{p+i-1} = t_i + \alpha x_{i-1} = x_i.$$

Hence $\{x_n\}$ is a solution of (5.53) with period p, as claimed. If p is the minimal period of $\{t_n\}$ and q is the minimal period of $\{x_n\}$ then by what we just established, $q \leq p$. However, by Part (a) $\{t_n\}$ has period $q \geq p$ since p is the minimal period for $\{t_n\}$. Therefore, $q = p$. ∎

Recall that the particular values $\alpha = \pm 1$ correspond to the inversion and identity form symmetries for the additive group of the field $\mathbb{C}$. The next example shows that for these special types of linear form symmetry the cofactor equations may not preserve the periodic structures of the factor equations. Note that 1 is a p-th root of unity for every positive integer p and -1 is a p-th root of unity for all even positive integers p.

Example 5.7
Consider the following second-order difference equation on $\mathbb{R}$:

$$x_{n+1} = \frac{a}{x_n - bx_{n-1}} + bx_n, \quad a, b \neq 0.$$

Rearranging terms and substituting $t_n = x_n - bx_{n-1}$ yields the factor equation

$$t_{n+1} = \frac{a}{t_n}.$$

Every solution of the above equation with initial value $t_0 \neq a$ has period $p = 2$:

$$\left\{ t_0, \frac{a}{t_0}, t_0, \frac{a}{t_0}, \ldots \right\}.$$

If $b = 1$ (corresponding to the inversion form symmetry) then the solution of the cofactor equation $x_{n+1} = t_{n+1} + x_n$ is

$$x_n = x_0 + \sum_{j=1}^{n} t_j$$

which is nonperiodic for all $t_0 \neq 0$.

If $b = -1$ (corresponding to the identity form symmetry) then the solution of the cofactor equation $x_{n+1} = t_{n+1} - x_n$ is

$$x_n = (-1)^n x_0 + \sum_{j=1}^{n} (-1)^{n-j} t_j.$$

Although periodic solutions are possible for some initial values $t_0 = x_0 + x_{-1}$ clearly not all solutions $\{x_n\}$ are periodic. ☐

The next result is an immediate consequence of Lemma 5.5 and Corollary 5.5.

COROLLARY 5.7

Assume that the quantity defined in (5.56) is independent of u_0 for all functions f_n. Then:

(a) Eq. (5.1) has the linear form symmetry with SC factorization given by (5.52) and (5.53).

(b) Let p be a positive integer and let $\alpha \in \mathbb{C}$ such that $\alpha \neq 0$ and α is not a p-th root of unity; i.e., $\alpha^p \neq 1$. If the factor equation (5.52) has a solution $\{t_n\}$ of minimal (or prime) period p then Eq. (5.1) has a solution $\{x_n\}$ with minimal period p whose cycles are given by equations (5.71).

Example 5.8

Let α be a real number, $\alpha \neq 0$ and consider the second order difference equation

$$x_{n+1} = \alpha x_n + \left| \frac{1}{x_n - \alpha x_{n-1}} - 1 \right| \qquad (5.72)$$

with real initial values x_0, x_{-1}. After subtracting the term αx_n from both sides, the substitution $t_n = x_n - \alpha x_{n-1}$ defines a linear form symmetry on $\mathbb{R}$ with a factor equation

$$t_{n+1} = \left| \frac{1}{t_n} - 1 \right|. \qquad (5.73)$$

In Theorem 4.10 of Section 4.4 it is established that Eq. (5.73) *has a solution of period p for every $p \neq 3$ with each cycle* $\{r_1, \ldots, r_p\}$ given by

$$r_1 = \frac{1 + \sqrt{5}}{2}, \ r_2 = \frac{\sqrt{5} - 1}{\sqrt{5} + 1} \quad (p = 2)$$

$$r_1 = \frac{1}{2} \left[y_{p-4} + \sqrt{y_{p-4}^2 + 4 y_{p-4} y_{p-1}} \right],$$

$$r_k = \frac{y_{k-4} r_1 - y_{k-2}}{y_{k-3} - y_{k-5} r_1}, \ 2 \leq k \leq p, \ (p \geq 4).$$

Here y_n is the n-th Fibonacci number; i.e., $y_{n+1} = y_n + y_{n-1}$ for $n \geq 0$ where we define

$$y_{-3} = -1, \ y_{-2} = 1 \ y_{-1} = 1 \ y_0 = 1.$$

If $\alpha \neq \pm 1$ then by Corollary 5.7, Eq. (5.72) also has periodic solutions of period p in $\mathbb{R}$ for all positive integers $p \neq 3$. A cycle $\{\xi_0, \ldots, \xi_{p-1}\}$ of $\{x_n\}$ for each p is given by formula (5.71). Since Eq. (5.73) does not have a solution of period three, Lemma 5.5(a) and Theorem 3.1 imply that the second-order equation (5.72) has no solutions of period three either. These statements would be more difficult to establish without using the reduction of order to the first order equation (5.73).

This example also illustrates an important difference between the solutions of Eq. (5.72) and those of the second-order absolute value equation studied in Section 4.4. Whereas the qualitative properties of solutions of the first-order equation (5.73) are shared by the solutions of Eq. (5.72) for $\alpha \neq \pm 1$, in the case of the absolute value equation the properties of the solutions of (5.73) are not shared by the solutions of the second-order equation. Instead, they are seen only in sequences of *ratios of consecutive terms* of each solution of the absolute value equation. Thus the *field* linear form symmetry of $\mathbb{R}$ preserves the aforementioned qualitative properties, but the *inversion* form symmetry of the absolute value equation does not. $\quad\square$

REMARK 5.3 The relationship between a solution of period p of the factor equation and the period-p solution of original difference equation that is noted in Corollary 5.7 is not direct. Lemma 5.5, which specifies this relationship quantitatively, does so through the cofactor equation not the original equation (5.1). *However, if* $|\alpha| < 1$ then we find in the next section that an attracting periodic solution of the factor equation yields the same type of solution for the cofactor (hence also the original) equation, even if the restrictions on initial values stated in Lemma 5.5(b) are not met. This feature is particularly useful to have in numerical simulations on computers. $\quad\blacksquare$

5.5.3 *Limit cycles

As further applications of linear form symmetries, in this section we use the associated reduction in order to discuss *convergence* to periodic solutions of equations that have such a form symmetry. These attracting periodic solutions are the ones that are preserved by the linear form symmetry, as noted in Remark 5.3. The material in this section is not essential to understanding semiconjugate factorization and reduction of order so readers who are not interested in convergence issues may omit it without loss of continuity.

As in the previous section, we assume that $\mathcal{F}$ is the field $\mathbb{C}$ of complex numbers or some subfield of $\mathbb{C}$ (such as $\mathbb{R}$). Also, for convenience, we limit our work to the second-order version of (5.1)

$$x_{n+1} = f_n(x_n, x_{n-1}). \tag{5.74}$$

By Corollary 5.5 this equation has the linear form symmetry $H(u_0, u_1) = u_0 - \alpha u_1$ for some $\alpha \neq 0$ if and only if

$$f_n\left(u_0, \frac{u_0 - v_1}{\alpha}\right) - \alpha u_0$$

is independent of u_0 for all v_1 and all n. In this case, if we define

$$g_n(v_1) \doteq f_n\left(u_0, \frac{u_0 - v_1}{\alpha}\right) - \alpha u_0$$

then the factor equation of (5.74) is the first-order equation

$$t_{n+1} = g_n(t_n). \tag{5.75}$$

THEOREM 5.5

(Limit cycles) Assume that $\alpha \in \mathbb{C}$ with $0 < |\alpha| < 1$ and that functions f_n, g_n are defined as in the preceding discussion for $n \geq 0$. If $\{t_n\}$ is a limit cycle or an attracting periodic solution of (5.75) in $\mathbb{C}$ then $\{x_n\}$ is a limit cycle of (5.74).

PROOF   Let $\{\tau_0, \ldots, \tau_{p-1}\}$ be an attracting cycle for (5.75) in $\mathbb{C}$ with

$$\lim_{n \to \infty} t_{pn+i} = \tau_{i-1}, \quad i = 1, 2, \ldots, p.$$

Let $s_n = \sum_{j=1}^{n} \alpha^{n-j} t_j$. Then by rearranging terms in the summation we find that

$$\begin{aligned}
s_{pn} = {} & \alpha^{pn-1} t_1 + \alpha^{pn-2} t_2 \cdots + \alpha^{pn-p} t_p \\
& + \alpha^{pn-p-1} t_{p+1} + \alpha^{pn-p-2} t_{p+2} \cdots + \alpha^{pn-2p} t_{2p} \\
& + \cdots \\
& + \alpha^{p-1} t_{p(n-1)+1} + \alpha^{p-2} t_{p(n-1)+2} \cdots + \alpha^{pn-pn} t_{p(n-1)+p}.
\end{aligned}$$

Factoring and rearranging various terms gives

$$\begin{aligned}
s_{pn} = {} & \alpha^{p-1} (\alpha^{pn-p} t_1 + \alpha^{pn-2p} t_{p+1} + \cdots + t_{pn-p+1}) \\
& + \alpha^{p-2} (\alpha^{pn-p} t_2 + \alpha^{pn-2p} t_{p+2} + \cdots + t_{pn-p+2}) \\
& + \cdots \\
& + \alpha^{pn-p} t_p + \alpha^{pn-2p} t_{2p} + \cdots + t_{pn} \\
= {} & \sum_{i=1}^{p} \alpha^{p-i} \sum_{k=0}^{n-1} (\alpha^p)^{n-k-1} t_{pk+i}.
\end{aligned}$$

Now for $i = 1, 2, \ldots, p$ define

$$\sigma_n^i = \sum_{k=0}^{n-1} (\alpha^p)^{n-k-1} t_{pk+i} \quad \text{and} \quad \gamma_n^i = \sum_{k=0}^{n-1} (\alpha^p)^{n-k-1} \tau_{i-1} = \tau_{i-1} \sum_{k=0}^{n-1} \alpha^{pk}.$$

Notice that

$$\begin{aligned}
\left| \sigma_n^i - \frac{\tau_{i-1}}{1 - \alpha^p} \right| &\leq \left| \sigma_n^i - \gamma_n^i \right| + \left| \gamma_n^i - \frac{\tau_{i-1}}{1 - \alpha^p} \right| \\
&\leq \sum_{k=0}^{n-1} |\alpha^p|^{n-k-1} |t_{pk+i} - \tau_{i-1}| + \left| \gamma_n^i - \frac{\tau_{i-1}}{1 - \alpha^p} \right|
\end{aligned}$$

Clearly the second term on the right-hand side approaches 0 as $n \to \infty$. As for the first term, let $m \geq 1$ and define

$$\delta = \max_{1 \leq i \leq p} \left\{ \sup_{k \geq 1} |t_{pk+i} - \tau_{i-1}| \right\} < \infty, \quad \delta_m^i = \sup_{k \geq m} |t_{pk+i} - \tau_{i-1}|$$

and observe that for $m < n$

$$\sum_{k=0}^{n-1} |\alpha^p|^{n-k-1} |t_{pk+i} - \tau_{i-1}| = \sum_{k=0}^{m-1} |\alpha^p|^{n-k-1} |t_{pk+i} - \tau_{i-1}|$$

$$+ \sum_{k=m}^{n} |\alpha^p|^{n-k-1} |t_{pk+i} - \tau_{i-1}|$$

$$\leq |\alpha^p|^{n-m} \delta \sum_{k=0}^{m-1} |\alpha^p|^k + \delta_m^i \sum_{k=m}^{n} |\alpha^p|^{n-k-1}.$$

By taking n and m sufficiently large, each of the last two terms above can be made arbitrarily small. Therefore,

$$\lim_{n\to\infty} \sigma_n^i = \frac{\tau_{i-1}}{1 - \alpha^p} \quad i = 1, \dots, p.$$

It follows that

$$\lim_{n\to\infty} x_{pn} = \lim_{n\to\infty} s_{pn} = \sum_{i=1}^{p} \frac{\alpha^{p-i} \tau_{i-1}}{1 - \alpha^p} = \frac{\alpha^{p-1}\tau_0 + \cdots + \alpha\tau_{p-2} + \tau_{p-1}}{1 - \alpha^p}.$$

Therefore, $x_{pn} \to \xi_0$ as $n \to \infty$ with ξ_0 as in Lemma 5.5. From this and (5.53) we obtain

$$\lim_{n\to\infty} x_{pn+1} = \lim_{n\to\infty} (t_{pn+1} + \alpha x_{pn}) = \tau_0 + \alpha\xi_0 = \xi_1.$$

Inductively, we find that $x_{pn+i} \to \xi_i$ for $i = 0, 1, \dots, p-1$. This implies that $\{x_n\}$ is an attracting periodic solution of (5.74). ∎

Example 5.9
Consider the following second-order equation on $\mathbb{R}$,

$$x_{n+1} = \alpha x_n + c_n(x_n - \alpha x_{n-1})^q \tag{5.76}$$

$$0 < |\alpha|, |q| < 1, \ c_{2m} = c_0 > 0, \ c_{2m+1} = c_1 > 0, \ m = 0, 1, \dots \tag{5.77}$$

In Eq. (5.76) it is possible to define a new variable $t_n = x_n - \alpha x_{n-1}$ which identifies a linear form symmetry $H(u_0, u_1) = u_0 - \alpha u_1$ with $\phi_n(t) = c_n t^q$ for $t > 0$.

Now consider the solutions of the first-order factor equation

$$t_{n+1} = c_n t_n^q. \tag{5.78}$$

This equation is nonautonomous but for each $j \geq 1$ two separate autonomous difference equations determine the even and odd terms of its solutions; specifically,

$$t_{2j} = c_{2j-1} t_{2j-1}^q = c_{2j-1} c_{2j-2}^q t_{2j-2}^{q^2} = c_1 c_0^q t_{2j-2}^{q^2},$$

$$t_{2j+1} = c_{2j} t_{2j}^q = c_{2j} c_{2j-1}^q t_{2j-1}^{q^2} = c_0 c_1^q t_{2j-1}^{q^2}.$$

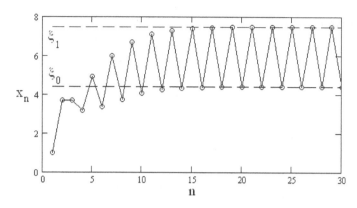

FIGURE 5.3

Convergence to a 2-cycle discussed in Example 5.9.

Using these simple first-order equations, it can be shown by straightforward induction that under conditions (5.77) all positive solutions of the factor equation (5.78) converge to a 2-cycle

$$t_{2n} \to c_0^{q/(1-q^2)} c_1^{1/(1-q^2)} = \tau_0,$$
$$t_{2n+1} \to c_0^{1/(1-q^2)} c_1^{q/(1-q^2)} = \tau_1.$$

Now Theorem 5.5 implies that every solution of (5.76) in the invariant region or set $\{(x, y) : x > \alpha y\}$ converges to the attracting cycle

$$\xi_0 = \frac{\alpha \tau_0 + \tau_1}{1 - \alpha^2}, \quad \xi_1 = \frac{\alpha \tau_1 + \tau_0}{1 - \alpha^2}.$$

Figures 5.3 and 5.4 illustrate the preceding results with the following parameter values:

$$\alpha = \frac{1}{2}, \ q = -\frac{2}{3}, \ c_0 = 4, \ c_1 = 2.$$

□

Example 5.9 can be extended to sequences c_n with any period p with slightly more calculating effort; see the Problems for this chapter.

We close this section with the next result concerning the uniform boundedness of solutions of Eq. (5.74) when linear form symmetry is present.

PROPOSITION 5.2

(Boundedness) Let $|\alpha| < 1$. If $\{t_n\}$ is a bounded sequence with $|t_n| \le B$ for some $B > 0$, then the corresponding solution $\{x_n\}$ of the cofactor (5.53) is also bounded and there is a positive integer N such that

$$|x_n| < |\alpha| + \frac{B}{1 - |\alpha|} \quad \text{for all } n \ge N. \tag{5.79}$$

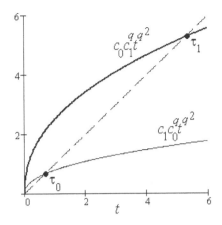

FIGURE 5.4

Constituent curves for the 2-cycle in Figure 5.3.

PROOF From (5.54) we obtain

$$|x_n| \le |\alpha|^n|x_0| + \sum_{j=1}^{n} |\alpha|^{n-j}B < |\alpha|^n|x_0| + B\sum_{k=0}^{\infty} |\alpha|^k = |\alpha|^n|x_0| + \frac{B}{1-|\alpha|}.$$

Now if n is large enough, then $|\alpha|^n|x_0| \le |\alpha|$ from which (5.79) follows.

5.6 Difference equations with linear arguments

The existence of a linear form symmetry leads to an interesting array of applications of the SC factorization method. In this section we examine the occurrence of linear form symmetry in difference equations of the following type:

$$x_{n+1} = \sum_{j=0}^{k} a_j x_{n-j} + g_n\left(\sum_{j=0}^{k} b_j x_{n-j}\right) \tag{5.80}$$

where the constants a_j, b_j, $j = 0, 1, 2, \ldots, k$ are given in a nontrivial field $\mathcal{F}$ and $g_n : \mathcal{F} \to \mathcal{F}$ are given functions for all $n = 0, 1, 2, \ldots$

We refer to Eq. (5.80) as a *difference equation with linear argument* of order $k+1$ on the field $\mathcal{F}$. These equations include linear nonhomogeneous difference equation with constant coefficients

$$x_{n+1} = \sum_{j=0}^{k} a_j x_{n-j} + \gamma_n$$

as special cases with $g_n(u) = \gamma_n$ constants in $\mathcal{F}$.

In this section we find necessary and sufficient conditions for Eq. (5.80) to have the linear form symmetry and therefore, a SC factorization. In the special case of linear equations these conditions are satisfied; see Corollary 5.11 below. Later in Chapter 6 we show that such linear equations also possess SC factorizations through a different form symmetry that leads to type-$(1, k)$ reductions. Finally, in Chapter 7, after defining time-dependent form symmetries, we derive SC factorizations of linear equations in the general case, i.e., with variable coefficients in arbitrary algebraic fields.

5.6.1   The reduction theorem

We now establish the necessary and sufficient conditions for the existence of a linear form symmetry for Eq. (5.80) and calculate the corresponding SC factorization. We state the following consequence of Corollary 5.5 as a theorem since it lays the foundation for the SC factorizations of equations with linear arguments. It may be noted that the conclusion of the theorem is valid for arbitrary functions $g_n : \mathcal{F} \to \mathcal{F}$ since the stated criteria for the existence of linear form symmetry do not depend on the choice of g_n.

THEOREM 5.6

Let $\mathcal{F}$ be a nontrivial field.

(a) Eq. (5.80) has the linear form symmetry (5.51) if and only if the following polynomials have a common, nonzero root $\alpha \in \mathcal{F}$

$$P(u) = u^{k+1} - a_0 u^k - a_1 u^{k-1} - \cdots a_{k-1} u - a_k$$
$$Q(u) = b_0 u^k + b_1 u^{k-1} + \cdots + b_{k-1} u + b_k.$$

(b) If α is a common nonzero root of P and Q in $\mathcal{F}$ then (5.80) has a type-$(k, 1)$ order reduction with the factor equation

$$t_{n+1} = -\sum_{i=0}^{k-1} p_i t_{n-i} + g_n \left(\sum_{i=0}^{k-1} q_i t_{n-i} \right) \tag{5.81}$$

where the numbers p_i and q_i are defined for $i = 0, \ldots, k - 1$ as

$$p_i = \alpha^{i+1} - a_0 \alpha^i - \cdots - a_i, \quad q_i = b_0 \alpha^i + b_1 \alpha^{i-1} + \cdots + b_i. \tag{5.82}$$

PROOF   (a) For each $n \geq 0$ define the function

$$f_n(u_0, \ldots, u_k) = \sum_{j=0}^{k} a_j u_j + g_n \left(\sum_{j=0}^{k} b_j u_j \right).$$

By Corollary 5.5 Eq. (5.80) has the linear form symmetry if and only if the quantity (5.56) is independent of u_0. For any nonzero $\alpha \in \mathcal{F}$ we find

$$f_n(u_0, \zeta_1, \ldots, \zeta_k) - \alpha u_0 = (a_0 - \alpha) u_0 + \sum_{j=1}^{k} a_j \left(\frac{u_0}{\alpha^j} - \sum_{i=1}^{j} \frac{v_i}{\alpha^{j-i+1}} \right) +$$

$$+ g_n \left(b_0 u_0 + \sum_{j=1}^{k} b_j \left[\frac{u_0}{\alpha^j} - \sum_{i=1}^{j} \frac{v_i}{\alpha^{j-i+1}} \right] \right)$$

$$= \left(a_0 - \alpha + \sum_{j=1}^{k} \frac{a_j}{\alpha^j} \right) u_0 - \sum_{j=1}^{k} a_j \sum_{i=1}^{j} \frac{v_i}{\alpha^{j-i+1}} +$$

$$+ g_n \left(\left[b_0 + \sum_{j=1}^{k} \frac{b_j}{\alpha^j} \right] u_0 - \sum_{j=1}^{k} b_j \sum_{i=1}^{j} \frac{v_i}{\alpha^{j-i+1}} \right).$$

From the last expression above it is evident that the linear form symmetry exists if and only if α can be chosen such that

$$a_0 - \alpha + \sum_{j=1}^{k} \frac{a_j}{\alpha^j} = 0 \quad \text{and} \quad b_0 + \sum_{j=1}^{k} \frac{b_j}{\alpha^j} = 0.$$

Multiplying the two equalities above by α^k and rearranging terms shows that α must be a common nonzero root of the polynomials P and Q.

(b) Let α be a common nonzero root of P and Q in $\mathcal{F}$. By Corollary 5.5 and the calculations in Part (a), the factor functions are

$$\phi_n(v_1, \ldots, v_k) = -\sum_{j=1}^{k} a_j \sum_{i=1}^{j} \frac{v_i}{\alpha^{j-i+1}} + g_n \left(-\sum_{j=1}^{k} b_j \sum_{i=1}^{j} \frac{v_i}{\alpha^{j-i+1}} \right)$$

$$= -\sum_{i=1}^{k} \sum_{j=i}^{k} \frac{a_j}{\alpha^{j-i+1}} v_i + g_n \left(-\sum_{i=1}^{k} \sum_{j=i}^{k} \frac{b_j}{\alpha^{j-i+1}} v_i \right).$$

For each $i = 1, \cdots, k$, since α is a root of the polynomial P it follows that

$$\sum_{j=i}^{k} \frac{a_j}{\alpha^{j-i+1}} = \frac{1}{\alpha^{k-i+1}} \left(a_i \alpha^{k-i} + a_{i+1} \alpha^{k-i-1} + \cdots + a_{k-1} \alpha + a_k \right)$$

$$= \frac{1}{\alpha^{k-i+1}} (\alpha^{k+1} - a_0 \alpha^k - \cdots - a_{i-1} \alpha^{k-i+1})$$

$$= \alpha^i - a_0 \alpha^{i-1} - \cdots - a_{i-1}$$

Similarly, since α is also a root of the polynomial Q it follows that

$$\sum_{j=i}^{k} \frac{b_j}{\alpha^{j-i+1}} = \frac{1}{\alpha^{k-i+1}} \left(b_i \alpha^{k-i} + b_{i+1} \alpha^{k-i-1} + \cdots + b_{k-1}\alpha + b_k \right)$$

$$= \frac{1}{\alpha^{k-i+1}} \left(-b_0 \alpha^k - b_1 \alpha^{k-1} - \cdots - b_{i-1}\alpha^{k-i+1} \right)$$

$$= -b_0 \alpha^{i-1} - b_1 \alpha^{i-2} - \cdots - b_{i-1}.$$

Now, if the quantities p_i and q_i are defined as in (5.82) then the preceding calculations show that

$$\sum_{j=i}^{k} \frac{a_j}{\alpha^{j-i+1}} = p_{i-1} \quad \text{and} \quad \sum_{j=i}^{k} \frac{b_j}{\alpha^{j-i+1}} = -q_{i-1}.$$

Using these quantities we write the factor functions as

$$\phi_n(v_1, \ldots, v_k) = -\sum_{i=1}^{k} p_{i-1} v_i + g_n \left(\sum_{i=1}^{k} q_{i-1} v_i \right).$$

Identifying v_i with t_{n-i+1} as usual gives the factor equation (5.81)

$$t_{n+1} = \phi_n(t_n, \ldots, t_{n-k+1})$$

$$= -\sum_{i=1}^{k} p_{i-1} t_{n-i+1} + g_n \left(\sum_{i=1}^{k} q_{i-1} t_{n-i+1} \right)$$

$$= -\sum_{i=0}^{k-1} p_i t_{n-i} + g_n \left(\sum_{i=0}^{k-1} q_i t_{n-i} \right).$$

■

For difference equations with linear arguments of order two ($k = 1$) we have the following corollary.

COROLLARY 5.8

Let $\mathcal{F}$ be a nontrivial field, $a_0, a_1, b_0, b_1 \in \mathcal{F}$ and let $\{g_n\}$ be a sequence of functions $g_n : \mathcal{F} \to \mathcal{F}$. If

$$b_1^2 + a_0 b_0 b_1 - a_1 b_0^2 = 0 \quad \text{with} \quad b_0 b_1 \neq 0 \tag{5.83}$$

then the equation

$$x_{n+1} = a_0 x_n + a_1 x_{n-1} + g_n(b_0 x_n + b_1 x_{n-1}) \tag{5.84}$$

has the SC factorization

$$t_{n+1} = \left(a_0 + \frac{b_1}{b_0} \right) t_n + g_n (b_0 t_n)$$

$$x_{n+1} = -\frac{b_1}{b_0} x_n + t_{n+1}.$$

PROOF   The polynomial Q in Theorem 5.6 is $b_0 u + b_1$ whose only root is $\lambda = -b_1/b_0$. Note that $\lambda \in \mathcal{F}$ and $\lambda \neq 0$ under conditions (5.83). Since λ is also a root of the polynomial P, Theorem 5.6 applies to Eq. (5.84) under conditions (5.83).   ∎

5.6.2   Economics: Modeling the business cycle

A special case of the equation in the next example with $j = k = 1$ has been used to model the Keynesian business cycle in macroeconomics; see the remarks following the next example.

Example 5.10

Let $a_0, a_1, \ldots, a_k \in \mathbb{C}$ and suppose that $g_n : \mathbb{C} \to \mathbb{C}$ is a given sequence of functions. Consider the difference equation

$$x_{n+1} = a_0 x_n + \cdots + a_k x_{n-k} + g_n(x_n - x_{n-j}), \quad 1 \leq j \leq k. \tag{5.85}$$

By Theorem 5.6, Eq. (5.85) has the linear form symmetry if and only if there is $\alpha \in \mathbb{C}$, $\alpha \neq 0$ such that

$$\alpha^{k+1} - a_0 \alpha^k - \cdots a_{k-1}\alpha - a_k = 0, \tag{5.86}$$

$$\alpha^k - \alpha^{k-j} = 0. \tag{5.87}$$

By Eq. (5.87) α must be a j-th root of unity, i.e., $\alpha^j = 1$. If $j = 1$ then $\alpha = 1$ and from (5.86) it follows that Eq. (5.85) has the linear form symmetry if and only if

$$a_0 + a_1 \cdots + a_k = 1, \quad (j = 1).$$

In this case Theorem 5.6 gives the factor equation of (5.85) as

$$t_{n+1} = g_n(t_n) - \sum_{i=0}^{k-1} (1 - a_0 \cdots - a_i) t_{n-i}. \tag{5.88}$$

If $\{t_n\}$ is a solution of (5.88) then the cofactor $x_{n+1} = x_n + t_{n+1}$ yields the corresponding solution $\{x_n\}$ of (5.85) as

$$x_n = x_0 + \sum_{i=1}^{n} t_i. \tag{5.89}$$

◻

The following special case of Eq. (5.85)

$$x_{n+1} = cx_n + sx_{n-1} + g_n(x_n - x_{n-1}), \quad c+s = 1, \ 0 < c \leq 1 \qquad (5.90)$$

has been used to model the business cycle in macroeconomics. In (5.90) the number c is the "marginal propensity to consume" (or MPC) and the number $s = 1 - c$ is the "marginal propensity to save" (or MPS). In this context, the variable x_n represents the "output" usually total income or gross domestic product in period n; a percentage c of the current period income x_n is consumed and affects the income in the next period. Also consumed is the amount sx_{n-1} saved in the preceding period. Thus Eq. (5.90) represents a case where all savings are fully consumed. The functions g_n in (5.90) usually denote "investment functions" but often include other factors such as government spending. They are independent of n in some models but need not be so in general.

Models of this type that are based on Keynesian concepts such as MPC, MPS, etc. were explored throughout the 20th century to explain the occurrence of "boom-bust" cycles in the absence of (and sometimes in spite of) external factors such as government policy or other types of interference. See the Notes section for references.

By Example 5.10 the first-order factor equation of (5.90) is

$$t_{n+1} = g_n(t_n) - st_n$$

from which either an explicit formula for the solutions of (5.90) or significant details about the solutions may be obtained via (5.89).

5.6.3   Repeated reductions of order

An important by-product of Theorem 5.6 is that the factor equation of (5.80), namely, (5.81) is of the same type as (5.80), but it has a lower order. This suggests that Theorem 5.6 can be applied repeatedly to each time reduce the order of the original equation with linear arguments by one, as long as *common*, nonzero polynomial roots, or eigenvalues, exist. For convenience we now give a name to such eigenvalues.

DEFINITION 5.1   *The two polynomials in Corollary 5.6 are the* **joint characteristic polynomials** *of Eq. (5.80) if they have a common, nonzero root. Each common, nonzero root of the polynomials P and Q is a* **joint eigenvalue** *of Eq. (5.80).*

The next result simplifies the task of searching for joint eigenvalues by limiting it to all common roots of the original pair of polynomials P and Q.

LEMMA 5.6

*Let $\mathcal{F}$ be a nontrivial field and let $\alpha \in \mathcal{F}$ be a joint eigenvalue of Eq. (5.80).
If $\alpha_1 \in \mathcal{F}$ is another joint eigenvalue of (5.80) then α_1 is a common nonzero
root of both of the polynomials*

$$P_1(u) = u^k + p_0 u^{k-1} + p_1 u^{k-2} + \cdots + p_{k-1}$$
$$Q_1(u) = q_0 u^{k-1} + q_1 u^{k-2} + \cdots + q_{k-1}$$

*where the coefficients p_i and q_i are defined in terms of α as in (5.82) for
$i = 0, \ldots, k-1$.*

PROOF By assumption,

$$P(\alpha_1) = Q(\alpha_1) = 0. \tag{5.91}$$

Now

$$(u - \alpha)P_1(u) = (u - \alpha)\left(u^k + \sum_{j=0}^{k-1} p_j u^{k-j-1}\right)$$

$$= u^{k+1} + \sum_{j=0}^{k-1}[p_j - \alpha p_{j-1}]u^{k-j} - \alpha p_{k-1}.$$

where we define $p_{-1} = 1$ to simplify the notation. From (5.82) for each
$j = 0, 1, \ldots, k-1$ we obtain

$$p_j - \alpha p_{j-1} = -a_j$$

and further, since $P(\alpha) = 0$,

$$\alpha p_{k-1} = \alpha\left(\alpha^k - a_0 \alpha^{k-1} - \cdots - a_{k-1}\right)$$
$$= P(\alpha) + a_k$$
$$= a_k.$$

Thus

$$(u - \alpha)P_1(u) = P(u)$$

and if $\alpha_1 \neq \alpha$ then $P_1(\alpha_1) = 0$ by (5.91). If $\alpha_1 = \alpha$ then α is a double root of
both P and Q so that their derivatives are zeros, i.e.,

$$P'(\alpha) = Q'(\alpha) = 0. \tag{5.92}$$

Now, using (5.82) we obtain

$$P_1(\alpha) = \alpha^k + \sum_{j=0}^{k-1}(\alpha^{j+1} - a_0\alpha^j - \cdots - a_{j-1}\alpha - a_j)\alpha^{k-j-1}$$

$$= (k+1)\alpha^k - \sum_{j=0}^{k-1}(k-j)a_j\alpha^{k-j-1}$$

$$= P'(\alpha).$$

In particular, if $\alpha_1 = \alpha$ then $P_1(\alpha_1) = 0$ by (5.92). Similar calculations (left to the reader) show that $Q_1(\alpha_1) = 0$, thus completing the proof. ∎

An interesting by-product of the above proof is the following result which we state as a proposition.

PROPOSITION 5.3
At each joint eigenvalue α of (5.80) the second-stage polynomials P_1 and Q_1 are related to the joint characteristic polynomials P and Q at α via the equations

$$P_1(\alpha) = P'(\alpha) \quad \text{and} \quad Q_1(\alpha) = Q'(\alpha).$$

Theorem 5.6 and Lemma 5.6 imply the following result.

COROLLARY 5.9
(a) Assume that Eq. (5.80) has two joint eigenvalues α, α_1 in a nontrivial field $\mathcal{F}$. Then the factor equation (5.81) has the linear form symmetry with the factor equation

$$r_{n+1} = -\sum_{i=0}^{k-2}p_{1,i}r_{n-i} + g_n\left(\sum_{i=0}^{k-2}q_{1,i}r_{n-i}\right) \tag{5.93}$$

where

$$p_{1,i} = \alpha_1^{i+1} + p_0\alpha_1^i + \cdots + p_i \text{ and } q_{1,i} = q_0\alpha_1^i + q_1\alpha_1^{i-1} + \cdots + q_i.$$

(b) The repeated SC factorization of (5.80) with joint eigenvalues α, α_1 consists of the factor equation (5.93) and the two cofactor equations

$$t_{n+1} = \alpha_1 t_n + r_{n+1}$$
$$x_{n+1} = \alpha x_n + t_{n+1}.$$

(c) If Eq. (5.80) has m joint eigenvalues in $\mathcal{F}$ (counting repeated roots) where $1 \leq m \leq k$ then it has a factor chain of length m and can be reduced in order repeatedly m times.

PROOF   (a) The proof applies the same arguments as in the proof of Theorem 5.6 to the factor equation (5.81) that was obtained using the joint eigenvalue α. We simply change a_i to $-p_i$ and b_i to q_i in the proof and make other minor modifications to complete the proof of this part.

(b) This follows from the general form of cofactors associated with the linear form symmetry.

(c) This follows by induction, using Part (a).   ∎

The preceding corollary leads to a natural question: Are there special cases of Eq. (5.80) with factor chains of length k? Or in different terminology, are there special cases of Eq. (5.80) with full triangular factorizations? The following result answers this question affirmatively.

COROLLARY 5.10
Let $\mathcal{F}$ be a nontrivial field and consider the following difference equation with linear arguments

$$x_{n+1} = a_0 x_n + \cdots + a_{k-1} x_{n-k+1} + g_n \left(b x_n - a_0 b x_{n-1} - \cdots - a_{k-1} b x_{n-k} \right) \tag{5.94}$$

where $b, a_j \in \mathcal{F}$, $j = 0, \ldots, k-1$. If $a_{k-1} \neq 0$ and (5.94) has k joint eigenvalues in $\mathcal{F}$ then (5.94) admits repeated SC factorizations into a triangular system of first-order equations over $\mathcal{F}$.

PROOF   By Definition 5.1 a nonzero $\alpha \in \mathcal{F}$ is a joint eigenvalue of (5.94) if $P(\alpha) = Q(\alpha) = 0$ i.e., if

$$\alpha^{k+1} - a_0 \alpha^k - a_1 \alpha^{k-1} - \cdots - a_{k-1} \alpha = 0 \tag{5.95}$$
$$b \alpha^k - a_0 b \alpha^{k-1} - \cdots - a_{k-2} b \alpha - a_{k-1} b = 0. \tag{5.96}$$

First, suppose that $b \neq 0$. Since $\alpha, a_{k-1} \neq 0$, after dividing (5.95) above by α and (5.96) by b both equations reduce to the following single equation

$$\alpha^k - a_0 \alpha^{k-1} - a_1 \alpha^{k-2} - \cdots a_{k-1} = 0. \tag{5.97}$$

Clearly, every common, nonzero root of P and Q is a root of the polynomial equation (5.97) and conversely, every root of (5.97) is nonzero and a root of both P and Q. Therefore, Eq. (5.94) has k joint eigenvalues in $\mathcal{F}$ and by Corollary 5.9 it has a factor chain of length k, or equivalently, it has a complete triangular SC factorization.

If $b = 0$ then (5.96) is true trivially for all α. Therefore, again the single equation (5.97) remains whose k nonzero roots are all joint eigenvalues. So by Corollary 5.9 Eq. (5.94) has a factor chain of length k and a full triangular SC factorization.   ∎

5.6.4   SC factorization of linear equations, factor chains

When $b = 0$ in (5.94) then the values $g_n(0) = \gamma_n$ represent constants in the field $\mathcal{F}$ so that (5.94) reduces to a linear nonhomogeneous equation. Therefore, the following result is a straightforward consequence of Corollary 5.10.

COROLLARY 5.11

Let $\mathcal{F}$ be a nontrivial field and consider the linear nonhomogeneous difference equation of order k

$$x_{n+1} = a_0 x_n + a_1 x_{n-1} + \cdots + a_{k-1} x_{n-k+1} + \gamma_n \qquad (5.98)$$

with $a_j, \gamma_n \in \mathcal{F}$ for $j = 0, \ldots, k-1$, $n \geq 0$ and $a_{k-1} \neq 0$. If $\mathcal{F}$ contains all eigenvalues of (5.98) then (5.98) has repeated SC factorizations into a triangular system over $\mathcal{F}$ via a factor chain as follows: If $\lambda_j \in \mathcal{F}$ are the eigenvalues for $j = 0, \ldots k-1$ then

$$y_{0,n+1} = \lambda_0 y_{0,n} + y_{1,n+1}, \quad y_{0,n} = x_n$$
$$y_{1,n+1} = \lambda_1 y_{1,n} + y_{2,n+1},$$
$$\vdots$$
$$y_{k-1,n+1} = \lambda_{k-1} y_{k-1,n} + \gamma_n.$$

PROOF   The existence of the full SC factorization is established in Corollary 5.10. Note that the condition $a_{k-1} \neq 0$ assures that all eigenvalues of (5.98) are nonzero. To obtain the triangular system, we use Theorem 5.6 to obtain the first SC factorization of (5.98) as

$$x_{n+1} = \lambda_0 x_n + t_{n+1}$$
$$t_{n+1} = -\sum_{i=0}^{k-1} p_i t_{n-i} + \gamma_n. \qquad (5.99)$$

For convenience, we re-label $x_n = y_{0,n}$ and $t_n = y_{1,n}$. Since the factor equation (5.99) is again linear and by Lemma 5.6 λ_1 is one of its eigenvalues, we obtain the further factorization

$$y_{0,n+1} = \lambda_0 y_{0,n} + y_{1,n+1}, \quad y_{0,n} = x_n$$
$$y_{1,n+1} = \lambda_1 y_{1,n} + y_{2,n+1},$$
$$y_{2,n+1} = -\sum_{i=0}^{k-2} p_{1,i} y_{2,n-i} + \gamma_n$$

where the numbers $p_{1,i}$ are defined in Corollary 5.9. Continuing the above argument we obtain the triangular system by induction.   ∎

REMARK 5.4   The triangular system in Corollary 5.11 is derived from a factor chain through a series of type-$(k,1)$ factorizations. The system is shown in ascending form, with the bottom equation being independent of the rest. For full triangular factorizations of linear equations over the complex numbers $\mathbb{C}$ via *cofactor chains* are obtained in Section 6.2.2. Clearly, the triangular system that results by either method is the same up to a rearrangement of eigenvalues. For additional important remarks about the use of the SC factorization see Section 6.2.2.   ∎

The results on difference equations with linear arguments apply to more general equations than (5.98). We now proceed to extend Corollary 5.11 to a class of nonhomogeneous linear equations with variable coefficients that properly includes all equations with constant coefficients as special cases. The significance of the particular type of variable coefficients encountered in the next corollary is explained in Section 7.4.3. The SC factorizations of *general* nonhomogeneous linear equations with variable coefficients require the introduction of time-dependent form symmetries and will be discussed in Section 7.4.

COROLLARY 5.12

Let $\mathcal{F}$ be a nontrivial field and let $\{\sigma_n\}$, $\{\gamma_n\}$ be sequences in $\mathcal{F}$. Consider the nonhomogeneous linear difference equation

$$x_{n+1} = \sum_{j=0}^{k}(a_j + b_j\sigma_n)x_{n-j} + \gamma_n \tag{5.100}$$

with constants $a_j, b_j \in \mathcal{F}$ for $j = 0, \ldots, k$ and $a_k \neq 0$.

(a) If the polynomials P and Q in Theorem 5.6 have a joint eigenvalue $\lambda_0 \in \mathcal{F}$ then Eq. (5.100) has a type-$(k,1)$ reduction with SC factorization

$$t_{n+1} = -\sum_{i=0}^{k-1}(p_i + q_i\sigma_n)t_{n-i} + \gamma_n \tag{5.101}$$

$$x_{n+1} = \lambda_0 x_n + t_{n+1} \tag{5.102}$$

where the constants p_i, q_i are as defined by (5.82) in Theorem 5.6.

(b) Since Eq. (5.101) is of the same type as (5.100) the reduction to factor can be repeated as long as joint eigenvalues exist in $\mathcal{F}$. In particular, if $\mathcal{F}$ is an algebraically closed field then the nonhomogeneous linear difference equation with variable coefficients

$$x_{n+1} = (a_0 + b\sigma_n)x_n + \cdots + (a_{k-1} - a_{k-2}b\sigma_n)x_{n-k+1} + a_{k-1}b\sigma_n x_{n-k} + \gamma_n \tag{5.103}$$

has a full SC factorization into a triangular system of $k+1$ first-order linear difference equations.

PROOF   If we define the functions $g_n : \mathcal{F} \to \mathcal{F}$ as $g_n(u) = \sigma_n u + \gamma_n$ then the proof of Theorem 5.6 can be repeated to obtain the SC factorization in (a) into (5.101) and (5.102). The first statement in (b) is obvious. Further, the repeated SC factorization of Eq. (5.103) into a triangular system is obtained directly using Corollary 5.10.   ∎

For second-order linear nonhomogeneous equations we have the following consequence of Corollary 5.12.

COROLLARY 5.13
Let $\mathcal{F}$ be a nontrivial field, $a_0, a_1, b_0, b_1 \in \mathcal{F}$ and let $\{\sigma_n\}$, $\{\gamma_n\}$ be given sequences in $\mathcal{F}$. If

$$b_1^2 + a_0 b_0 b_1 - a_1 b_0^2 = 0 \quad \text{with} \quad b_0 b_1 \neq 0 \tag{5.104}$$

then the linear nonhomogeneous difference equation

$$x_{n+1} = (a_0 + b_0 \sigma_n) x_n + (a_1 + b_1 \sigma_n) x_{n-1} + \gamma_n \tag{5.105}$$

has the SC factorization

$$t_{n+1} = \left(a_0 + \frac{b_1}{b_0} + b_0 \sigma_n \right) t_n + \gamma_n, \quad t_0 = x_0 - a_0 x_{-1}$$

$$x_{n+1} = -\frac{b_1}{b_0} x_n + t_{n+1}.$$

Example 5.11
Consider the nonautonomous linear equation

$$x_{n+1} = \left(\alpha + \frac{\beta}{n+1} \right) x_n - \frac{\alpha\beta}{n+1} x_{n-1}, \quad n = 0, 1, 2, \ldots \tag{5.106}$$

where $\alpha, \beta \in \mathbb{R}$ and $\alpha\beta \neq 0$. Eq. (5.106) is a special case of (5.105) with

$$a_0 = \alpha, \ b_0 = \beta, \ a_1 = 0, \ b_1 = -\alpha\beta, \ \sigma_n = \frac{1}{n+1}.$$

Therefore, Conditions (5.104) are satisfied and the following SC factorization is obtained

$$t_{n+1} = \frac{\beta}{n+1} t_n, \quad x_{n+1} = \alpha x_n + t_{n+1}. \tag{5.107}$$

The solution of the factor equation in (5.107) is readily found as

$$t_{n+1} = \frac{\beta^{n+1} t_0}{(n+1)!} \quad \text{where } n \geq 0, \ t_0 = x_0 - \alpha x_{-1}.$$

Inserting this into the cofactor equation in (5.107) and solving the resulting equation gives the explicit solution of (5.106) as

$$x_n = \alpha^n \left[x_0 + t_0 \sum_{j=1}^{n} \frac{(\beta/\alpha)^j}{j!} \right].$$

Since for all nonzero, real α, β

$$\lim_{n \to \infty} \sum_{j=1}^{n} \frac{(\beta/\alpha)^j}{j!} = e^{\beta/\alpha} - 1$$

the following conclusions are easily arrived at:

(i) If $\alpha > 1$ then $\lim_{n\to\infty} x_n = \infty$ or $\lim_{n\to\infty} x_n = -\infty$ depending on the initial values x_0, x_{-1} while if $\alpha = 1$ then

$$\lim_{n \to \infty} x_n = x_0 + t_0(e^\beta - 1);$$

(ii) If $|\alpha| < 1$ then $\lim_{n\to\infty} x_n = 0$;

(iii) If $\alpha < -1$ then each solution $\{x_n\}$ of (5.106) is unbounded and oscillatory while if $\alpha = -1$ then $\{x_n\}$ converges to the 2-cycle

$$(-1)^n \left[x_0 + t_0(e^{-\beta} - 1) \right].$$

□

5.6.5 *A rational equation with chaotic solutions

In this section we show that the solutions of a rational recursive equation exhibit chaotic behavior in the field of real numbers. This fact is difficult to prove without reduction of order and it is significant for showing that difference equations with linear arguments are capable of generating complex behavior. Rational equations of the type studied in this section are also special types of quadratic difference equations that we study systematically in Chapter 8. The material in this section is not essential to understanding reduction of order by semiconjugate factorization and may be omitted without loss of continuity.

Consider the following second-order difference equation with linear arguments

$$x_{n+1} = cx_n + g(x_n - cx_{n-1}) \tag{5.108}$$

where g is the one-dimensional rational map

$$g(t) = pt + q + \frac{s}{t}.$$

It is evident that Eq. (5.108) has the linear form symmetry $-cu$ with the SC factorization

$$t_{n+1} = g(t_n) \tag{5.109}$$
$$x_{n+1} = t_{n+1} + cx_n. \tag{5.110}$$

Rearranging terms in (5.108) using the above g we obtain the second-order, rational recursive equation

$$x_{n+1} = \frac{(p+c)x_n^2 - c(2p+c)x_n x_{n-1} + pc^2 x_{n-1}^2 + qx_n - qcx_{n-1} + s}{x_n - cx_{n-1}}. \tag{5.111}$$

We point out that Eq. (5.111) is a special case of Eq. (5.57) that satisfies all of the conditions (5.58)–(5.60) in Corollary 5.6 as may be readily verified.

REMARK 5.5   If $t_n > 0$ for all $n \geq n_0$ where n_0 is a positive integer, then $x_{n+1} > cx_n$ for all $n \geq n_0$. This is evident from (5.110):

$$x_{n_0+1} = cx_{n_0} + t_{n_0+1} > cx_{n_0}.$$

This conclusion extends by induction to all points (x_n, x_{n+1}) in the plane with $n \geq n_0$. Therefore, if $g(t) > 0$ for all $t > 0$ and we start with $t_0 = x_0 - cx_{-1} > 0$ then the entire orbit (x_n, x_{n+1}) of (5.108) stays in the region $\{(u, v) : v > cu\}$. To avoid singularities, we may assume that the initial values for (5.111) satisfy

$$x_0 > cx_{-1}. \tag{5.112}$$

∎

Note that if $c \leq 0$ then in particular every pair of positive initial values satisfies (5.112). In the next lemma we obtain, among other information, conditions that imply $g(t) > 0$ for all $t > 0$. Note that if $\bar{t}$ is a fixed point of (5.109) then by direct calculation, the second-order equation (5.108) has a corresponding fixed point

$$\bar{x} = \frac{\bar{t}}{1-c}. \tag{5.113}$$

LEMMA 5.7
Assume that $s > 0, p \geq 0$ and consider

$$g(t) = pt + q + \frac{s}{t}, \quad t > 0. \tag{5.114}$$

(a) If $p > 0$ then the function g attains its global minimum value $2\sqrt{sp} + q$ on $(0, \infty)$ at $t = \sqrt{s/p}$. Hence for $p \geq 0$, $g(0, \infty) \subset (0, \infty)$ if and only if

$$q > -2\sqrt{sp}.$$

(b) *The positive fixed points of g, when they exist, are solutions of the equation* $\phi(t) = 0$ *where*

$$\phi(t) = (p-1)t^2 + qt + s.$$

There are three possible cases:

(i) *If* $p = 1$ *and* $q < 0$ *then* g *has a unique fixed point*

$$\bar{t} = \frac{s}{|q|} > 0.$$

(ii) *If* $p < 1$ *then* g *has a unique positive fixed point*

$$\bar{t} = \frac{q + \sqrt{q^2 + 4s(1-p)}}{2(1-p)}.$$

(iii) *If* $p > 1$ *then* g *has up to two positive fixed points*

$$\bar{t}_1 = \frac{|q| - \sqrt{q^2 - 4s(p-1)}}{2(p-1)} \leq \bar{t}_2 = \frac{|q| + \sqrt{q^2 - 4s(p-1)}}{2(p-1)}.$$

(c) *Let* $q > -2\sqrt{sp}$ *and* $g^2(t) = g(g(t))$. *Then*

$$g^2(t) = t \quad \text{iff} \quad \phi(t)\psi(t) = 0$$

where

$$\psi(t) = [p(p+1)t^2 + q(p+1)t + sp]. \tag{5.115}$$

PROOF (a) $\lim_{t \to 0^+} g(t) = \lim_{t \to \infty} g(t) = \infty$ and

$$g'(t) = p - \frac{s}{t^2}, \quad g''(t) = \frac{2s}{t^3}.$$

Hence, g has a global minimum value $2\sqrt{sp}$ on $(0, \infty)$ at $t = \sqrt{s/p}$.
(b) The fixed points of g are solutions of the equation $g(t) = t$, or equivalently of

$$0 = g(t) - t = tg(t) - t^2 = \phi(t).$$

Solving this quadratic equation by routine calculation readily gives the roots structure as claimed in (i)–(iii).
(c) The equation $g^2(t) = t$ written explicitly is

$$p^2t + q(p+1) + \frac{sp}{t} + \frac{st}{pt^2 + qt + s} = t. \tag{5.116}$$

Since $pt^2 + qt + s$ has no real roots when $q > -2\sqrt{sp}$, the real solutions of (5.116) in $(0, \infty)$ are precisely those of

$$t(pt^2 + qt + s)g^2(t) = t^2(pt^2 + qt + s).$$

After multiplying out the terms and rearranging them, the following is obtained:

$$p(p^2 - 1)t^4 + q(p+1)(2p-1)t^3 + [q^2(p+1) + 2sp^2]t^2 - sp(2p+1)t + s^2p = 0. \tag{5.117}$$

The roots of the quartic polynomial on the left-hand side of (5.117) are precisely the solutions of the equation $g^2(t) = t$. The solutions of the latter equation obviously include the fixed points, or the roots of $\phi(t)$ so the quartic is divisible by $\phi(t)$. Dividing in the straightforward fashion yields the polynomial $\psi(t)$ as the quotient. ∎

As an initial value problem, we consider (5.111) and (5.112) to be semiconjugate to the first-order discrete initial value problem

$$t_{n+1} = pt_n + q + \frac{s}{t_n}, \quad t_0 = x_0 - cx_{-1} > 0. \tag{5.118}$$

When $c = 0$ then Eq. (5.111) reduces to (5.118). On the other hand, as $|c|$ moves away from 0, approaching 1, the behavior of solutions of (5.111) becomes completely different from the first-order case and generally, it becomes more complex. In this sense, c acts as a *focusing parameter* that relates the first-order equation to the second order one; however, c does not affect the first-order equation (5.118).

The next result on the global attractivity of the positive fixed point shows that the coefficients of the linear terms in the numerator of Eq. (5.111) can play a significant role in the behavior of solutions of (5.111).

THEOREM 5.7
(Global attractivity) Assume that $s > 0$, $0 \le p \le 1$ and $0 < |c| < 1$.
   (a) Let $p < 1$ and assume that

$$q \ge -2p\sqrt{\frac{s}{p+1}}. \tag{5.119}$$

Then the positive fixed point

$$\bar{x} = \frac{q + \sqrt{q^2 + 4s(1-p)}}{2(1-p)(1-c)}$$

attracts all solutions of (5.111) satisfying (5.112).
   (b) Let $p = 1$. If

$$-\sqrt{2s} \le q < 0 \tag{5.120}$$

then all solutions of (5.111) satisfying (5.112) converge to the positive fixed point

$$\bar{x} = \frac{s}{q(c-1)}.$$

If

$$q \geq 0 \tag{5.121}$$

then every solution of (5.111) satisfying (5.112) is unbounded.

PROOF   (a) Because of Theorem 5.5 it is sufficient to prove that the fixed point $\bar{t} = \bar{x}(1-c)$ is a global attractor for (5.118). To show that $\bar{t}$ is attracting on $(0, \infty)$ we prove that

$$g^2(t) > t \text{ if } t < \bar{t} \quad \text{and} \quad g^2(t) < t \text{ if } t > \bar{t}, \quad t > 0 \tag{5.122}$$

where $g^2(t) = g(g(t))$; see the Appendix. By Lemma 5.7(c) the positive solutions of $g^2(t) = t$ are the positive zeros of $\phi(t)\psi(t)$. The only positive zero of $\phi(t)$ is the fixed point $\bar{t}$. Also ψ has no positive real roots; for $p > 0$ its discriminant is negative by (5.119):

$$q^2(p+1)^2 - 4sp^2(p+1) < 0.$$

Therefore, the only zero of $g^2(t) = t$ in $(0, \infty)$ is $\bar{t}$. From its explicit form in (5.116) we see that $g^2(t) \to \infty$ as $t \to 0^+$ so it must be that $g^2(t) > t$ if $t < \bar{t}$. Also, for t sufficiently large the last two terms of $g^2(t)$ are negligible and bounded above by $\varepsilon > 0$ such that $\varepsilon < (1 - p^2)t - q(p+1)$ and so

$$g^2(t) \leq p^2 t + q(p+1) + \varepsilon = t - (1 - p^2)t + q(p+1) + \varepsilon < t.$$

Thus (5.122) holds and it follows that $\bar{t}$ is a global attractor for (5.118). Therefore, $\bar{x}$ attracts all solutions of (5.111) satisfying (5.112).

(b) If $p = 1$ then the first inequality in (5.120) is (5.119) and the second is needed to obtain the unique fixed point $\bar{t} = -s/q$ for (5.118). Repeating an argument similar to the proof of Part (a) shows that (5.122) holds and thus $\bar{t}$ is a global attractor in this case too.

Finally, if (5.121) holds then $g(t) = t + q + s/t > t$ for all $t > 0$. Hence, for every $t_0 > 0$, $t_n = g^n(t_0) \to \infty$ monotonically as $n \to \infty$. Hence, (5.110) implies that $x_n \to \infty$. ∎

THEOREM 5.8
(2-cycles) Assume that

$$s, p > 0, \ 0 < |c| < 1, \ q < -2p\sqrt{\frac{s}{p+1}}. \tag{5.123}$$

Then Eq. (5.111) has a positive periodic solution with minimal (or prime) period 2 (or a 2-cycle)

$$\xi_0 = \frac{c\tau_0 + \tau_1}{1 - c^2}, \quad \xi_1 = \frac{c\tau_1 + \tau_0}{1 - c^2}$$

where

$$\tau_0 = \frac{|q| - \sqrt{q^2 - 4sp^2/(p+1)}}{2p}, \quad \tau_1 = \frac{|q| + \sqrt{q^2 - 4sp^2/(p+1)}}{2p}. \quad (5.124)$$

This 2-cycle is asymptotically stable if in addition to (5.123) the following holds:

$$q > -\frac{\sqrt{2sp(2p^2 + 2p + 1)}}{p+1} = -2p\sqrt{\frac{s}{p+1}}\sqrt{1 + \frac{1}{2p(p+1)}}. \quad (5.125)$$

PROOF   Conditions (5.123) imply that the polynomial $\psi(t)$ in (5.115) has two positive roots which are readily computed as the numbers τ_0, τ_1 in (5.124). Since these are the non-fixed point solutions of the equation $g^2(t) = t$ it follows that $\{\tau_0, \tau_1\}$ is a 2-cycle of (5.118). Lemma 5.5(a) now gives the 2-cycle $\{\xi_0, \xi_1\}$ for (5.111).

By Theorem 5.5 the 2-cycle $\{\xi_0, \xi_1\}$ is attracting if $\{\tau_0, \tau_1\}$ is an attracting 2-cycle of (5.118); i.e. if

$$|g'(\tau_0)g'(\tau_1)| < 1.$$

Calculating the values of the derivative g' we obtain

$$\left| \left(p - \frac{s}{\tau_0^2} \right) \left(p - \frac{s}{\tau_1^2} \right) \right| < 1$$

$$\left| p^2 + \frac{s^2}{\tau_0^2 \tau_1^2} - \frac{sp(\tau_0^2 + \tau_1^2)}{\tau_0^2 \tau_1^2} \right| < 1. \quad (5.126)$$

Note that

$$\tau_0 + \tau_1 = \frac{q}{p} \text{ and } \tau_0 \tau_1 = \frac{s}{p+1}$$

imply that

$$\tau_0^2 + \tau_1^2 = (\tau_0 + \tau_1)^2 - 2\tau_0 \tau_1 = \frac{q^2(p+1) - 2sp^2}{p^2(p+1)}.$$

After inserting these values in (5.126) and doing some routine calculations we obtain

$$\frac{4sp^2}{p+1} < q^2 < \frac{2sp(2p^2 + 2p + 1)}{(p+1)^2}.$$

This completes the proof.   ∎

REMARK 5.6   If $s = p = 1$ then Theorems 5.7 and 5.8 indicate that a (globally) attracting fixed point $\bar{x} = 1/(c-1)q$ exists when

$$-\sqrt{2} \leq q < 0.$$

The fixed point $\bar{x}$ becomes unstable as the value of q goes below $-\sqrt{2}$; thereafter, the stable 2-cycle $\{\xi_0, \xi_1\}$ in Theorem 5.8 emerges over the range

$$-\sqrt{\frac{5}{2}} < q < -\sqrt{2}.$$

Thus a period-doubling bifurcation occurs for the second-order equation (5.111) as the bifurcation parameter q decreases and crosses $-\sqrt{2}$. Then as q crosses $-\sqrt{5/2}$ a second period-doubling bifurcation destabilizes the 2-cycle and creates a stable 4-cycle. The mapping

$$g(t) = t + q + \frac{1}{t} \tag{5.127}$$

exhibits the usual bifurcations to higher periods that follow the Sharkovski ordering of cycles as the parameter q continues to decrease further. The requirement that g in (5.127) be positive puts a lower bound on q; in fact, by Lemma 5.7(a) it is necessary that $q > -2\sqrt{sp} = -2$. ∎

The next result marks the emergence of a period 3 solution (through a tangent bifurcation rather than a period-doubling one).

LEMMA 5.8
If $q = -\sqrt{3}$ then the function g in (5.127) has a unique set of positive period 3 points given by

$$\tau_0 = \frac{2}{\sqrt{3}}\left(1 + \cos\frac{\pi}{9}\right), \quad \tau_1 = g(\tau_0), \quad \tau_2 = g(\tau_1). \tag{5.128}$$

PROOF A period-3 point is a solution of the equation $g^3(t) = t$, which is equivalent to

$$\frac{1}{t} + \frac{t}{t^2 - \sqrt{3}t + 1} + \frac{t(t^2 - \sqrt{3}t + 1)}{t(t - 2\sqrt{3})(t^2 - \sqrt{3}t + 1) + 2t^2 - \sqrt{3}t + 1} = 3\sqrt{3}.$$

Multiplying out and rearranging various terms, the above equation may be written as the polynomial equation

$$P(t) = 3\sqrt{3}t^7 - 39t^6 + 66\sqrt{3}t^5 - 168t^4 + 77\sqrt{3}t^3 - 57t^2 + 7\sqrt{3}t - 1 = 0.$$

By Lemma 5.7(b)(i) g has a unique positive fixed point at $1/\sqrt{3}$ and no negative fixed points so this is the only fixed point root of $P(t)$. Factoring it out yields

$$P(t) = 3\sqrt{3}\left(t - 1/\sqrt{3}\right)Q(t)$$

where

$$Q(t) = t^6 - 4\sqrt{3}t^5 + 18t^4 - \frac{38\sqrt{3}}{3}t^3 + 13t^2 - 2\sqrt{3}t + \frac{1}{3}.$$

Since all the fixed points are accounted for, the 6 roots of Q give two sets of period 3 points (if they are all real). These two sets are identical if Q is a perfect square, i.e.,

$$Q(t) = (t^3 + \lambda t^2 + \omega t + \sigma)^2. \tag{5.129}$$

We show that this is in fact the case. Indeed, by matching coefficients on both sides of (5.129) we find a set of numbers

$$\lambda = -2\sqrt{3}, \quad \omega = 3, \quad \sigma = -\frac{1}{\sqrt{3}}$$

for which (5.129) holds for all $t > 0$. Therefore,

$$P(t) = 3(\sqrt{3}t - 1)\left(t^3 - 2\sqrt{3}t^2 + 3t - \frac{1}{\sqrt{3}}\right)^2.$$

The roots of the cubic polynomial above can be found using the standard formula with radicals; one root using this formula is found to be

$$\tau_0 = \frac{1}{\sqrt{3}}\left(2 + \sqrt[3]{\frac{z}{2}} + \sqrt[3]{\frac{\bar{z}}{2}}\right) \quad \text{where } z = 1 + i\sqrt{3} = 2e^{i\pi/3}.$$

Therefore,

$$\tau_0 = \frac{1}{\sqrt{3}}\left(2 + e^{i\pi/9} + e^{-i\pi/9}\right) = \frac{1}{\sqrt{3}}\left(2 + 2\cos\frac{\pi}{9}\right)$$

as in (5.128).   ∎

LEMMA 5.9
Assume that $|c| < 1$. If the first order equation (5.109) has an invariant interval $[\mu, \nu]$ then every solution of the second-order equation (5.108) is eventually contained in the planar compact, convex set:

$$S_{\mu,\nu} = \{(x, y) : cx + \mu \leq y \leq cx + \nu\} \cap \left[-|c| - \frac{\max\{|\mu|, |\nu|\}}{1 - |c|}, |c| + \frac{\max\{|\mu|, |\nu|\}}{1 - |c|}\right]^2.$$

If (5.109) is chaotic in $[\mu, \nu]$ (e.g. if f has a period-3 point) then (5.108) is chaotic in $S_{\mu,\nu}$.

It is clear from Theorem 5.5 and Lemma 5.9 that as long as $|c| < 1$ the various properties of the first-order equation (5.109) directly lead to corresponding properties of (5.111).

The next theorem follows readily from the preceding results.

THEOREM 5.9
Let $p = s = 1$ in (5.111).

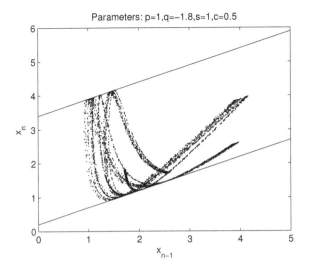

FIGURE 5.5

A state-space orbit of Eq. (5.111).

(a) If $q = -\sqrt{3}$ then (5.111) has a unique 3-cycle

$$\xi_0 = \frac{c^2\tau_0 + c\tau_1 + \tau_2}{1 - c^3}, \quad \xi_1 = \frac{c^2\tau_1 + c\tau_2 + \tau_0}{1 - c^3}, \quad \xi_2 = \frac{c^2\tau_2 + c\tau_0 + \tau_1}{1 - c^3}$$

where τ_0, τ_1, τ_2 *are given by (5.128).*

(b) If $-2 < q \le -\sqrt{3}$ then Eq. (5.111) has periodic solutions of all possible periods.

(c) For $-2 < q < -\sqrt{3}$ solutions of (5.111) exhibit chaotic behavior in the sense of Li and Yorke.

(d) If $q > -2\sqrt{1+c}$ then all solutions of (5.111) with positive initial values are positive; i.e. $(0, \infty)$ is invariant. In particular, this is the case for all $-2 < q < 0$ if $c > 0$.

Figures 5.5, 5.6, and 5.7 illustrate Theorem 5.9. The straight lines in Figure 5.5 have equations

$$y = cx + \mu, \quad y = cx + \nu$$

where $\mu = g(1) = 0.2$ is the minimum value of g in (5.127) when $q = -1.8$ and $\nu = g(0.2) = 3.4$; refer to Lemmas 5.9 and 5.7(a). Figure 5.7 shows the emergence of the 3-cycle at $q = -\sqrt{3}$ as well as the initial period-doubling bifurcations whose beginning is described in Theorem 5.8.

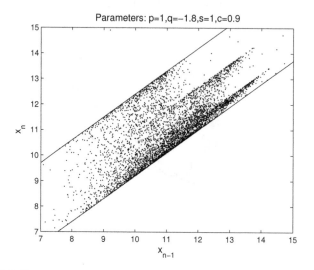

FIGURE 5.6

Another state-space orbit of Eq. (5.111).

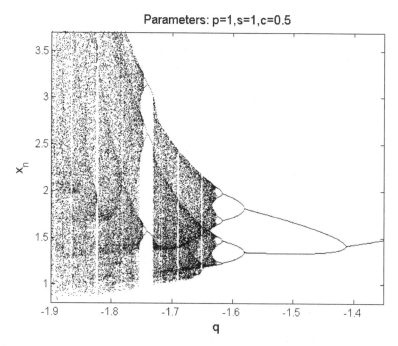

FIGURE 5.7

A bifurcation diagram for Eq. (5.111).

5.7   Field-inverse form symmetry

Let $\mathcal{F}$ be a nontrivial field and define $h : \mathcal{F} \to \mathcal{F}$ as $h(u) = -\alpha/u$ where α is a fixed nonzero element in $\mathcal{F}$. As before we use the reciprocal notation for the multiplicative inverse. The mapping h defines the form symmetry

$$H(u_0, u_1, \ldots, u_k) = \left[u_0 - \frac{\alpha}{u_1}, u_1 - \frac{\alpha}{u_2}, \ldots, u_{k-1} - \frac{\alpha}{u_k} \right].$$

Because H depends on the multiplicative inverse in a field we call it the *field-inverse form symmetry*. This form symmetry is characterized by a change of variables to

$$t_n = x_n - \frac{\alpha}{x_{n-1}}.$$

The SC factorization has the following factor and cofactor equations

$$t_{n+1} = \phi_n(t_n, \ldots, t_{n-k+1}),$$
$$x_{n+1} = t_{n+1} + \frac{\alpha}{x_n}.$$

Notice that h is a self-inverse mapping, i.e., $h = h^{-1}$. The following is a straightforward consequence of Theorem 5.1.

COROLLARY 5.14
For every $u_0, v_1, \ldots, v_k \in \mathcal{F}$ let $\zeta_0 = u_0$ and define

$$\zeta_j = \frac{\alpha}{\zeta_{j-1} - v_j}, \quad j = 1, \ldots, k.$$

Then Eq. (2.3) has the field-inverse form symmetry if and only if the quantity

$$f_n(u_0, \zeta_1, \ldots, \zeta_k) - \frac{\alpha}{u_0}$$

is independent of u_0.

The next example gives an application of this corollary.

Example 5.12
Consider the third-order equation

$$x_{n+1} = a_n + \frac{b}{x_n} + \frac{c x_{n-2}}{x_{n-1} x_{n-2} + d} \qquad (5.130)$$

where a_n is a sequence of real numbers and b, c, d are nonzero real numbers. For this equation we calculate

$$f_n(u_0, \zeta_1, \zeta_2) - \frac{\alpha}{u_0} = a_n + \frac{b}{u_0} + \frac{c\zeta_2}{\zeta_1 \zeta_2 + d} - \frac{\alpha}{u_0}$$

$$= a_n + \frac{b - \alpha}{u_0} + \frac{c\alpha/(\zeta_1 - v_2)}{\alpha^2/[(u_0 - v_1)(\zeta_1 - v_2)] + d}.$$

As this expression is to be independent of u_0 we set $\alpha = b$ and $\zeta_1 = \alpha/(u_0 - v_1)$ to obtain

$$f_n(u_0, \zeta_1, \zeta_2) - \frac{b}{u_0} = a_n + \frac{bc(u_0 - v_1)}{b^2 + bd - d(u_0 - v_1)v_2}$$

If $d = -b$ then the quantity on the right-hand side above reduces to

$$a_n + \frac{c}{v_2}$$

which is independent of u_0. In this case, Corollary 5.14 implies that Eq. (5.130) has the field inverse form symmetry $-b/u$ and the following SC factorization:

$$t_{n+1} = a_n + \frac{c}{t_{n-1}},$$

$$x_{n+1} = t_{n+1} + \frac{b}{x_n}.$$

□

5.8   Notes

References on the inversion form symmetry appear in the Notes for Chapter 4. Linear form symmetries appeared in Dehghan et al. (2008a) and Sedaghat (2009c) to simplify the study of solutions of certain second-order difference equations. In particular, the material in Section 5.6.5 is taken from Dehghan et al. (2008a) while more general results are studied in Sedaghat (2009c), including the results in Sections 5.5.2 and 5.5.3. The results in Section 5.6 above in turn generalize those in Sedaghat (2009c); Corollary 5.8 provides the specific link.

Difference equations with linear arguments appear in a variety of pure and applied contexts. For applications involving biology see, e.g., Fisher and Goh (1984), Kocic and Ladas (1993), and Hamaya (2005). Economic models involving the business cycle that use difference equations with linear arguments appear in theoretical publications by prominent economists in the 20th century, e.g., Samuelson (1939), Hicks (1965), and Puu (1993). These models and mathematical issues pertaining to them have been discussed in detail in Sedaghat (1997, 1998, 2003, 2004b), and Kent and Sedaghat (2004b) which collectively contain extensive bibliographies.

Discovering the invertible map criterion led to identifying more form symmetries, including the identity and the field-inverse as is indicated in the text of this chapter; see also Sedaghat (2009d, 2009e). The material on the discrete Riccati equation is from Dehghan et al. (2010). For a discussion of the discrete Riccati equation in one dimension see Kulenovic and Ladas (2002). That discussion is based on a detailed study of the solutions of a second-order linear difference equation.

5.9 Problems

5.1 Consider the following variation of Eq. (5.16) in Example 5.2:

$$x_{n+1} = \frac{x_n x_{n-1}}{a_n x_{n-1} + b_n x_{n-2}}. \tag{5.131}$$

(a) Use Corollary 5.1 to show that this equation does *not* have the identity form symmetry.

(b) Show that Eq. (5.131) is HD1 and use the inversion form symmetry to find a SC factorization of (5.131).

(c) Show that the reciprocal form symmetry (order-preserving) transforms the second-order factor equation in Part (b) into a linear equation.

5.2 Show that the following variation of Eq. (5.16) in Example 5.2:

$$x_{n+1} = \frac{x_{n-1} x_{n-2}}{a_n x_n + b_n x_{n-2}}$$

has the identity form symmetry and determine its SC factorization relative to this form symmetry.

5.3 Prove Corollary 5.5.

5.4 Let β_n, γ_n be sequences of complex numbers such that β_n is not eventually zero. For each of the following difference equations, find the value of α that gives the equation the linear form symmetry, then find the corresponding SC factorization:

$$x_{n+1} = \alpha x_n + \frac{\beta_n}{x_n + x_{n-2} + \gamma_n}, \quad x_{n+1} = 2i x_n + \frac{\beta_n}{x_n + \alpha x_{n-1} + \gamma_n}.$$

5.5 Let $a, b, c, d \in \mathbb{R}$ and consider the second-order difference equation

$$x_{n+1} = ax_n + bx_{n-1} + c(x_n + dx_{n-1})^2. \qquad (5.132)$$

(a) Find the value of a in terms of the other parameters such that Eq. (5.132) has the linear form symmetry and determine its SC factorization.

(b) Assume that $0 < |d| < 1$, $b, c \neq 0$ and a has the value in (a). Prove that for certain values of parameters b, c, d Eq. (5.132) has periodic solutions of all possible periods within an invariant region in $\mathbb{R}^2$. Determine the invariant region and the appropriate parameter ranges.

Hint. Show that the factor equation for (5.132) is conjugate to the logistic equation $s_{n+1} = rs_n(1 - s_n)$ for some r.

5.6 Let $a, b, c, p \in \mathbb{R}$, $p, c \neq 0$ and let $\{\gamma_n\}$ be a given sequence in $\mathbb{R}$. For each of the following difference equations find conditions on the parameters that imply the existence of a linear form symmetry. In each case also determine the SC factorization.

$$(a) \quad x_{n+1} = a(x_n + bx_{n-1})^p + cx_{n-k} + \gamma_n,$$
$$(b) \quad x_{n+1} = ax_n + b(x_{n-j} + cx_{n-k})^p, \quad 0 \le j < k.$$

5.7 Consider the difference equation

$$x_{n+1} = ax_n + \frac{\delta_n}{x_n + bx_{n-k} + w_n} \qquad (5.133)$$

where a, b and the sequences $\{\delta_n\}$ and $\{w_n\}$ are complex. Show that if $b = -a^k$ then (5.133) has the linear form symmetry and its reduction to factor is

$$t_{n+1} = \frac{\delta_n}{t_n + at_{n-1} + \cdots + a^{k-1}t_{n-k+1} + w_n}.$$

5.8 Consider the following variation of Eq. (5.133)

$$x_{n+1} = ax_n + \frac{b_n}{x_{n-k+1} - ax_{n-k} + c_n} \qquad (5.134)$$

where a and the sequences $\{b_n\}, \{c_n\}$ are real.

(a) Show that Eq. (5.134) has the linear form symmetry and determine its SC factorization.

(b) Let $k = 2$, $0 < |a| < 1$ and let $b_n = b > 0$ and $c_n = c > 0$ be constant sequences. Prove that if $x_0 \ge ax_{-1}$, $x_{-1} \ge ax_{-2}$ then the corresponding solution $\{x_n\}$ of (5.134) converges to its positive fixed point.

(c) Assume the hypotheses of Part (b) except that $c = 0$ and $x_0 > ax_{-1}$, $x_{-1} > ax_{-2}$. Determine the global behavior of the corresponding solution $\{x_n\}$ of (5.134) in this case.

(d) Assume the hypotheses of Part (b) except that $a = 1$ (this makes (5.134) an HD1 equation). Determine the global behavior of the corresponding solution $\{x_n\}$ in this case.

(e) Assume the hypotheses of Part (b) except that $a = -1$. Determine the global behavior of the corresponding solution $\{x_n\}$ in this case.

5.9 (a) Show that the following third-order difference equation has the linear form symmetry and find its factor equation:

$$x_{n+1} = x_n + 2x_{n-1} + \rho_n + e^{\sigma_n - bx_n - 3bx_{n-1} - 2bx_{n-2}},$$
$$b, \rho_n, \sigma_n \in \mathbb{R} \text{ for all } n = 0, 1, 2, \cdots, \quad b \neq 0.$$

(b) Repeat Part (a) for the following equation

$$x_{n+1} = x_n + 2x_{n-1} + \rho_n + e^{\sigma_n + bx_n - bx_{n-1} - 2bx_{n-2}}.$$

(c) Show that the second-order factor equation that is obtained in Part (b) again has the linear form symmetry and use this fact to its first-order factor equation.

5.10 Let $\{\alpha_n\}$, $\{\beta_n\}$ and $\{\gamma_n\}$ be sequences of complex numbers. For all $a, b \in \mathbb{C}$ such that $|a| + |b| > 0$ find all values of δ for which the third-order equation

$$x_{n+1} = ax_n + bx_{n-1} + \frac{\alpha_n(x_{n-1} + \delta x_{n-2})}{x_n + \delta x_{n-1} + \beta_n} + \gamma_n$$

has the linear form symmetry and find its factor equation for each δ.

5.11 Let $\{\sigma_n\}$ be a sequence of complex numbers. Determine necessary and sufficient conditions on parameters $a, b, c, \gamma, \delta \in \mathbb{C}$ so that the third-order difference equation

$$x_{n+1} = ax_n + bx_{n-1} + (cx_n + x_{n-2})e^{\sigma_n - \gamma x_n - \delta x_{n-1}}$$

has the linear form symmetry. Then find the second-order factor equation. (*Hint.* A suitable g_n can be used in Example 5.6.)

5.12 Let a_i, b_i, c_i be constants in a nontrivial field $\mathcal{F}$ for $i = 0, 1, \ldots, k$. By extending the ideas presented for difference equations with linear arguments and Example 5.6, show that a necessary and sufficient condition for the existence of linear form symmetry for the difference equation

$$x_{n+1} = \sum_{i=0}^{k} a_i x_{n-i} + g_n \left(\sum_{i=0}^{k} b_i x_{n-i}, \sum_{i=0}^{k} c_i x_{n-i} \right)$$

is that the following three polynomials have a common, nonzero root $\alpha \in \mathcal{F}$

$$u^{k+1} - a_0 u^k - a_1 u^{k-1} - \cdots - a_{k-1} u - a_k = 0,$$
$$b_0 u^k + b_1 u^{k-1} + \cdots + b_{k-1} u + b_k = 0,$$
$$c_0 u^k + c_1 u^{k-1} + \cdots + c_{k-1} u + c_k = 0.$$

5.13 Let $j = 2$ in Eq. (5.85) of Example 5.10; i.e., consider the difference equation with linear arguments

$$x_{n+1} = a_0 x_n + \cdots + a_k x_{n-k} + g_n(x_n - x_{n-2}), \quad k \geq 2. \tag{5.135}$$

(a) Determine necessary and sufficient conditions for Eq. (5.135) to have the linear form symmetry $-\alpha u$.

(b) Find the SC factorization of (5.135) corresponding to a value $\alpha \neq 1$.

5.14 Let $\{c_n\}$ be a periodic sequence of real numbers with period $p \geq 1$ and consider the second-order difference equation

$$x_{n+1} = \alpha x_n + c_n(x_n - \alpha x_{n-1})^q$$
$$0 < |\alpha|, |q| < 1, \ c_n \geq 0 \text{ for all } n.$$

(a) Let $\{c_n\}$ be a periodic sequence with period $p \geq 1$. If $c_j > 0$ for all $j = 1, \ldots, p$ then use Theorem 5.5 and the idea presented in Example 5.9 to show that every solution of the above difference equation that originates in the invariant region $\{(x, y) : x > \alpha y\}$ converges to a p-cycle. Also determine this p-cycle.

(b) Use the results of Part (a) to determine the asymptotic behaviors of all solutions of the following difference equation with nonnegative initial values:

$$x_{n+1} = \frac{2^{\cos(n\pi/3)+1/2}}{\sqrt{2x_n + x_{n-1}}} - \frac{1}{2}x_n.$$

(c) If $c_j = 0$ in Part (a) for some $j \in \{1, \ldots, p\}$ then prove that all solutions converge to zero.

5.15 Complete the proof of Lemma 5.6 by doing the calculations for the polynomial Q_1.

5.16 Consider the following higher-order version of Example 5.12

$$x_{n+1} = a_n + \frac{b}{x_n} + \frac{cx_{n-k}}{x_{n-k+1}x_{n-k} - b}$$

where k is a fixed positive integer.

(a) Show that the above equation has the field-inverse form symmetry and find its SC factorization.

Hint: Consider a slight modification of the factor equation in Example 5.12.

(b) Let $a_n = a > 0$ for all n, $b, c > 0$ and assume that the initial values x_j satisfy

$$x_{-k+1} > 0, \quad x_j x_{j-1} > b, \ j = -k + 2, \ldots, 0.$$

Prove that the corresponding solution $\{x_n\}$ of the above difference equation converges to a positive fixed point.

(c) Determine the behavior of $\{x_n\}$ in (b) if $a = 0$.

5.17 (a) Let $p > 0$. Find the explicit solution of the nonautonomous linear equation

$$x_{n+1} = \left[\alpha + \frac{\beta}{(n+1)^p}\right] x_n - \frac{\alpha\beta}{(n+1)^p} x_{n-1}$$

and discuss the asymptotic behaviors of its solutions.

Hint: See Example 5.11.

(b) Find the explicit solution of the nonautonomous linear equation

$$x_{n+1} = \left(\alpha + \frac{\beta n}{n+1}\right) x_n - \frac{\alpha\beta n}{n+1} x_{n-1}$$

and discuss the asymptotic behaviors of its solutions.

6

Type-(1,k) Reductions

From Section 3.3.3 recall that for a type-$(1, k)$ reduction, $m = 1$ so that $h : G^k \to G$ and the form symmetry is a scalar function

$$H(u_0, \ldots, u_k) = u_0 * h(u_1, \ldots, u_k). \tag{6.1}$$

This form symmetry is characterized by the change of variables

$$t_n = h(x_{n-1}, \ldots, x_{n-k})$$

in Eq. (2.3), i.e.,

$$x_{n+1} = f_n(x_n, x_{n-1}, \ldots, x_{n-k}) \tag{6.2}$$

which yields the SC factorization

$$t_{n+1} = g_n(t_n) \tag{6.3}$$
$$x_{n+1} = t_{n+1} * h(x_n, \ldots, x_{n-k+1})^{-1}. \tag{6.4}$$

We note that the cofactor equation plays a role that is similar to that played by the factor equation in Chapter 5; i.e., the roles of factor and cofactor equations are switched. Because the form symmetry component function h is now multivariable and encodes more information, it may be more difficult to identify. On the other hand, since H is a scalar function more technical tools may be available for handling it.

In this chapter we discuss classes of difference equations that have the scalar form symmetry (6.1). The classes discussed here involve separable difference equations that are defined below. We obtain conditions on separable equations that imply the existence of a form symmetry of type (6.1) that we call the *separable form symmetry*. In particular, we find that linear equations possess the separable form symmetry and the corresponding semiconjugate factorizations as equations of type 6.3 and 6.4. Mirroring Section 5.6 above, we also discover that every linear nonhomogeneous equation with constant coefficients has a complete SC factorization into a triangular system of first-order equations. However, this time the system is reached by a *cofactor chain* rather than a factor chain.

6.1   Linear form symmetry revisited

We first consider the case where $h(u_1, \ldots, u_k) = -\alpha_1 u_1 - \cdots - \alpha_k u_k$ is a linear function on a nontrivial field $\mathcal{F}$ with $\alpha_j \in \mathcal{F}$ for $j = 1, \ldots, k$. In this case,

$$H(u_0, \ldots, u_k) = u_0 - \alpha_1 u_1 - \cdots - \alpha_k u_k \qquad (6.5)$$

yields a type-$(1, k)$ linear form symmetry. Difference equations possessing the form symmetry (6.5) have been encountered in Chapter 5 in a different context. Eq. (6.2) has the form symmetry (6.5) if the unfoldings F_n satisfy the semiconjugate relation. In this case, Eq. (3.11) in Section 3.2 takes the form

$$f_n(u_0, \ldots, u_k) - \alpha_1 u_0 - \cdots - \alpha_k u_{k-1} = g_n(u_0 - \alpha_1 u_1 - \cdots - \alpha_k u_k)$$

or equivalently,

$$f_n(u_0, \ldots, u_k) = \alpha_1 u_0 + \cdots + \alpha_k u_{k-1} + g_n(u_0 - \alpha_1 u_1 - \cdots - \alpha_k u_k).$$

Therefore, equations having the form symmetry (6.5) are special types of equations with linear arguments, specifically of type (5.94) seen in Corollary 5.10. We refer to Section 5.6.3 for further details.

6.2   Separable difference equations

In this section we consider reduction of order in a class of difference equations that are distinct from all others that we studied in previous chapters. This class of equations is characterized by the following property.

DEFINITION 6.1   *Let G be a commutative group. A function $\phi : G^{k+1} \to G$ is **separable** relative to G if there are $k + 1$ functions $\phi_j : G \to G$, $j = 0, 1, \ldots, k$ such that for all $u_0, \ldots, u_k \in G$,*

$$\phi(u_0, \ldots, u_k) = \phi_0(u_0) * \cdots * \phi_k(u_k).$$

*If f_n in (6.2) is a separable function for every n then we call (6.2) a **separable difference equation** over G.*

The concept of separable in Definition 6.1 may be extended to arbitray groups with a consideration of the *ordering of terms* in the factorization being important; e.g.,

$$\phi_0(u_0) * \phi_1(u_1) \neq \phi_1(u_1) * \phi_0(u_0).$$

We do not consider the general case in this book.

Example 6.1
Let $\mathbb{R}$ be the set of real numbers which is a field under its familiar operations of addition and multiplication. Each of the following functions is separable with respect to one or the other of these two operations in $\mathbb{R}$:

$$\phi(u_0, u_1) = au_0 + bu_1 + c, \quad a, b, c \in \mathbb{R}$$

$$\psi(u_0, u_1) = au_0 + \frac{b}{u_1 + c},$$

$$\sigma(u_0, u_1) = \frac{au_0 + b}{u_1 + c} = (au_0 + b)\frac{1}{u_1 + c},$$

$$\eta(u_0, u_1) = u_1 e^{au_0 + bu_1 + c} = (e^{au_0 + c})(u_1 e^{bu_1})$$

$$\xi(u_0, u_1, u_2) = u_0 u_1^a e^{bu_0 + cu_2} = (u_0 e^{bu_0})(u_1^a)(e^{cu_2}).$$

⬜

6.2.1 Form symmetries of additive forms over fields

Consider a separable difference equation of the following type (written in additive notation)

$$x_{n+1} = \alpha_n + \phi_0(x_n) + \phi_1(x_{n-1}) + \cdots \phi_k(x_{n-k}). \tag{6.6}$$

We assume here that (6.6) is defined over a nontrivial field $\mathcal{F}$ where G represents the additive group structure of $\mathcal{F}$. Specifically, we assume that

$$x_{-j}, \alpha_n \in \mathcal{F}, \quad \phi_j : \mathcal{F} \to \mathcal{F}, \quad j = 0, 1, \ldots, k, \quad n = 0, 1, 2, \ldots \tag{6.7}$$

The results in this section require the field structure. As usual, a commonly encountered field is the complex numbers $\mathbb{C}$ under ordinary addition and multiplication with its subfield of all real numbers $\mathbb{R}$. On the other hand, the group $i\mathbb{R}$ of all imaginary numbers under addition fails to be a subfield of $\mathbb{C}$ since $i\mathbb{R}$ is not closed under the ordinary multiplication of complex numbers.
The unfolding of (6.6) is the mapping

$$F_n(u_0, u_1, \ldots, u_k) = (\alpha_n + \phi_0(u_0) + \cdots + \phi_k(u_k), u_0, \ldots, u_{k-1}).$$

Based on the additive nature of Eq. (6.6) we look for a separable form symmetry of type

$$H(u_0, u_1, \ldots, u_k) = u_0 + h_1(u_1) + \cdots + h_k(u_k) \tag{6.8}$$

where $h_j : \mathcal{F} \to \mathcal{F}$ for each $j = 1, \ldots, k$. With these assumptions, equality (3.11) in Section 3.2 takes the form

$$\alpha_n + \phi_0(u_0) + \cdots + \phi_k(u_k) + h_1(u_0) + \cdots + h_k(u_{k-1}) = g_n(u_0 + h_1(u_1) + \cdots + h_k(u_k)) \tag{6.9}$$

where $g_n : \mathcal{F} \to \mathcal{F}$. Our aim is to determine functions h_j and g_n that satisfy the functional equation (6.9) and use these to obtain a form symmetry of type (6.8). We do this in two stages as indicated below:

(I) A derivation in the field of real numbers

To simplify our task and make tools from calculus available, we start our calculations with the field $\mathbb{R}$ of real numbers and assume at first that the functions ϕ_j and h_j are all differentiable on $\mathbb{R}$ for $j = 0, \ldots, k$.

Taking partial derivatives of both sides of (6.9) for each of the $k+1$ variables $u_0, u_1, \ldots, u_k \in \mathbb{R}$ yields the following system of $k + 1$ partial differential equations:

$$\phi_0'(u_0) + h_1'(u_0) = g_n'(u_0 + h_1(u_1) + \cdots + h_k(u_k)) \qquad (6.10)$$

$$\phi_1'(u_1) + h_2'(u_1) = g_n'(u_0 + h_1(u_1) + \cdots + h_k(u_k))h_1'(u_1) \qquad (6.11)$$

$$\vdots$$

$$\phi_{k-1}'(u_{k-1}) + h_k'(u_{k-1}) = g_n'(u_0 + h_1(u_1) + \cdots + h_k(u_k))h_{k-1}'(u_{k-1})$$

$$\phi_k'(u_k) = g_n'(u_0 + h_1(u_1) + \cdots + h_k(u_k))h_k'(u_k). \qquad (6.12)$$

Taking equations (6.10) and (6.11) together, we may eliminate g_n' to obtain

$$\phi_0'(u_0) + h_1'(u_0) = \frac{\phi_1'(u_1) + h_2'(u_1)}{h_1'(u_1)}. \qquad (6.13)$$

The variables u_0 and u_1 are separated on different sides of PDE (6.13) so each side must be equal to a nonzero constant $c \in \mathbb{R}$. The left-hand side of (6.13) gives (in single-variable notation u)

$$\phi_0'(u) + h_1'(u) = c \Rightarrow h_1(u) = cu - \phi_0(u). \qquad (6.14)$$

Similarly, from the right-hand side of (6.13) we get

$$\frac{\phi_1'(u) + h_2'(u)}{h_1'(u)} = c \Rightarrow h_2(u) = ch_1(u) - \phi_1(u).$$

Note that from (6.10) and (6.14) it follows that

$$g_n'(z_0 + h_1(z_1) + \cdots + h_k(z_k)) = c. \qquad (6.15)$$

Next, using (6.15) in the third equation in the above system of PDE's gives

$$\frac{\phi_2'(u) + h_3'(u)}{h_2'(u)} = c \Rightarrow h_3(u) = ch_2(u) - \phi_2(u).$$

It is evident that this calculation can be repeated for equations 4 to $k-1$ in the above system of PDE's and the following functional recursion is established by induction

$$h_j = ch_{j-1} - \phi_{j-1}, \quad j = 1, \ldots, k, \ h_0(u) \doteq u. \qquad (6.16)$$

Finally, the last PDE (6.12) gives the following equality of functions

$$ch_k = \phi_k. \tag{6.17}$$

On the other hand, setting $j = k$ in (6.16) gives $h_k = ch_{k-1} - \phi_{k-1}$. Therefore, equality (6.17) implies the existence of a restriction on the functions ϕ_j. To determine this restriction, first notice that by (6.17) and (6.16)

$$c^2 h_{k-1}(u) - c\phi_{k-1}(u) - \phi_k(u) = 0 \quad \text{for all } u.$$

Now, applying (6.16) repeatedly $k - 1$ more times eliminates the functions h_j to give the following identity

$$c^{k+1}u - c^k \phi_0(u) - c^{k-1}\phi_1(u) - \cdots - c\phi_{k-1}(u) - \phi_k(u) = 0 \quad \text{for all } z. \tag{6.18}$$

This equality shows that the existence of form symmetries of type (6.8) requires that the given maps $\phi_0, \phi_1, \ldots, \phi_k$ plus the identity function form a linearly dependent set. Equivalently, (6.18) determines any one of the functions $\phi_0, \phi_1, \ldots, \phi_k$ as a linear combination of the identity function and the other k functions; e.g.,

$$\phi_k(u) = c^{k+1}u - c^k \phi_0(u) - c^{k-1}\phi_1(u) - \cdots - c\phi_{k-1}(u). \tag{6.19}$$

Thus if there is $c \in \mathbb{R}$ such that (6.18) or (6.19) hold then Eq. (6.6) has a form symmetry of type (6.8). This form symmetry H can now be calculated as follows: Using the recursion (6.16) repeatedly gives

$$h_j(u) = c^j u - c^{j-1}\phi_0(u) - \cdots - \phi_{j-1}(u), \quad j = 1, \ldots k. \tag{6.20}$$

(II) Derivation of a separable form symmetry over any field

At this stage, note that the equations of interest in the derivation of form symmetries, i.e., (6.18) and (6.20) are meaningful for arbitrary functions ϕ_j over *any nontrivial field* $\mathcal{F}$ although their derivations required stronger assumptions. If there is a nonzero element $c \in \mathcal{F}$ for which (6.18) is satisfied for every $u \in \mathcal{F}$ then we may use the functions h_j as defined by (6.20) to define the following separable function

$$H(u_0, u_1, \ldots, u_k) = u_0 + h_1(u_1) + \cdots + h_k(u_k)$$

$$= u_0 + \sum_{j=1}^{k} [c^j u_j - c^{j-1}\phi_0(u_j) - \cdots - \phi_{j-1}(u_j)].$$

For H to be a form symmetry, it must be surjective, i.e., $H(\mathcal{F}^{k+1}) = \mathcal{F}$. To show this, first note that since the constant c is in the field $\mathcal{F}$, by (6.20) $\mathcal{F}$ is invariant under all h_j. Now, for each $u \in \mathcal{F}$ and arbitrary $u_1, \ldots, u_k \in \mathcal{F}$ we may define

$$u_0 = u - [h_1(u_1) + \cdots + h_k(u_k)] \in \mathcal{F}$$

so that $u = H(u_0, u_1, \ldots, u_k)$. Therefore, H is surjective, hence a separable form symmetry for Eq. (6.6).

With the form symmetry calculated, we now proceed with the SC factorization of Eq. (6.6). To determine the first-order factor equation, the formula for g_n can be determined as follows: Note that since $\mathcal{F}$ is invariant under all h_j it is also invariant under all g_n. Since H is onto $\mathcal{F}$, every element $u \in \mathcal{F}$ can be expressed as the image $H(u_0, u_1, \ldots, u_k)$ of certain other elements in $u_0, u_1, \ldots, u_k \in \mathcal{F}$. Thus, by (6.9), (6.16) and (6.17):

$$g_n(u) = g_n(u_0 + h_1(u_1) + \cdots + h_k(u_k))$$

$$= \alpha_n + \sum_{j=1}^{k} [\phi_{j-1}(u_{j-1}) + h_j(u_{j-1})] + \phi_k(u_k)$$

$$= \alpha_n + c[u_0 + h_1(u_1) + \cdots + h_k(u_k)]$$

$$= \alpha_n + cu.$$

This determines the factor equation (3.31) for (6.6) and the cofactor is given by Eq. (3.32); i.e., the SC factorization of Eq. (6.6) is

$$t_{n+1} = \alpha_n + ct_n, \quad t_0 = x_0 + h_1(x_{-1}) + \cdots + h_k(x_{-k}) \tag{6.21}$$

$$x_{n+1} = t_{n+1} - h_1(x_n) - \cdots - h_k(x_{n-k+1}). \tag{6.22}$$

The following result summarizes the preceding discussion.

THEOREM 6.1
Let $\mathcal{F}$ be a nontrivial field. Eq. (6.6) has a separable form symmetry (6.8) if there is $c \in \mathcal{F}$ such that (6.18) holds. In this case, (6.6) has a type-$(1, k)$ reduction to the triangular system of equations (6.21) and (6.22) of lower orders with functions h_j given by (6.20).

Example 6.2
Let $\mathcal{F}$ be a nontrivial field and consider the difference equation

$$x_{n+1} = \alpha_n + ax_n + bx_{n-1} + \psi(x_{n-k+1}) + \psi(x_{n-k}) \tag{6.23}$$

where $\alpha_n, a, b, \in \mathcal{F}$ for all $n \geq 0$ and $\psi : \mathcal{F} \to \mathcal{F}$ is a given nonconstant function. To determine conditions that imply the existence of a separable form symmetry we use Theorem 6.1. If $k = 1, 2$ then indices overlap and result in three separate cases: $k = 1$, $k = 2$ and $k \geq 3$.

First suppose that $k \geq 3$. If we define

$$\phi_0(u) = au, \quad \phi_1(u) = bu, \quad \phi_{k-1}(u) = \phi_k(u) = \psi(u),$$

$$\phi_j(u) = 0 \quad \text{for all other values of } j \leq k$$

then identity (6.18) yields

$$0 = c^{k+1}u - c^k au - c^{k-1}bu - c\psi(u) - \psi(u)$$
$$= c^{k-1}(c^2 - ac - b)u - (c+1)\psi(u)$$

which must hold for every $u \in \mathcal{F}$. Since ψ is not constant, it follows that $c = -1 \in \mathcal{F}$. With this value of c,

$$(-1)^2 - (-1)a - b = 0,$$

or equivalently,

$$a = b - 1. \tag{6.24}$$

If (6.24) holds then Eq. (6.23) has a separable form symmetry of type (6.8). We determine the SC factorization in this case as follows: By (6.20) and (6.24)

$$h_1(u) = -u - au = -bu.$$

If $2 \leq j \leq k - 1$ then

$$h_j(u) = (-1)^j u - (-1)^{j-1}au - (-1)^{j-2}bu$$
$$= (-1)^{j-2}(1 + a - b)u$$
$$= 0$$

whereas

$$h_k(u) = (-1)^k u - (-1)^{k-1}au - (-1)^{k-2}bu - \psi(u)$$
$$= (-1)^{k-2}(1 + a - b)u - \psi(u)$$
$$= -\psi(u).$$

Now (6.21) and (6.22) take the forms

$$t_{n+1} = \alpha_n - t_n, \quad t_0 = x_0 - bx_{-1} - \psi(x_{-k}), \tag{6.25}$$
$$x_{n+1} = t_{n+1} + bx_n + \psi(x_{n-k+1}). \tag{6.26}$$

Next, assume that $k = 2$. Then for the third-order difference equation

$$x_{n+1} = \alpha_n + (b-1)x_n + bx_{n-1} + \psi(x_{n-1}) + \psi(x_{n-2}) \tag{6.27}$$

we have

$$\phi_0(u) = (b-1)u, \quad \phi_1(u) = bu + \psi(u), \quad \phi_2(u) = \psi(u)$$

and again by (6.20) and (6.24)

$$h_1(u) = -u - (b-1)u = -bu,$$
$$h_2(u) = (-1)^2 u - (-1)(b-1)u - bu - \psi(u) = -\psi(u).$$

Thus the SC factorization of (6.27) is

$$t_{n+1} = \alpha_n - t_n, \quad t_0 = x_0 - bx_{-1} - \psi(x_{-2}),$$
$$x_{n+1} = t_{n+1} + bx_n + \psi(x_{n-1}).$$

Finally, let $k = 1$ to obtain the the second-order difference equation

$$x_{n+1} = \alpha_n + (b-1)x_n + bx_{n-1} + \psi(x_n) + \psi(x_{n-1}) \qquad (6.28)$$

with

$$\phi_0(u) = (b-1)u + \psi(u), \quad \phi_1(u) = bu + \psi(u).$$

By (6.20) and (6.24)

$$h_1(u) = -u - (b-1)u - \psi(u) = -bu - \psi(u).$$

Thus (6.28) has a SC factorization

$$t_{n+1} = \alpha_n - t_n, \quad t_0 = x_0 - bx_{-1} - \psi(x_{-1})$$
$$x_{n+1} = t_{n+1} + bx_n + \psi(x_n).$$

Although calculated differently, it is evident that in the three cases above, the SC factorization of Eq. (6.23) with $a = b - 1$ is given by the same system of equations (6.25) and (6.26) for all $k \geq 1$. ⬜

The following result gives an application of Theorem 6.1 to second-order equations.

COROLLARY 6.1
Let $\mathcal{F}$ be a nontrivial field, $c \in \mathcal{F}$, α_n a sequence in $\mathcal{F}$ and let $\psi : \mathcal{F} \to \mathcal{F}$ a given nonconstant function.

(a) The separable, second-order equation

$$x_{n+1} = \alpha_n + \psi(x_n) + c^2 x_{n-1} - c\psi(x_{n-1}) \qquad (6.29)$$

has a SC factorization on $\mathcal{F}$ into the following pair of first order equations

$$t_{n+1} = \alpha_n + ct_n, \quad t_0 = x_0 + cx_{-1} - \psi(x_{-1})$$
$$x_{n+1} = t_{n+1} - cx_n + \psi(x_n).$$

(b) The separable, second-order equation

$$x_{n+1} = \alpha_n + cx_n - \frac{1}{c}\psi(x_n) + \psi(x_{n-1}) \qquad (6.30)$$

has a SC factorization on $\mathcal{F}$ into the following pair of first order equations

$$t_{n+1} = \alpha_n + ct_n, \quad t_0 = x_0 + \frac{1}{c}\psi(x_{-1})$$
$$x_{n+1} = t_{n+1} - \frac{1}{c}\psi(x_n).$$

PROOF To prove (a) let $\phi_0 = \psi$ and note that identity (6.19) holds with $k = 1$. Therefore, Theorem 6.1 applies with

$$H(u_0, u_1) = u_0 + h_1(u_1),$$
$$h_1(u) = cu - \psi(u)$$

Part (b) is proved similarly since identity (6.19) is satisfied with $\phi_0(u) = cu - \psi(u)/c$. ∎

A generalization of the preceding corollary to higher-order equations appears in the Problems section of this chapter. The next example concerns a finite field.

Example 6.3

Consider the second-order equation

$$x_{n+1} = \alpha_n + \psi(x_n) + 4x_{n-1} + 3\psi(x_{n-1}) \qquad (6.31)$$

on the finite field $\mathbb{Z}_5$. In Eq. (6.29) set $c^2 = 4$ or $c = 2$. Then

$$3 = 5 - 2 = -2 = -c$$

so by Corollary 6.1(a), Eq. (6.31) has a separable form symmetry on $\mathbb{Z}_5$ with a SC factorization

$$t_{n+1} = \alpha_n + 2t_n, \quad t_0 = x_0 + 2x_{-1} + 4\psi(x_{-1})$$
$$x_{n+1} = t_{n+1} + 3x_n + \psi(x_n).$$

∎

Example 6.4

Consider the separable, second-order difference equation

$$x_{n+1} = \alpha_n + Ax_n + \frac{B}{x_n} + ax_{n-1} + \frac{b}{x_{n-1}}, \qquad (6.32)$$
$$A, B, a, b, \alpha_n \in \mathbb{R}, \quad n = 0, 1, 2, \ldots$$

Define $\psi(t) = at + b/t$ for nonzero $t \in \mathbb{R}$ in Eq. (6.30) to obtain

$$x_{n+1} = \alpha_n + cx_n - \frac{a}{c}x_n - \frac{b}{cx_n} + ax_{n-1} + \frac{b}{x_{n-1}}.$$

This equation is the same as (6.32) if there is a nonzero, real value of c such that

$$c - \frac{a}{c} = A \text{ and } -\frac{b}{c} = B. \qquad (6.33)$$

Eliminating c from the above pair of equations yields the following quadratic identity among the coefficients of (6.32)

$$b^2 + bAB - aB^2 = 0. \tag{6.34}$$

If identity (6.34) holds then with $c = -b/B$ obtained from (6.33), Corollary 6.1 implies that Eq. (6.32) has the SC factorization

$$t_{n+1} = \alpha_n - \frac{b}{B}t_n, \quad t_0 = x_0 - \frac{Ba}{b}x_{-1} - \frac{B}{x_{-1}} \tag{6.35}$$

$$x_{n+1} = t_{n+1} + \frac{Ba}{b}x_n + \frac{B}{x_n}. \tag{6.36}$$

We note that $A = aB/b - b/B$ from (6.34). To illustrate the usefulness of the SC factorization, we consider the behavior of solutions of Eq. (6.32) in a special case where $\alpha_n = 0$ for all n and $B = -b$. Then (6.32) reduces to

$$x_{n+1} = (1-a)x_n - \frac{b}{x_n} + ax_{n-1} + \frac{b}{x_{n-1}}. \tag{6.37}$$

This equation has a trivial factor equation $t_{n+1} = t_n$; its cofactor is the equation

$$x_{n+1} = t_0 - ax_n - \frac{b}{x_n}, \quad t_0 = x_0 + ax_{-1} + \frac{b}{x_{-1}}. \tag{6.38}$$

Changes in the initial values x_0, x_{-1} lead (through t_0) to the occurrence of bifurcations in (6.38) even for fixed values of the coefficients a, b, a fact that is not readily apparent from a direct examination of Eq. (6.37).

To expose the source of bifurcations, suppose for brevity that

$$a, b > 0$$

and define the one parameter function

$$\zeta(t; t_0) = t_0 - at - \frac{b}{t}.$$

Then ζ has a fixed point $\bar{t}$ which is a solution of the equation

$$t = t_0 - at - \frac{b}{t}$$

or equivalently, a root of the quadratic equation

$$(1+a)t^2 - t_0 t + b = 0.$$

Thus,

$$\bar{t} = \frac{t_0 \pm \sqrt{t_0^2 - 4b(1+a)}}{2(1+a)}.$$

The dependence of $\bar{t}$ on the initial values x_0, x_{-1} via t_0 is noteworthy. Positive real values of $\bar{t}$ are obtained if $t_0 \geq 2\sqrt{b(1+a)}$. In particular, a tangent bifurcation occurs when the value of t_0 crosses the number $2\sqrt{b(1+a)}$. For values of t_0 near $2\sqrt{b(1+a)}$ it can be shown that $\bar{t}$ is an asymptotically stable fixed point. As t_0 increases further, we come across a familiar pattern of period-doubling bifurcations leading to stable aperiodic oscillations (eventually becoming chaotic after $\bar{t}$ becomes a snap-back repeller and a homoclinic orbit appears).

These bifurcations result in qualitative changes in the behaviors of solutions of Eq. (6.37) depending on the values of x_0, x_{-1}. Thus, using the factor equation (6.38) we conclude that (6.37) has qualitatively different, coexisting solutions that range from stable convergent to stable aperiodic to unbounded oscillatory even with fixed, positive values of a and b. We leave a further exploration of Eq. (6.32) to the interested reader. ☐

6.2.2   SC factorization of linear equations, cofactor chains

An important feature of Eq. (6.22) is that it is of the same type as (6.6). Thus if the functions $h_1, \ldots, h_k$ satisfy the analog of (6.18) for some constant $c' \in \mathcal{F}$ then Theorem 6.1 can be applied to (6.22). In the important case of linear difference equations this process can be continued until we are left with a triangular system of first-order linear equations. In Section 5.6.4 we showed that each linear nonhomogeneous equation with constant coefficients in a field $\mathcal{F}$ has a complete SC factorization by a factor chain if $\mathcal{F}$ contains all of the eigenvalues of the homogeneous part of the original linear equation (e.g., if $\mathcal{F}$ is algebraically closed). The resulting triangular system consists of first-order equations where the coefficients are the eigenvalues of the homogeneous part of the original linear equation. The next result, which may be compared with Corollary 5.11, shows that for the same types of linear equations, the triangular system can be also arrived at via a cofactor chain through a series of type$(1, k)$ factorizations.

COROLLARY 6.2

Consider the linear nonhomogeneous difference equation of order $k + 1$

$$x_{n+1} + b_0 x_n + b_1 x_{n-1} + \cdots + b_k x_{n-1} = \alpha_n \tag{6.39}$$

with constant coefficients $b_0, \ldots, b_k$ and the sequence $\{\alpha_n\}$ in a nontrivial field $\mathcal{F}$. If $\mathcal{F}$ contains every eigenvalue of the homogeneous part, then Eq. (6.39) is equivalent to the following system of $k + 1$ first-order linear nonhomogeneous *equations*

$$z_{0,n+1} = \alpha_n + c_0 z_{0,n},$$
$$z_{1,n+1} = z_{0,n+1} + c_1 z_{1,n}$$

$$\vdots$$

$$z_{k,n+1} = z_{k-1,n+1} + c_k z_{k,n}$$

in which $z_{k,n} = x_n$ is the solution of Eq. (6.39) and the constants $c_0, c_1, \ldots, c_k$ are the eigenvalues of the homogeneous part of (6.39), i.e., they are the roots of the characteristic polynomial

$$P(z) = z^{k+1} + b_0 z^k + b_1 z^{k-1} + \cdots + b_{k-1} z + b_k. \qquad (6.40)$$

PROOF   Define $\phi_j(z) = -b_j z$ for $j = 1, \ldots k$ and for all $z \in \mathcal{F}$. Applying Theorem 6.1 yields the linear system

$$z_{0,n+1} = \alpha_n + c_0 z_{0,n}$$
$$x_{n+1} = z_{0,n+1} - \beta_{1,0} x_n - \cdots - \beta_{1,k-1} x_{n-k+1} \qquad (6.41)$$

where c_0 satisfies (6.18)

$$c_0^{k+1} z + c_0^k b_0 z + c_0^{k-1} b_1 z + \cdots + c_0 b_{k-1} z + b_k z = 0$$

for all $z \in \mathcal{F}$, i.e. c_0 is a root of the characteristic polynomial P in (7.41). Further, the numbers $\beta_{1,j}$ are given via the functions h_j in (6.17) and (6.20) as

$$h_j(z) = \beta_{1,j-1} z, \quad \beta_{1,j-1} = c_0^j + c_0^{j-1} b_0 + \cdots + b_{j-1}, \ c_0 \beta_{1,k-1} = -b_{k-1}.$$

Alternatively, the numbers $\beta_{1,j}$ may be calculated from the recursion

$$\beta_{1,j} = c_0 \beta_{1,j-1} + b_j, \quad j = 1, \ldots k - 1, \qquad (6.42)$$
$$\beta_{1,0} = c_0 + b_0, \ c_0 \beta_{1,k-1} = -b_{k-1}.$$

Now, since Eq. (6.41) is also of the same type as (6.39), Theorem 6.1 can be applied to it to obtain the SC factorization

$$z_{1,n+1} = z_{0,n+1} + c_1 z_{1,n}$$
$$x_{n+1} = z_{1,n+1} - \beta_{2,0} x_n - \cdots - \beta_{2,k-2} x_{n-k+2}$$

in which c_1 satisfies (6.18) for (6.41), i.e., with the power reduced by 1 and coefficients adjusted appropriately; i.e.,

$$c_1^k + \beta_{1,0} c_1^{k-1} + \beta_{1,1} c_1^{k-2} + \cdots + \beta_{1,k-2} c_1 + \beta_{1,k-1} = 0. \qquad (6.43)$$

Now we show that c_1 is also a root of P in (7.41). Define the new polynomial

$$P_1(z) = z^k + \beta_{1,0}z^{k-1} + \cdots + \beta_{1,k-2}z + \beta_{1,k-1}$$

so that by (6.43) c_1 is a root of P_1. Direct calculation using (6.42) shows that

$$
\begin{aligned}
(z - c_0)P_1(z) &= z^{k+1} + \beta_{1,0}z^k + \beta_{1,1}z^k + \cdots + \beta_{1,k-1}z \\
&\quad - c_0z^k - c_0\beta_{1,0}z^{k-1} - \cdots - c_0\beta_{1,k-2}z - c_0\beta_{1,k-1} \\
&= z^{k+1} + (c_0 + b_0)z^k + (c_0\beta_{1,0} + b_1)z^{k-1} \cdots + (c_0\beta_{1,k-2} + b_{k-2})z \\
&\quad - c_0z^k - c_0\beta_{1,0}z^{k-1} - \cdots - c_0\beta_{1,k-2}z - c_0\beta_{1,k-1} \\
&= P(z).
\end{aligned}
$$

Thus, if $c_1 \neq c_0$ then $P(c_1) = 0$, i.e., c_1 is a root of P as claimed. The case $c_1 = c_0$ is handled by considering the derivative of P as in the proof of Lemma 5.6. We leave the details to the reader as an exercise. Now the above process inductively generates the system in the statement of this corollary. ∎

REMARK 6.1 *(Operator factorization, complementary and particular solutions)*
1. Let $\mathcal{F} = \mathbb{C}$, the algebraically closed field of complex numbers. We show that the SC factorizations in Corollaries 5.11 and 6.2 are essentially what is obtained through operator factorization. If $Ex_n = x_{n+1}$ represents the forward shift operator then as is well known from the classical theory, the eigenvalues "factor" the operator $P(E)$ with P defined by (7.41); i.e., (6.39) can be written as

$$(E - c_0)(E - c_1) \cdots (E - c_k)x_{n-k} = \alpha_n. \tag{6.44}$$

Now if we define

$$(E - c_1) \cdots (E - c_k)x_{n-k} = y_{0,n} \tag{6.45}$$

then (6.44) can be written as

$$y_{k-1,n+1} - c_0 y_{0,n} = \alpha_n$$

which is the first equation in the triangular system of Corollary 6.2 with $y_{0,n} = z_{0,n}$. We may continue in this fashion by applying the same idea to (6.45); we set

$$(E - c_2) \cdots (E - c_k)x_{n-k} = y_{1,n}$$

and write (6.45) as $y_{1,n+1} - c_1 y_{1,n} = z_{0,n}$ which is the second equation in the triangular system if $y_{1,n} = z_{1,n-1}$. The reduction in the time index n here is due to the removal of one occurrence of E. Proceeding in this fashion, setting $y_{j,n} = z_{j,n-j}$ at each step, we eventually arrive at

$$(E - c_k)x_{n-k} = y_{k-1,n} \Rightarrow x_{n+1-k} = y_{k-1,n} + c_k x_{n-k}.$$

Thus, with $y_{k-1,n} = z_{k-1,n-k+1}$ the preceding equation is the same as the last equation in the system of Corollary 6.2.

2. In Corollaries 5.11 and 6.2 the eigenvalues and both the particular solution and the solution of the homogeneous part of (6.39) are simultaneously obtained without needing to *guess* linearly independent exponential solutions as is done in the classical theory. We indicate how this is done in the second-order case $k = 1$ which is also representative of the higher-order cases. First, for a given sequence $s = \{s_n\}$ of complex numbers and for each $c \in \mathbb{C}$, let us define the quantity

$$\sigma_n(s;c) = \sum_{j=1}^{n} c^{j-1} s_{n-j}$$

and note that for sequences s, t and numbers $a, b \in \mathbb{C}$, $\sigma_n(as + bt; c) = a\sigma_n(s;c) + b\sigma_n(t;c)$, i.e., $\sigma_n(\cdot, c)$ is a linear operator on the space of complex sequences for each $n \geq 1$ and each $c \in \mathbb{C}$. Further, if $s_n = ab^n$ then it is easy to see that

$$\sigma_n(s;c) = \begin{cases} a(b^n - c^n)/(b - c), & c \neq b \\ anb^{n-1}, & c = b \end{cases} \tag{6.46}$$

Now, if $k = 1$ then the semiconjugate factorization of (6.39) into first order equations is

$$z_{n+1} = \alpha_n + c_0 z_n, \quad z_0 = x_0 + (c_0 - b_0)x_{-1} \tag{6.47}$$

$$x_{n+1} = z_{n+1} + c_1 x_n. \tag{6.48}$$

A straightforward inductive argument gives the solution of (6.47) as

$$z_n = z_0 c_0^n + \sigma_n(\alpha; c_0) \tag{6.49}$$

where $\alpha = \{\alpha_n\}$. Next, insert (6.49) into (6.48), set $\gamma_n = z_{n+1}$ and repeat the above argument to obtain the general solution of (6.39) for $k = 1$, i.e., $x_n = x_0 c_1^n + \sigma_n(\gamma; c_1)$. If $c_1 \neq c_0$ then from (6.46) we obtain after combining some terms and noting that $\gamma_0 = z_1 = \alpha_0 + c_0 z_0$,

$$x_n = \left(\frac{\alpha_0 + c_0 z_0}{c_0 - c_1}\right) c_0^n + \left(x_0 - \frac{\alpha_0 + c_0 z_0}{c_0 - c_1}\right) c_1^n + \sigma_n(\sigma'(\alpha; c_0); c_1)$$

where $\sigma' = \{\sigma_{n+1}(\alpha; c_0)\}$. We recognize the first two terms of the above sum as giving the solution of the homogeneous part of (6.39) and the last term as giving the particular solution. In the case of repeat eigenvalues, i.e., $c_1 = c_0$ again from (6.46) we get

$$x_n = [x_0 c_0 + (\alpha_0 + c_0 z_0)n]c_0^{n-1} + \sigma_n(\sigma'(\alpha; c_0); c_0).$$

6.2.3 Form symmetries of multiplicative forms over subgroups of $\mathbb{C}\backslash\{0\}$

In this section we study a class of separable difference equations whose form symmetries involve exponential and power functions. Therefore, we limit our focus to the set $\mathbb{C}$ of complex numbers. On the other hand, a *field structure is not required in this section* so we stay with groups.

Let G be a nontrivial subgroup of the group $\mathbb{C}_0 = \mathbb{C}\backslash\{0\}$ of nonzero complex numbers under multiplication. Obvious examples of G include the sets $\mathbb{R}\backslash\{0\}$ and $(0,\infty)$ of nonzero real numbers and the positive real numbers, respectively, each under ordinary multiplication. Another important subgroup of $\mathbb{C}_0$ is the circle group $\mathbb{T}$. Aside from $\mathbb{R}$ and $\mathbb{T}$ and their subgroups, there are also countable multiplicative subgroups of $\mathbb{C}$ such as the cyclic groups

$$G_a = \{a^n : n \in \mathbb{Z}, \ |a| \neq 0, 1\}$$

and the dense subgroup

$$\mathbb{C}_r = \{z : \operatorname{Re} z, \operatorname{Im} z \in \mathbb{Q}, \ z \neq 0\}$$

of complex numbers with rational real and imaginary parts. See the Problems section for this chapter for examples of nontrivial difference equations having these groups as invariant sets.

Consider the following difference equation

$$x_{n+1} = \alpha_n \psi_0(x_n)\psi_1(x_{n-1})\cdots\psi_k(x_{n-k}), \qquad (6.50)$$
$$\alpha_n, x_{-j} \in G, \ \psi_j : G \to G, \ j = 0,\ldots k. \qquad (6.51)$$

This difference equation is clearly separable on G in the sense of Definition 6.1. If $G = (0,\infty)$ then Eq. (6.50) can be transformed into the additively separable equation (6.6) by taking logarithms to change its multiplicative form into an additive one. Specifically by defining

$$x'_n = \ln x_n, \ x_n = e^{x'_n}, \ \ln \alpha_n = \alpha'_n, \ \phi_j(r) = \ln \psi_j(e^r), \ j = 0,\ldots k, \ r \in \mathbb{R}$$

we can transform (6.50) into (6.6), which is additively separable. For this special situation, Theorem 6.1 gives the form symmetry H that can be transformed into a multiplicative form by applying the exponential function. An important feature of this multiplicative version of the form symmetry for $(0,\infty)$ is that it does not involve the exponential or logarithmic functions. Therefore, it is a possible form symmetry of the multiplicative equation 6.50) for all subgroups of $\mathbb{C}_0$ under suitable conditions. This is established in the next theorem which is proved directly, without using the additive Theorem 6.1. This proof does not use complex logarithms so its conclusion is valid for all nontrivial subgroups of $\mathbb{C}_0$.

THEOREM 6.2

Let G be a nontrivial subgroup of $\mathbb{C}_0$ and let α_n, x_{-j}, ψ_j be as in (6.51). If there is $c \in \mathbb{C}_0$ such that the following is true for all $z \in G$,

$$\psi_0(z)^{c^k}\psi_1(z)^{c^{k-1}}\cdots\psi_k(z) = z^{c^{k+1}} \tag{6.52}$$

then Eq. (6.50) has the form symmetry

$$H(u_0, u_1, \ldots, u_k) = u_0 h_1(u_1)\cdots h_k(u_k), \quad u_0, u_1, \ldots, u_k \in G \tag{6.53}$$

where the functions $h_j : G \to \mathbb{C}$ in (6.53) are given as

$$h_j(z) = z^{c^j}\psi_0(z)^{-c^{j-1}}\cdots\psi_{j-1}(z)^{-1}, \quad j = 1, \ldots k, \; z \in G \tag{6.54}$$

This form symmetry yields a type-$(1,k)$ order reduction with the SC factorization

$$t_{n+1} = \alpha_n t_n^c, \quad t_0 = x_0 h_1(x_{-1})\cdots h_k(x_{-k}) \tag{6.55}$$

$$x_{n+1} = \frac{t_{n+1}}{h_1(x_n)\cdots h_k(x_{n-k+1})}. \tag{6.56}$$

PROOF   The unfoldings or associated vector functions F_n of Eq. (6.50) are

$$F_n(z_0, z_1, \ldots, z_k) = [\alpha_n\psi_0(z_0)\psi_1(z_1)\cdots\psi_k(z_k), z_0, z_1, \ldots, z_{k-1}]$$

where $(z_0, z_1, \ldots, z_k) \in G^{k+1}$. Therefore, if H is defined by (6.53) and (6.54) then

$$\begin{aligned}
H(F_n(z_0, z_1, \ldots, z_k)) &= \alpha_n\psi_0(z_0)\cdots\psi_k(z_k)h_1(z_0)h_2(z_1)\cdots h_k(z_{k-1})\\
&= \alpha_n\psi_0(z_0)\psi_1(z_1)\cdots\psi_{k-1}(z_{k-1})\psi_k(z_k)[z_0^c\psi_0(z_0)^{-1}]\\
&\quad [z_1^{c^2}\psi_0(z_1)^{-c}\psi_1(z_1)^{-1}]\cdots[z_{k-1}^{c^k}\psi_0(z_{k-1})^{-c^{k-1}}\\
&\quad \cdots\psi_{k-1}(z_{k-1})^{-1}]\\
&= \alpha_n\psi_k(z_k)z_0^c z_1^{c^2}\psi_0(z_1)^{-c}\cdots z_{k-1}^{c^k}\psi_0(z_{k-1})^{-c^{k-1}}\cdots\\
&\quad \cdots\psi_{k-1}(z_{k-1})^{-c}.
\end{aligned}$$

On the other hand, if we define $\phi_n(z) = \alpha_n z^c$ as in (6.55) for all $z \in G$ and all integers $n \geq 0$ then

$$\begin{aligned}
\phi_n(H(z_0, z_1, \ldots, z_k)) &= \alpha_n[z_0 h_1(z_1)h_2(z_2)\cdots h_k(z_k)]^c\\
&= \alpha_n z_0^c[z_1^c\psi_0(z_1)^{-1}]^c\cdots[z_{k-1}^{c^{k-1}}\psi_0(z_{k-1})^{-c^{k-2}}\cdots\\
&\quad \cdots\psi_{k-1}(z_{k-1})^{-1}]^c[h_k(z_k)]^c\\
&= \alpha_n z_0^c z_1^{c^2}\psi_0(z_1)^{-c}\cdots z_{k-1}^{c^k}\psi_0(z_{k-1})^{-c^{k-1}}\cdots\\
&\quad \cdots\psi_{k-1}(z_{k-1})^{-c}\psi_k(z_k)
\end{aligned}$$

where the last equality follows by (6.52) because

$$[h_k(z_k)]^c = z_k^{c^{k+1}} \psi_0(z_k)^{-c^k} \cdots \psi_{k-1}(z_k)^{-c} = z_k^{c^{k+1}} z_k^{-c^{k+1}} \psi_k(z_k) = \psi_k(z_k).$$

Hence the semiconjugate relation holds for each n.

To establish the equivalence of (6.50) to the system of equations (6.55) and (6.56) we show that every solution $\{x_n\}$ of (6.50) corresponds uniquely to a solution $\{(t_n, y_n)\}$ of the system and vice versa. First, let $\{x_n\}$ be the unique solution of (6.50) generated by a given set of initial values $x_0, x_{-1}, \ldots, x_{-k}$ in G and define the sequence

$$t_n = H(x_n, \ldots, x_{n-k}).$$

Then the semiconjugate relation implies that

$$\begin{aligned}
t_{n+1} &= H(x_{n+1}, x_n, \ldots, x_{n-k+1}) \\
&= H(\alpha_n \psi_0(x_n) \cdots \psi_k(x_{n-k}), x_n, x_{n-1}, \ldots, x_{n-k+1}) \\
&= H(F_n(x_n, \ldots, x_{n-k})) \\
&= \phi_n(H(x_n, \ldots, x_{n-k})) \\
&= \phi_n(t_n).
\end{aligned}$$

Therefore, $\{t_n\}$ is the solution of equation (6.55) that is uniquely defined by the initial value

$$t_0 = x_0 h_1(x_{-1}) \cdots h_k(x_{-k}).$$

Further, notice that

$$x_{n+1} h_1(x_n) \cdots h_k(x_{n-k+1}) = H(x_{n+1}, x_n, \ldots, x_{n-k+1}) = t_{n+1}$$

so that (6.56) holds with $y_n = x_n$ for all $n \geq -k$.

Conversely, let $\{(t_n, y_n)\}$ be the unique solution of the system of equations (6.55) and (6.56) with a given set of initial values

$$t_0, y_{-1}, y_{-2}, \ldots, y_{-k} \in G. \tag{6.57}$$

We note that t_0 generates the sequence $\{t_n\}$, which satisfies (6.55) independently of (6.56). These values t_n then contribute to the calculation of the sequence $\{y_n\}$, which satisfies (6.56). By the latter equation,

$$\begin{aligned}
t_{n+1} &= y_{n+1} h_1(y_n) \cdots h_k(y_{n-k+1}) \\
&= H(y_{n+1}, \ldots, y_{n-k+1}).
\end{aligned}$$

It follows that for all $n \geq 0$, $t_n = H(y_n, \ldots, y_{n-k})$. Now by (6.55), (6.56) and the definition of H,

$$
\begin{aligned}
y_{n+1} &= \frac{\phi_n(t_n)}{h_1(y_n) \cdots h_k(y_{n-k+1})} \\
&= \frac{\phi_n(H(y_n, \ldots, y_{n-k}))}{h_1(y_n) \cdots h_k(y_{n-k+1})} \\
&= \frac{H(F_n(y_n, \ldots, y_{n-k}))}{h_1(y_n) \cdots h_k(y_{n-k+1})} \\
&= \frac{H(\alpha_n \psi_0(y_n) \cdots \psi_k(y_{n-k}), y_n, y_{n-1}, \ldots, y_{n-k+1})}{h_1(y_n) \cdots h_k(y_{n-k+1})} \\
&= \alpha_n \psi_0(y_n) \cdots \psi_k(y_{n-k}).
\end{aligned}
$$

It follows that $\{y_n\}$ is a solution of (6.50) with initial values (6.57). ∎

Example 6.5
Let G be a nontrivial subgroup of $\mathbb{C}_0$ and let $\alpha_n \in G$ for all $n \geq 1$. Consider the multiplicatively separable difference equation

$$
x_{n+1} = \alpha_n x_n x_{n-1}, \quad x_0, x_{-1} \in G. \tag{6.58}
$$

In Eq. (6.58) we may define $\psi_0(z) = \psi_1(z) = z$ for all $z \in G$. Condition (6.52) holds if there is $c \in \mathbb{C}$ such that

$$
z^c z = z^{c^2} \quad \text{for all } z \in G.
$$

Thus c must be a root of the polynomial

$$
c^2 - c - 1 = 0, \tag{6.59}
$$

e.g., the positive root

$$
c = \frac{1 + \sqrt{5}}{2}.
$$

The form symmetry is $H(u_0, u_1) = u_0 h_1(u_1)$ where $h_1(z) = z^c z^{-1} = z^{c-1}$. The SC factorization of (6.58) is thus

$$
r_{n+1} = \alpha_n r_n^c, \quad r_0 = x_0 x_{-1}^{c-1} \tag{6.60}
$$

$$
x_{n+1} = r_{n+1} x_n^{1-c}. \tag{6.61}
$$

Note that the number $1 - c$ is the other root of the polynomial (6.59); therefore, the factor and cofactor equations of (6.58) involve the two roots of the same polynomial (6.59). Not surprisingly, if $G = (0, \infty)$ then the transformation of (6.58) via logarithms into a linear nonhomogeneous equation on the real line also shows the roots of (6.59) to be the eigenvalues of the linear equation. In this case, the behavior of solutions of (6.58) reflect the linear behavior since the real exponential and logarithm are homeomorphisms and isomorphisms of $(0, \infty)$ and $\mathbb{R}$. However, as noted in Chapter 2, solutions of (6.58) may exhibit a different behavior in a different subgroup of $\mathbb{C}_0$.

For instance, if $\mathbb{T}$ is the circle group and $\alpha_n = e^{ia_n}$ for $a_n \in \mathbb{R}$ then $\mathbb{T}$ is invariant under (6.58) and its SC factorization. The factor equation (6.60) can be written as

$$\theta_{n+1} = (a_n + c\theta_n) \bmod 2\pi, \quad e^{i\theta_n} = r_n$$

which leads to a dense orbit in $\mathbb{T}$ in most cases, even if a_n is a constant sequence. Similarly, the cofactor equation (6.61) can be written as

$$\xi_{n+1} = [\theta_{n+1} + (1 - c)\xi_n] \bmod 2\pi, \quad e^{i\xi_n} = x_n$$

whose orbits are generally dense in $\mathbb{T}$. $\quad\square$

The preceding example highlights a feature of the SC factorization in Theorem 6.2 that is reminiscent of linear equations, namely, that even if G is invariant under Eq. (6.58) then its SC factorization need not be (since G need not be invariant under the functions h_j). For instance, if $\alpha_n, x_0, x_{-1} \in \mathbb{R}$, the numbers r_n may be complex; e.g., if $x_0 = -1$, $x_{-1} = 1$ and $\alpha_n = 1$ for all n then $r_0 = -1$ so that

$$r_1 = (-1)^c = \cos \pi c + i \sin \pi c$$

is not real for the above value of c. Of course, the numbers x_n are real since the set of nonzero real numbers is invariant under (6.58); for instance,

$$x_1 = r_1 x_0^{1-c} = (-1)^c (-1)^{1-c} = -1.$$

This phenomenon is similar to what is seen in linear equations having real coefficients but complex eigenvalues.

Example 6.6

Let G be a nontrivial subgroup of $\mathbb{C}_0$ and let $\psi : G \to G$ be a given function. For positive integers k, m with $k > m \geq 0$, consider the separable difference equation

$$x_{n+1} = \alpha_n \psi(x_{n-m})\psi(x_{n-k})x_{n-k}, \tag{6.62}$$

$$x_0, x_{-1}, \ldots, x_{-k}, \alpha_n \in G \text{ for all } n \geq 0.$$

Special cases of Eq. (6.62) include the following that are obtained by specializing the function ψ:

$$x_{n+1} = \alpha_n(x_{n-m} + \beta)(x_{n-k} + \beta)x_{n-k}, \quad \beta \in G,$$

$$x_{n+1} = \frac{\alpha_n x_{n-k}}{(x_{n-m} + \beta)(x_{n-k} + \beta)}, \quad \beta \in G,$$

$$x_{n+1} = x_{n-k}e^{\gamma_n - \beta x_{n-m} - \beta x_{n-k}}, \quad \beta, \gamma_n \in G.$$

Eq. (6.62) is of type (6.50) with

$$\psi_j(z) = \begin{cases} \psi(z), & \text{if } j = m \\ z\psi(z), & \text{if } j = k \\ 1, & \text{if } j \neq m, k \end{cases}.$$

Identity (6.52) is satisfied if there is $c \in \mathbb{C}_0$ such that

$$z^{c^{k+1}-1} = \psi(z)^{c^{k-m}+1} \text{ for all } z \in G. \tag{6.63}$$

For arbitrary ψ, identity (6.63) holds only if both exponents are zeros, i.e., if c is both a $(k+1)$-st root of unity and a $(k-m)$-th root of -1. This value of c is determined by the delay pattern (the values of m and k). Some special cases are listed below:

$$x_{n+1} = \alpha_n \psi(x_n)\psi(x_{n-1})x_{n-1}, \quad c = -1; \tag{6.64}$$

$$x_{n+1} = \alpha_n \psi(x_{n-1})\psi(x_{n-3})x_{n-3}, \quad c = i; \tag{6.65}$$

$$x_{n+1} = \alpha_n \psi(x_{n-3})\psi(x_{n-7})x_{n-7}, \quad c = e^{i\pi/4} = (1+i)/\sqrt{2}. \tag{6.66}$$

The SC factorization of Eq. (6.64) is

$$t_{n+1} = \frac{\alpha_n}{t_n}, \quad x_{n+1} = t_{n+1}x_n\psi(x_n).$$

For the SC factorizations of the other two equations see the Problems for this chapter. ▯

The difference equation (6.56) has order k, one less than (6.50), and by Theorem 3.1 its solutions subject to the conditions of Lemma 6.2 are identical to the solutions $\{x_n\}$ of (6.50). In this sense, we may think of the cofactor equation (6.56) as a *reduction of order* of (6.50) and to distinguish it from (6.50), we may write it as an independent equation in a new variable

$$y_{n+1} = t_{n+1}h_1(y_n)^{-1} \cdots h_k(y_{n-k+1})^{-1}. \tag{6.67}$$

The variable coefficient t_{n+1} is calculated independently from the simple first-order equation (6.55) by induction as stated in the next lemma whose straightforward proof is left as an exercise.

LEMMA 6.1
The general solution of Eq. (6.55) is

$$t_{n+1} = t_0^{c^{n+1}} \prod_{j=0}^{n} \alpha_{n-j}^{c^j}. \tag{6.68}$$

If $\alpha_n = \alpha$ is constant for all n (i.e., (6.50) is an autonomous equation) then

$$t_{n+1} = \begin{cases} \rho(t_0/\rho)^{c^{n+1}} \text{ with } \rho = \alpha^{1/(1-c)}, & \text{if } c \neq 1, \\ t_0\alpha^{n+1}, & \text{if } c = 1. \end{cases}.$$

6.3 Equations with exponential and power functions

In this section we use Theorem 6.2 to explore multiplicatively separable difference equations of the following type

$$x_{n+1} = x_n^{a_0} x_{n-1}^{a_1} \cdots x_{n-k}^{a_k} e^{\alpha_n - b_0 x_n - b_1 x_{n-1} - \cdots - b_k x_{n-k}} \qquad (6.69)$$

that consist only of complex exponential and power functions. We generally assume that the parameters are complex because, like the case of linear equations, the SC factorizations of (6.69) often yield complex parameters even if the original parameters are real.

Positive solutions of (6.69) are guaranteed to exist if

$$a_j, b_j, \beta_n \in \mathbb{R}, \ x_{-j} > 0, \ j = 0, 1, \ldots, k, \ n = 0, 1, 2, \ldots \qquad (6.70)$$

Under these conditions Eq. (6.69) is separable on the invariant multiplicative group $(0, \infty)$. Conditions (6.70) hold for equations that appear in applied contexts; under these conditions, Eq. (6.69) can have a wide variety of positive solutions as we will see in some of the examples in this section. A significant and helpful aspect of the positive solutions under conditions (6.70) is that they do not involve the technical complications due to multiple-valued complex power functions. Further, most applied models involve real parameters and positive solutions. However, even in the case of positive solutions on $(0, \infty)$ under conditions (6.70) the SC factorization of (6.69) may involve complex parameters in its factor or cofactor equation.

Because Eq. (6.69) involves only power and exponential functions we refer to it as an *expow equation* for short. Note that (6.69) can also be written in the equivalent though more stylized form

$$x_{n+1} = \gamma_n x_n^{a_0} x_{n-1}^{a_1} \cdots x_{n-k}^{a_k} \beta_0^{x_n} \beta_1^{x_{n-1}} \cdots \beta_k^{x_{n-k}}$$

where $\beta_j = e^{-b_j}$ for $j = 0, 1, \ldots, k$ and $\gamma_n = e^{\alpha_n}$ for all n. As is customary, if $a_j = 0$ for some j then we set $x_{n-j}^{a_j} = 1$.

6.3.1 The reduction theorem

To find sufficient conditions on the parameters a_j, b_j that allow a reduction of order of (6.69) via Theorem 6.2 and Lemma 6.1, the functions

$$\psi_j(z) = z^{a_j} e^{-b_j z}, \quad j = 0, 1, \ldots, k$$

must satisfy identity (6.52); i.e., for some $c \in \mathbb{C}_0$ the following must hold for all $z \in G$

$$z^{a_0 c^k} e^{-b_0 c^k z} z^{a_1 c^{k-1}} e^{-b_1 c^{k-1} z} \cdots z^{a_k} e^{-b_k z} = z^{c^{k+1}}. \qquad (6.71)$$

After some rearranging of terms, we see that identity (6.71) holds for all z if c is a common nonzero root of the polynomials

$$P_0(z) = z^{k+1} - a_0 z^k - a_1 z^{k-1} - \cdots - a_k,$$
$$Q_0(z) = b_0 z^k + b_1 z^{k-1} + \cdots + b_k.$$

For such $c \in \mathbb{C}_0$ the form symmetry and semiconjugate factorization are determined using the functions

$$h_j(z) = z^{c^j} \psi_0(z)^{-c^{j-1}} \cdots \psi_{j-1}(z)^{-1} \tag{6.72}$$
$$= z^{c^j - a_0 c^{j-1} - \cdots - a_{j-1}} e^{(b_0 c^{j-1} + b_1 c^{j-2} + \cdots + b_{j-1})z}$$

as shown in the following result.

THEOREM 6.3

Assume that the polynomials P_0 and Q_0 have a common root $c_0 \in \mathbb{C}_0$. Then the following statements are true:

(a) Eq. (6.69) has a reduction of order to the expow equation

$$y_{n+1} = t_{n+1} y_n^{-p_{0,0}} y_{n-1}^{-p_{0,1}} \cdots y_{n-k+1}^{-p_{0,k-1}} e^{-q_{0,0} y_n - q_{0,1} y_{n-1} - \cdots - q_{0,k-1} y_{n-k+1}},$$
$$\tag{6.73}$$

where for $j = 0, 1, \ldots, k-1$,

$$p_{0,j} = c_0^{j+1} - a_0 c_0^j - a_1 c_0^{j-1} - \cdots - a_{j-1} c_0 - a_j, \tag{6.74}$$
$$q_{0,j} = b_0 c_0^j + b_1 c_0^{j-1} + \cdots + b_{j-1} c_0 + b_j \tag{6.75}$$

and

$$t_{n+1} = t_0^{c_0^{n+1}} e^{\sigma_n}, \quad \sigma_n = \sum_{j=0}^{n} \alpha_{n-j} c_0^j, \quad t_0 = x_0 h_1(x_{-1}) \cdots h_k(x_{-k}).$$

(b) Let $c_1 \in \mathbb{C}_0$ be also a common root of both P_0 and Q_0. Then c_1 is a root of both of the following polynomials

$$P_1(z) = z^k + p_{0,0} z^{k-1} + p_{0,1} z^{k-2} + \cdots + p_{0,k-1},$$
$$Q_1(z) = q_{0,0} z^{k-1} + q_{0,1} z^{k-2} + \cdots + q_{0,k-1}$$

so by Part (a) the expow equation (6.73) has a reduction of order to

$$z_{n+1} = r_{n+1} z_n^{-p_{1,0}} z_{n-1}^{-p_{1,1}} \cdots z_{n-k+2}^{-p_{1,k-2}} e^{-q_{1,0} z_n - q_{1,1} z_{n-1} - \cdots - q_{1,k-2} z_{n-k+2}}, \tag{6.76}$$

where for $i = 0, 1, \ldots, k-2$,

$$p_{1,i} = c_1^{i+1} + p_{0,0} c_1^i + p_{0,1} c_1^{i-1} + \cdots + p_{0,i-1} c_1 + p_{0,i},$$
$$q_{1,i} = q_{0,0} c_1^i + q_{0,1} c_1^{i-1} + \cdots + q_{0,i-1} c_1 + q_{0,i}$$

and

$$r_{n+1} = r_0^{c_1^{n+1}} \prod_{j=0}^{n} t_{n+1-j}^{c_1^j}.$$

PROOF (a) The proof of this part is straightforward by induction using (6.72) in (6.67), then combining various exponents and finally using Lemma 6.1 (with $\beta_n = e^{\alpha_n}$) for the numbers t_{n+1}. We leave the details as an exercise.

(b) By assumption,

$$P_0(c_1) = Q_0(c_1) = 0. \tag{6.77}$$

Now

$$(z - c_0)P_1(z) = (z - c_0)\left(z^k + \sum_{j=0}^{k-1} p_{0,j} z^{k-j-1}\right)$$

$$= z^{k+1} + \sum_{j=0}^{k-1} [p_{0,j} - c_0 p_{0,j-1}] z^{k-j} - c_0 p_{0,k-1}$$

where we define $p_{0,-1} = 1$. From (6.74) for each $j = 0, 1, \ldots, k-1$ we obtain

$$p_{0,j} - c_0 p_{0,j-1} = -a_j$$

and further, since $P_0(c_0) = 0$,

$$c_0 p_{0,k-1} = c_0 \left(c_0^k - a_0 c_0^{k-1} - \cdots - a_{k-1}\right)$$
$$= P_0(c_0) + a_k$$
$$= a_k.$$

Thus

$$(z - c_0)P_1(z) = P_0(z)$$

and if $c_1 \neq c_0$ then $P_1(c_1) = 0$ by (6.77). If $c_1 = c_0$ then c_0 is a double root of both P_0 and Q_0 so that their derivatives are zeros, i.e.,

$$P_0'(c_0) = Q_0'(c_0) = 0. \tag{6.78}$$

In this case, using (6.74) we obtain

$$P_1(c_0) = c_0^k + \sum_{j=0}^{k-1} (c_0^{j+1} - a_0 c_0^j - \cdots - a_{j-1} c_0 - a_j) c_0^{k-j-1}$$

$$= (k+1)c_0^k - \sum_{j=0}^{k-1} (k-j) a_j c_0^{k-j-1}$$

$$= P_0'(c_0).$$

Therefore, if $c_1 = c_0$ then $P_1(c_1) = 0$ by (6.78). Similar calculations show that $Q_1(c_1) = 0$; thus by Part (a) Eq. (6.73) has a semiconjugate factorization.

The factorization is obtained as in Part (a). For Eq. (6.76) we need only change a_j to $-p_{0,j}$ and b_j to $q_{0,j}$ in our hypotheses for each $j = 0, 1, \ldots, k-1$ to obtain the new numbers $p_{1,i}$ and $q_{1,i}$ for (6.76) as stated in the statement of the theorem. ∎

6.3.2   Reductions in orders of expow equations

The next result applies Theorem 6.3 to second-order expow equations. The similarity of hypotheses between this result and Corollary 5.8 is rather interesting.

COROLLARY 6.3

Let $a_0, a_1, b_0, b_1, \alpha_n \in \mathbb{C}$ for $n \geq 1$ such that

$$b_1^2 + a_0 b_0 b_1 - a_1 b_0^2 = 0, \quad b_1 \neq 0. \tag{6.79}$$

Then the second-order expow equation

$$x_{n+1} = x_n^{a_0} x_{n-1}^{a_1} e^{\alpha_n - b_0 x_n - b_1 x_{n-1}} \tag{6.80}$$

has a reduction of order to the first-order equation

$$y_{n+1} = t_0^{c^{n+1}} y_n^{a_0 - c} e^{\sigma_n - b_0 y_n}, \quad c = -b_1/b_0, \quad y_0 = x_0, \tag{6.81}$$

$$\sigma_n = \sum_{j=0}^{n} \alpha_{n-j} c^j, \quad t_0 = x_0 x_{-1}^{c-a_0} e^{b_0 x_{-1}}.$$

In particular, if $a_0, a_1, b_0, b_1, \alpha_n \in \mathbb{R}$ for $n \geq 1$ then $c \in \mathbb{R}$ and reduction of order is defined on $\mathbb{R}$.

PROOF   First we note that conditions (6.79) imply that b_0 and at least one of a_0, a_1 are nonzero. Now $c = -b_1/b_0$ is the unique nonzero root of Q_0 and by the equality in (6.79),

$$c^2 - a_0 c - a_1 = \frac{b_1^2}{b_0^2} + a_0 \frac{b_1}{b_0} - a_1 = 0.$$

So the above value of c is also a root of P_0. Now the proof is completed by applying Theorem 6.3. The last assertion is obivous. ∎

In the next corollary, like Corollary 6.3, the reductions of order are defined over the real numbers if the parameters are real.

COROLLARY 6.4

Let $k \geq 1$, $b_0, b_k \in \mathbb{C}$ with $b_k \neq 0$ and $\alpha_n \in \mathbb{C}$ for $n \geq 0$. *Consider the expow delay equation:*

$$x_{n+1} = x_{n-1} e^{\alpha_n - b_0 x_n - b_k x_{n-k}}. \tag{6.82}$$

(a) Eq. (6.82) has two possible reductions of order: (i) If $b_k = -b_0$ then

$$y_{n+1} = \frac{t_0}{y_n} e^{\sigma_n - b_0 y_n - b_0 y_{n-1} - \cdots - b_0 y_{n-k+1}}, \tag{6.83}$$

$$t_{n+1} = t_0 e^{\sigma_n}, \quad \sigma_n = \sum_{j=0}^{n} \alpha_j, \tag{6.84}$$

$$t_0 = x_0 x_{-1} e^{b_0 (x_{-1} + x_{-2} + \cdots + x_{-k+1})}.$$

(ii) If k is odd and $b_k = b_0$ then

$$y_{n+1} = t_{n+1} y_n e^{-b_0 y_n + b_0 y_{n-1} - \cdots + (-1)^{k-1} b_0 y_{n-k+1}}, \tag{6.85}$$

$$t_{n+1} = t_0^{(-1)^{n+1}} e^{\sigma_n}, \quad \sigma_n = \sum_{j=0}^{n} (-1)^j \alpha_{n-j}, \tag{6.86}$$

$$t_0 = \frac{x_0}{x_{-1}} e^{b_0 (x_{-1} - x_{-2} + \cdots + (-1)^{k-1} x_{-k+1})}.$$

(b) If k is even then the expow equation (6.83) has a further reduction of order to

$$z_{n+1} = r_{n+1} e^{-b_0 z_n + b_0 z_{n-2} - b_0 z_{n-4} + \cdots + (-1)^{k-1} b_0 z_{n-k+2}},$$

$$r_{n+1} = r_0^{(-1)^{n+1}} \prod_{j=0}^{n} t_{n+1-j}^{(-1)^j}, \quad r_0 = x_0 e^{b_0 (x_0 - x_{-2} + \cdots + (-1)^k x_{-k+2})}.$$

(c) If $b_0, b_k \in \mathbb{R}$ with $b_k \neq 0$ and $\alpha_n \in \mathbb{R}$ for all n then the reductions of order in Parts (a) and (b) are defined on $\mathbb{R}$.

PROOF (a) Since $a_1 = 1$ and $a_j = 0$ for $j \neq 1$, we have

$$P_0(c) = c^{k+1} - c^{k-1} = 0 \Rightarrow c^{k-1}(c^2 - 1) = 0 \Rightarrow c = \pm 1.$$

If $c = 1$ then $Q_0(c) = b_0 + b_k = 0$ or $b_k = -b_0$. Applying Theorem 6.3 with $c = 1$ yields the claimed reduction of order. The other order reduction for odd k is obtained similarly with $c = -1$.

(b) In Eq. (6.83) which has order k the new parameter values are $a_0 = -1$ and $a_j = 0$ for $j > 0$. Now $P_0(c) = c^k + c^{k-1} = 0$ yields a nonzero value $c = -1$. This must also be a root of Q_0; since the new values of the parameters b_j are $b_j = b_0$ for $j = 0, \ldots, k-1$ we must have

$$(-1)^{k-1} b_0 + (-1)^{k-2} b_0 + \cdots + (-1) b_0 + b_0 = 0. \tag{6.87}$$

With $b_0 \neq 0$ (6.87) holds if and only if k is even, in which case applying Theorem 6.3 with $c = -1$ yields the stated reduction of order for (6.83).

(c) This is clear since $c = \pm 1$ is real in Parts (a) and (b).   ∎

The following result involves a more general delay pattern than that in Part (a)(i) of Corollary 6.4, i.e., when the parameters b_0, b_k have equal magnitudes but opposite signs.

COROLLARY 6.5

For $k, m \geq 1$, $\beta \in \mathbb{C}_0$ and $\alpha_n \in \mathbb{C}$ for $n \geq 0$ the expow delay equation

$$x_{n+1} = x_{n-m} e^{\alpha_n + \beta x_n - \beta x_{n-k}} \tag{6.88}$$

has a reduction of order to

$$y_{n+1} = \frac{t_0 e^{\sigma_n - \beta(y_n + y_{n-1} + \cdots + y_{n-k+1})}}{y_n y_{n-1} \cdots y_{n-m+1}},$$

$$\sigma_n = \sum_{j=0}^{n} \alpha_j, \quad t_0 = x_0 x_{-1} \cdots x_{-m+1} e^{\beta(x_{-1} + x_{-2} + \cdots + x_{-k+1})}.$$

For $m = 0$ Eq. (6.88) has a reduction of order to

$$y_{n+1} = t_0 e^{\sigma_n - \beta(y_n + y_{n-1} + \cdots + y_{n-k+1})}.$$

PROOF   First assume that $k \geq m \geq 1$. Then

$$P_0(c) = c^{k+1} - a_m c^{k-m} = c^{k-m}(c^{m+1} - 1)$$
$$Q_0(c) = -\beta c^k + \beta.$$

Clearly $c = 1$ is a common nonzero root of P_0 and Q_0 so Theorem 6.3 can be applied to obtain the order reduction by calculating

$$p_{0,j} = \begin{cases} c^{j+1} = 1, \text{ if } j < m \\ c^{j+1} - a_m c^{j-m} = 0 \text{ if } j \geq m \end{cases}$$

and

$$q_{0,j} = \begin{cases} b_0 c^j = -\beta, \text{ if } j < k \\ b_0 c^j + b_k = -\beta + \beta = 0, \text{ if } j = k. \end{cases}$$

Next, if $m > k \geq 1$ then

$$P_0(c) = c^{m+1} - a_m = c^{m+1} - 1$$
$$Q_0(c) = -\beta c^m + \beta c^{m-k} = -\beta c^{m-k}(c^k - 1).$$

Again $c = 1$ is a common nonzero root of P_0 and Q_0 so Theorem 6.3 can be applied similarly to the preceding case.

In the case $m = 0$ we have

$$P_0(c) = c^{k+1} - a_0 c^k = c^k(c-1), \quad Q_0(c) = -\beta c^k + \beta.$$

Thus $c = 1$ is a common nonzero root and the proof is completed by applying Theorem 6.3. ∎

The following result generalizes the delay pattern in Part (a)(ii) of Corollary 6.4, i.e., when the parameters b_0, b_k are equal. The proof follows the same argument as in the preceding corollary by showing that $c = -1$ is a common root of P_0 and Q_0. We omit the details of this proof.

COROLLARY 6.6

For $k, m \geq 1$, $\beta \in \mathbb{C}_0$ and $\alpha_n \in \mathbb{C}$ for $n \geq 0$ the expow delay equation

$$x_{n+1} = x_{n-2m+1} e^{\alpha_n - \beta x_n - \beta x_{n-2k+1}} \tag{6.89}$$

has a reduction of order to

$$y_{n+1} = \frac{t_0^{(-1)^{n+1}} y_n y_{n-2} \cdots y_{n-2m+2}}{y_{n-1} y_{n-3} \cdots y_{n-2m+3}} e^{\sigma_n - \beta[y_n - y_{n-1} + \cdots - y_{n-2k+2}]},$$

$$\sigma_n = \sum_{j=0}^{n} (-1)^j \alpha_{n-j}, \quad t_0 = \frac{x_0 x_{-2} \cdots x_{-2m+2}}{x_{-1} x_{-3} \cdots x_{-2m+3}} e^{\beta(x_{-1} - x_{-2} + \cdots + x_{-2k+1})}.$$

Example 6.10 below uses the order reduction in Corollary 6.6 to study the positive solutions of a special case of Eq. (6.89).

6.3.3 Repeated reductions of order

It is of interest that Eq. (6.73) in Theorem 6.3 is again an expow equation similar to (6.69). This suggests that the methods of the preceding section can be applied to (6.73). Indeed, under the conditions in Part (b) of Theorem 6.3 a further reductions of order was obtained by repeating the semiconjugate factorization process. A similar situation was encountered in Part (b) of Corollary 6.4.

The main impediment to repeatedly using Theorem 6.3 in this way is the requirement that the polynomials P_0 and Q_0 have *common* nonzero roots. The next result illustrates a special case in which P_0 and Q_0 reduce to a single polynomial and thus, the factorization process continues until a triangular system of first-order equations is obtained. In this way the original equation of order $k + 1$ is reduced to an equation of order one.

COROLLARY 6.7

The expow equation

$$x_{n+1} = x_n^{a_0} x_{n-1}^{a_1} \cdots x_{n-k+1}^{a_{k-1}} e^{\alpha_n + b x_n - a_0 b x_{n-1} - \cdots - a_{k-1} b x_{n-k}} \tag{6.90}$$

where for $j = 0, 1, \ldots, k - 1$ and all $n \geq 0$,

$$a_j, \alpha_n, b \in \mathbb{C}, \quad a_{k-1} \neq 0$$

has a complete semiconjugate factorization into a triangular system of first order equations over $\mathbb{C}_0$.

PROOF   First assume that $b \neq 0$. For Eq. (6.90) setting $P_0(c) = Q_0(c) = 0$ gives

$$c^{k+1} - a_0 c^k - a_1 c^{k-1} - \cdots - a_{k-1} c = 0,$$
$$b c^k - b a_0 c^{k-1} - \cdots - b a_{k-1} = 0. \tag{6.91}$$

Since $a_{k-1} \neq 0$ the nonzero roots of the above polynomials are identical to the zeros of the following polynomial

$$P(c) = c^k - a_0 c^{k-1} - \cdots - a_{k-1}.$$

Note that *every* root of $P(c)$ is nonzero and a root of *both* P_0 and Q_0. Therefore, by Theorem 6.3, not only Eq. (6.90) has a semiconjugate factorization, but also if $k > 1$ then the factorization process continues until the order of the cofactor equation is reduced to one.

Finally, if $b = 0$ (no exponential functions) then Eq. (6.91) reduces to a trivial identity so once again only one polynomial $P(c)$ remains.   ∎

We point out that by denoting $e^{\alpha_n} = \gamma_n$ and $e^{-b} = \beta$, Eq. (6.90) can be written in the following more symmetric form

$$x_{n+1} = \gamma_n x_n^{a_0} x_{n-1}^{a_1} \cdots x_{n-k+1}^{a_{k-1}} \beta^{-x_n} \beta^{a_0 x_{n-1}} \cdots \beta^{a_{k-1} x_{n-k}}.$$

Example 6.7

For each positive integer k, the expow delay equation

$$x_{n+1} = x_{n-k+1}^a e^{\alpha_n + b x_n - a b x_{n-k}}, \quad a, b, \alpha_n \in \mathbb{R}, \ a, b \neq 0 \tag{6.92}$$

is clearly of type (6.90), so by Corollary 6.7 it has a complete semiconjugate factorization into a triangular system of first-order equations. The polynomial $P(c) = c^k - a$ in this case so the common nonzero roots of P_0 and Q_0 are just the k-th roots of a.

In particular, if $k = 2$ then the polynomial $P(c) = c^2 - a$ has roots $\pm\sqrt{a}$. Using Theorem 6.3 we obtain the following sysetm of first order equations

$$t_{n+1} = e^{\alpha_n} t_n^{\sqrt{a}},$$
$$r_{n+1} = t_{n+1} r_n^{-\sqrt{a}},$$
$$y_{n+1} = r_{n+1} e^{b y_n}.$$

Note that if $a < 0$ then the above system is defined over the complex numbers even though all parameters in Eq. (6.92) are real. This situation is analogous to the previously discussed case of a linear difference equation with real coefficients having complex eigenvalues. □

6.3.4 Solutions of equations containing only power functions

In the preceding sections of this chapter the SC factorization of expow equations was of primary interest rather than their solutions. Expow equations are capable of generating interesting and complex solutions. Later in this chapter we discuss an expow equation that is derived from a population biology model whose real solutions exhibit multistable, chaotic behavior. In this section we examine the solutions of a different special type of expow equation in the complex plane.

The case $b = 0$ in Corollary 6.7 corresponds to the pure power-functions case

$$x_{n+1} = \gamma_n x_n^{a_0} x_{n-1}^{a_1} \cdots x_{n-k+1}^{a_{k-1}}, \quad a_{k-1} \neq 0. \tag{6.93}$$

As Corollary 6.7 shows, this equation admits repeated reductions of order so the methods of preceding sections apply. The equations of lower order that we obtain through a semiconjugate factorization can be readily solved in some cases to yield useful information about the nature of the solutions of Eq. (6.93); see the Problems for this chapter. It is also possible to obtain a wealth of information about (6.93) by studying the *moduli* of terms of each solution. We explore this idea next.

Eq. (6.93) bears a similarity to the *linear* nonhomogeneous difference equation of order $k \geq 1$. If $a_j \in \mathbb{R}$ for $j = 0, 1, \ldots, k - 1$ and $x_{-j}, \gamma_n \in (0, \infty)$ for $0 \leq j \leq k - 1$ and $n \geq 0$ then $x_n \in (0, \infty)$ for all n. Defining $y_n = \ln x_n$ and taking the logarithm of (6.93) gives the linear equation

$$y_{n+1} = a_0 y_n + a_1 y_{n-1} + \cdots a_{k-1} y_{n-k+1} + \delta_n, \quad \delta_n = \ln \gamma_n. \tag{6.94}$$

The characteristic polynomial of (6.94) is precisely P in the proof of Corollary 6.7. Therefore, the eigenvalues of the homogeneous part of (6.94) completely determine the SC factorization of (6.93). Further, for every solution $\{y_n\}$ of (6.94), the sequence $\{e^{y_n}\}$ is a positive solution of (6.93).

The preceding observations do not reduce the study of (6.93) to (6.94) since the linear equation only generates the *positive* solutions of (6.93). However, a more judicious use of the linear equation does yield substantial information about an important special case of (6.93) in the complex plane.

Consider the case where *all exponents are real numbers*; i.e., $a_j \in \mathbb{R}$ for $j = 0, 1, \ldots, k - 1$ but allow $x_{-j}, \gamma_n \in \mathbb{C}_0$ for $0 \leq j \leq k - 1$ and $n \geq 0$. Unless every a_j is an integer, the powers $x_{n-j}^{a_j}$ are multiple-valued for each n and solutions of (6.93) are not uniquely defined in $\mathbb{C}_0$. However, since the exponents a_j are real, the *moduli* of these powers are uniquely defined.

Specifically, if $x_{n-j} = \rho e^{i(\theta + 2\pi m)}$ for all integer values of m then the modulus of the power

$$|x_{n-j}^{a_j}| = \rho^{a_j} = |x_{n-j}|^{a_j}$$

does not depend on m. Further, since

$$|x_{n+1}| = |\gamma_n||x_n|^{a_0}|x_{n-1}|^{a_1}\cdots|x_{n-k+1}|^{a_{k-1}} \tag{6.95}$$

it follows that the moduli of terms $|x_n|$ in each solution of (6.93) in $\mathbb{C}_0$ yield a solution of the *real* difference equation (6.95) in $(0, \infty)$. Now, if we define

$$y_n = \ln|x_n|, \quad \delta_n = \ln|\gamma_n|$$

then *every* solution of (6.95) is of the form $\{e^{y_n}\}$ where $\{y_n\}$ is a solution of (6.94). Since the exponential function defines a conjugate mapping between the solutions of equations (6.94) and (6.95) we call (6.94) the *linear conjugate* of (6.95). It follows that solutions of (6.94) yield important information about the moduli of terms in a general solution of (6.93) in $\mathbb{C}_0$; in particular, by tracking the variations in modulus it is possible to determine which solutions of (6.93) are bounded without converging to zero.

Let us call a solution $\{x_n\}$ of (6.93) *interesting* if it is bounded and stays away from zero; i.e., if there are real numbers r_1, r_2 such that

$$0 < r_1 \leq |x_n| \leq r_2 < \infty. \tag{6.96}$$

We also call $(x_0, x_{-1}, \ldots, x_{-k+1}) \in \mathbb{C}_0^k$ an *interesting initial point* if the corresponding solution of (6.93) is an interesting solution, and define the *interesting set $I_0 \subset \mathbb{C}_0^k$* of (6.93) to be the collection of all initial points of (6.93). The set I_0 may be empty as in the case

$$x_{n+1} = \gamma_n x_n^\alpha \text{ where } 0 < |\alpha| \leq 1 \text{ and } |\gamma_n| \leq \rho < 1. \tag{6.97}$$

In this case, $|x_n| \leq \rho^n |x_0|^{\alpha^n}$ so $x_n \to 0$ as $n \to \infty$ regardless of the choice of initial value x_0. On the other hand, if $\{\gamma_n\}$ is any sequence in the unit circle $\mathbb{T}$ (i.e., $\gamma_n = e^{i\theta_n}$ where $\{\theta_n\}$ is an arbitrary sequence of real numbers) then by choosing all initial values $x_{-j} \in \mathbb{T}$ for $0 \leq j \leq k - 1$ we see from (6.95) that $\mathbb{T}$ is invariant. Therefore,

$$\gamma_n \in \mathbb{T} \text{ for all } n \Rightarrow \mathbb{T}^k \subset I_0.$$

We examine the interesting sets for a special case of (6.93) with order two; i.e., the second-order difference equation

$$x_{n+1} = \gamma_n x_n^\alpha x_{n-1}^\beta, \quad \gamma_n \in \mathbb{T}, \ n \geq 0 \tag{6.98}$$

where

$$\alpha, \beta \in \mathbb{R}, \beta \neq 0, \ x_{-1}, x_0 \in \mathbb{C}_0.$$

The linear conjugate of the modulus equation of (6.98) is

$$y_{n+1} = \alpha y_n + \beta y_{n-1} \tag{6.99}$$

which has eigenvalues

$$\lambda_{\pm} = \frac{\alpha \pm \sqrt{\alpha^2 + 4\beta}}{2}.$$

A number of different cases occur depending on the values of $\lambda_{\pm}$.

Case 1: $|\lambda_-|, |\lambda_+| > 1$. In this case all nonzero real solutions of (6.99) have unbounded magnitude; i.e., $\lim_{n \to \infty} |y_n| = \infty$. It follows that $|x_n| = e^{y_n}$ does not statisfy (6.96) and therefore, the only interesting solutions are the ones within the unit circle $\mathbb{T}$. Thus

$$I_0 = \mathbb{T} \times \mathbb{T}. \tag{6.100}$$

This also shows that the unit circle $\mathbb{T}$ is repelling since every uninteresting solution has a subsequence that approaches either 0 or ∞. Ranges of values for α, β that imply this case are easy to obtain from the above formula for $\lambda_{\pm}$. In particular, (6.100) is true with

$$\alpha > 0 \text{ and } \beta > \alpha + 1 \Rightarrow \lambda_- < -1 \text{ and } \lambda_+ > 1$$

for real eigenvalues. If the eigenvalues are complex with $\alpha^2 < -4\beta$ (or $|\alpha| < 2\sqrt{-\beta}$) then (6.100) holds if and only if

$$\beta < -1 \text{ and } \alpha^2 < -4\beta \Rightarrow |\lambda_-| = |\lambda_+| = \sqrt{-\beta} > 1.$$

Case 2: $|\lambda_-|, |\lambda_+| < 1$. This is the opposite extreme of Case 1 since in this case all real solutions of (6.99) converge to zero. Hence $\lim_{n \to \infty} |x_n| = e^0 = 1$ and we conclude that all solutions of (6.98) $\mathbb{C}_0$ in converge to the unit circle $\mathbb{T}$. In particular, all solutions of (6.98) in $\mathbb{C}_0$ are interesting; i.e,

$$I_0 = \mathbb{C}_0 \times \mathbb{C}_0. \tag{6.101}$$

For complex eigenvalues, (6.101) holds if and only if

$$-1 < \beta < 0 \text{ and } \alpha^2 < -4\beta.$$

For real eigenvalues, it is not hard to verify that (6.101) holds if

$$|\alpha| < 2 \text{ and } -1 < \beta < 1 - |\alpha|.$$

It is worth noting that *neither of the above conditions hold if α, β are integers with $\beta \neq 0$.*

Case 3. $|\lambda_-| \leq 1 \leq |\lambda_+|$ or $|\lambda_+| \leq 1 \leq |\lambda_-|$. In this case the solutions of (6.98) are neither universally attracted to the unit circle $\mathbb{T}$ as in Case 2 nor universally repelled by it as in Case 1. The interesting solutions are mixed with generic solutions that have 0 or ∞ as their limit points. Therefore, there is a less predictable (hence, more appealing) range of possibilities. Let us explore some of these possibilities next.

For real eigenvalues the following range is complementary to that stated in Case 1:

$$\alpha > 0 \text{ and } 0 < \beta \leq \alpha + 1 \Rightarrow -1 \leq \lambda_- < 0, \ \lambda_+ > 1$$

with $\lambda_- = -1$ if and only if $\beta = \alpha + 1$. When all inequalities above are strict, the interesting initial values are those which result in a zero coefficient for the dominant eigenvalue λ_+. From the formula

$$y_n = c_1 \lambda_+^n + c_2 \lambda_-^n$$

we obtain

$$y_{-1} = \frac{c_1}{\lambda_+} + \frac{c_2}{\lambda_-}, \quad y_0 = c_1 + c_2 \tag{6.102}$$

which yield

$$c_1 \left(\frac{1}{\lambda_+} - \frac{1}{\lambda_-} \right) = y_{-1} - \frac{y_0}{\lambda_-}.$$

Thus $c_1 = 0$ if and only if

$$y_0 = y_{-1}\lambda_- \Leftrightarrow e^{y_0} = e^{\lambda_- \ln |x_{-1}|} \Leftrightarrow |x_0| = |x_{-1}|^{\lambda_-}.$$

From this we conclude that if $\alpha > 0$ and $0 < \beta < \alpha + 1$ then

$$I_0 = \left\{ (x_0, x_{-1}) : |x_0| = |x_{-1}|^{\lambda_-} \right\}.$$

Further, with these initial values

$$\lim_{n \to \infty} |x_n| = \lim_{n \to \infty} e^{y_n} = \lim_{n \to \infty} e^{c_2 \lambda_-^n} = 1.$$

It follows that all interesting solutions of (6.98) in this case converge to the unit circle $\mathbb{T}$; i.e., if $\alpha > 0$ and $0 < \beta < \alpha + 1$ then for every $x_{-1} \in \mathbb{C}_0$ and each x_0 on the circle $|z| = |x_{-1}|^{\lambda_-}$, the solution that is generated is interesting and converges to an orbit in $\mathbb{T}$. No other solutions of (6.98) are interesting or attracted to $\mathbb{T}$, a situations that is markedly different from both Cases 1 and 2. As a specific example, the quadratic difference equation

$$x_{n+1} = x_n x_{n-1}$$

that we encountered earlier in Section 2.2 in this book. This equation generates uniquely defined trajectories in $\mathbb{C}$ due to its integer exponents and has a nontrivial interesting set

$$I_0 = \left\{ (x_0, x_{-1}) : |x_0| = |x_{-1}|^{1-\gamma} \right\}, \text{ where } \gamma = \frac{\sqrt{5}+1}{2} \text{ (the golden mean)}.$$

In the boundary case $\beta = \alpha + 1$, $\lambda_- = -1$ and $\lambda_+ = \alpha + 1$ so $(x_0, x_{-1}) \in I_0$ if and only if $|x_0| = |x_{-1}|^{-1}$. Further, from (6.102) we obtain

$$c_2 = \left(y_{-1} - \frac{y_0}{\lambda_+} \right) / \left(\frac{1}{\lambda_-} - \frac{1}{\lambda_+} \right) = \frac{y_0 - y_{-1}(\alpha + 1)}{\alpha + 2} = \frac{1}{\alpha + 2} \ln \left(\frac{|x_0|}{|x_{-1}|^{\alpha+1}} \right).$$

Therefore

$$|x_n| = e^{y_n} = e^{c_2(-1)^n} = \left(\frac{|x_0|}{|x_{-1}|^{\alpha+1}} \right)^{(-1)^n/(\alpha+2)} = |x_{-1}|^{(-1)^{n+1}}.$$

Evidently, for each given $x_{-1} \in C_0$, an interesting solution $\{x_n\}$ jumps between the circle of radius $|x_{-1}|^{-1}$ and the circle of radius $|x_{-1}|$ depending on whether n is even or odd. Therefore, the unit circle $\mathbb{T}$ neither attracts nor repels the interesting solutions in this case.

Finally, the eigenvalues are complex only when

$$|\lambda_-| = |\lambda_+| = 1 \quad \text{with } \beta = -1 \text{ and } |\alpha| < 2.$$

Solutions of (6.99) are as follows

$$y_n = c_1 \cos n\theta + c_2 \sin n\theta, \quad \cos\theta = \frac{\alpha}{2}, \quad \sin\theta = \sqrt{1 - \left(\frac{\alpha}{2} \right)^2}$$

where c_1, c_2 are functions of the initial values $y_0 = |x_0|$ and $y_{-1} = |x_{-1}|$. Hence,

$$|x_n| = e^{c_1 \cos n\theta + c_2 \sin n\theta}$$

It is clear that the modulus of every solution is almost periodic (or quasi-periodic) in this case, leading to a similar behavior for the actual solution in the complex plane. Figures 6.1 and 6.2 show plots of solutions of

$$x_{n+1} = \frac{\gamma_n \sqrt{x_n}}{x_{n-1}}$$

where $\alpha = 1/2$. The plotted points represent principal values only (argument index zero). We emphasize that these are *base-space plots* in the complex plane $\mathbb{C}$, not state-space orbits in $\mathbb{C}^2$.

If θ is a rational multiple of π then the modulus sequence is periodic; for example, if $\alpha = 1$ then $\theta = \pi/3$ and the modulus sequence is periodic with period 6. The difference equation in this case is

$$x_{n+1} = \frac{\gamma_n x_n}{x_{n-1}}$$

which we also encountered in Section 2.3. Exploration of additional cases of (6.98) that are not considered above (including a consideration of γ_n that is not in $\mathbb{T}$) is left to the reader, who may also wish to consider studying the solutions of (6.93) with order 3 or greater in the complex plane.

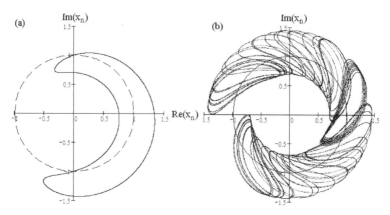

FIGURE 6.1
Plots of interesting solutions of Eq. (6.98) in the base-space $\mathbb{C}$ with $\alpha = 1/2$, $\beta = -1$, $x_{-1} = i$, $x_0 = 0.7i$ and: (a) $\gamma_n = 1$; (b) $\gamma_n = exp(n\pi i/1.1)$.

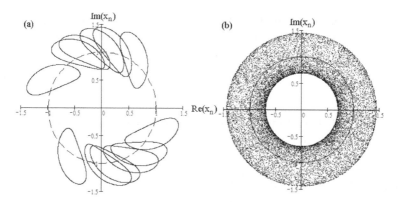

FIGURE 6.2
Plots of interesting solutions of Eq. (6.98) in the base-space $\mathbb{C}$ with $\alpha = 1/2$, $\beta = -1$, $x_{-1} = i$, $x_0 = 0.7i$ and: (a) $\gamma_n = exp(n\pi i/1.2)$, (b) $\gamma_n = exp(n\pi i/2)$.

6.3.5 Biology: Modeling populations

This section applies some of the results obtained in previous sections to study the positive solutions of certain expow equations. We discuss the asymptotic behaviors of solutions in two examples below where direct analysis (without order reduction) is more difficult.

The equation discussed in the next example appears in a discrete-time population model (see the Notes for this chapter for references). This example illustrates that solutions of Eq. (6.80) may exhibit a wide variety of coexisting, qualitatively different behaviors even in an applied context. Since these solutions are also stable, the equation discussed below is an example of a *multistable* equation.

Example 6.8

Consider the autonomous expow difference equation

$$x_{n+1} = x_{n-1}e^{a-x_n-x_{n-1}}, \quad a, x_{-1}, x_0 > 0. \tag{6.103}$$

It is easy to check that Eq. (6.103) has up to two isolated fixed points. One is the origin, which is repelling if $a > 0$ (eigenvalues of linearization are $\pm e^{a/2}$) and the other fixed point is $\bar{x} = a/2$. If $a > 4$ then $\bar{x}$ is unstable and nonhyperbolic because the eigenvalues of the linearization of (6.103) are -1 and $1 - a/2$. The computer-generated diagram in Figure 6.3 shows the variety of stable periodic and nonperiodic solutions that occur with $a = 4.6$ and one initial value $x_{-1} = 2.3$ fixed and the other initial value x_0 changing from 2.3 to 4.8; i.e., approaching (or moving away from) the fixed point $\bar{x}$ on a straight line segment in the plane.

In Figure 6.3, for every grid value of x_0 in the range 2.3–4.8, the last 200 (of 300) points of the solution $\{x_n\}$ are plotted vertically. In this figure, stable solutions with periods 2, 4, 8, 12, and 16 can be easily identified. All of the solutions that appear in Figure 6.3 represent *coexisting stable orbits* of Eq. (6.103). There are also periodic and nonperiodic solutions which do not appear in Figure 6.3 because they are unstable (e.g., the fixed point $\bar{x} = 2.3$). Additional bifurcations of both periodic and nonperiodic solutions occur outside the range 2.3–4.8 which are not shown in Figure 6.3.

Understanding the behavior for solutions of Eq. (6.103) is made easier when we look at its SC factorization as given by Corollary 6.3. Here $a_0 = 0$, $a_1 = 1$ and $b_0 = b_1 = b$. With these parameter values it is readily verified that condition (6.79) is satisfied and $c = -1$. Thus

$$t_{n+1} = \frac{e^a}{t_n}, \quad t_0 = \frac{x_0}{x_{-1}e^{-x_{-1}}} \tag{6.104}$$

$$x_{n+1} = t_{n+1}x_n e^{-x_n}. \tag{6.105}$$

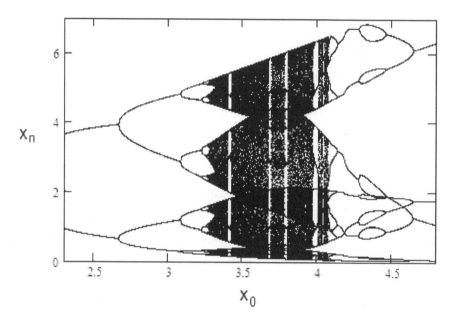

FIGURE 6.3
Bifurcations of solutions of Eq. (6.103) with a changing initial value; $a = 4.6$ is fixed.

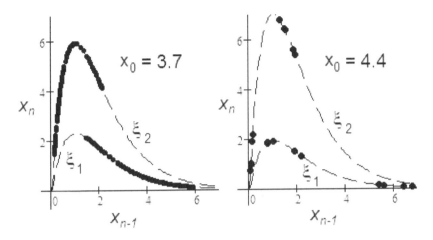

FIGURE 6.4
Two of the orbits in Figure 6.3 shown here on their loci of two curves ξ_1, ξ_2 in the state-space.

All solutions of (6.104) with $t_0 \neq e^{a/2}$ are periodic with period 2:

$$\left\{ t_0, \frac{e^a}{t_0} \right\} = \left\{ \frac{x_0}{x_{-1}e^{-x_{-1}}}, \frac{x_{-1}e^{a-x_{-1}}}{x_0} \right\}.$$

Hence the orbit of each nontrivial solution $\{x_n\}$ of (6.103) in the plane is restricted to the pair of curves

$$\xi_1(t) = \frac{e^a}{t_0} t e^{-t} \quad \text{and} \quad \xi_2(t) = t_0 t e^{-t}. \tag{6.106}$$

Now, if x_{-1} is fixed and x_0 changes, then t_0 changes proportionately to x_0. These changes in initial values are reflected as changes in *parameters* in (6.105). The orbits of the one dimensional map bte^{-t} where $b = t_0$ or e^a/t_0 exhibit a variety of behaviors as the parameter b changes according to well-known rules such as the fundamental ordering of cycles and the occurrence of chaotic behavior with the appearance of period-3 orbits when b is large enough. Eq. (6.105) splits these behaviors evenly over the pair of curves (6.106) as the initial value x_0 changes; see Figure 6.4, which shows the orbits of (6.103) for two different initial values x_0 with $a = 4.6$; the first 100 points of each orbit are discarded in these images so as to highlight the asymptotic behavior of each orbit. The splitting over the pair of curves ξ_1, ξ_2 also explains why odd periods do not appear in Figure 6.3.
□

The next example examines a slightly modified version of Eq. (6.103) that nevertheless shows a markedly different behavior. The semiconjugate factorization makes transparent the root causes of this substantial change in behavior that might be difficult to explain otherwise.

Example 6.9

By way of comparison with Eq. (6.103) consider the following expow equation

$$x_{n+1} = x_{n-1}e^{a+bx_n-bx_{n-1}}, \quad a \in \mathbb{R}, \ b > 0 \tag{6.107}$$

which is similar to (6.103) except for a sign change in the exponent. By Corollary 6.4, Eq. (6.107) reduces to the first-order equation:

$$y_{n+1} = \frac{t_0}{y_n}e^{(n+1)a+by_n}, \quad t_0 = x_0x_{-1}e^{-bx_{-1}}. \tag{6.108}$$

Since t_0 and $y_0 = x_0$ are positive, if $a > 0$ then evidently each solution of (6.108) is unbounded. If $a = 0$ then because the function e^{bu}/u is unimodal with a single minimum, the equation

$$y_{n+1} = \frac{t_0}{y_n}e^{by_n} = \frac{x_0x_{-1}e^{by_n-bx_{-1}}}{y_n} \tag{6.109}$$

has two, one, or no fixed points depending on how large the initial value t_0 is; in particular, for sufficiently small values of t_0 there are two fixed points.

This bifurcation is not so transparent in a direct examination of the second-order equation (6.107) but it has a significant effect on the asymptotic behaviors of the solutions of that equation because the appearance of the fixed point in (6.109) yields bounded solutions for Eq. (6.107) when $a = 0$. In this case, using straightforward analysis it is possible to determine the regions of initial points (x_{-1}, x_0) in the Euclidean plane that imply the occurrence of a particular aysmptotic behavior. In all cases, solutions are either unbounded or they converge to a positive fixed point. Finally, if $a < 0$ then we may in addition have convergence to zero. In no case are the complex behaviors exhibited by the solutions of Eq. (6.103) observed. ▯

Next we consider a more general, nonautonomous version of (6.103) with α replaced by a periodic sequence α_n. We first present a lemma to facilitate the discussion of the nonautonomous equation below.

LEMMA 6.2

Let $\{\sigma_n\}$ be a periodic sequence of positive numbers with minimal period $p \geq 1$ and let $\{t_n\}$ be a solution of the difference equation

$$t_{n+1} = \frac{\sigma_n}{t_n} \tag{6.110}$$

for a given initial value $t_0 > 0$. Then the following are true:

(a) If p is odd then $\{t_n\}$ is periodic with period $2p$ (not necessarily minimal).
(b) If p is even and

$$\sigma_0\sigma_2\cdots\sigma_{p-2} = \sigma_1\sigma_3\cdots\sigma_{p-1} \tag{6.111}$$

then $\{t_n\}$ is periodic with period p.
(c) If p is even but (6.111) does not hold then $\{t_n\}$ has a subsequence that decreases to zero and another subsequence that increases to infinity.

PROOF (a) Using straightforward induction we find that

$$t_n = \frac{t_0\sigma_1\sigma_3\cdots\sigma_{n-3}\sigma_{n-1}}{\sigma_0\sigma_2\cdots\sigma_{n-4}\sigma_{n-2}}, \text{ if } n \text{ is even and:} \tag{6.112}$$

$$t_n = \frac{\sigma_0\sigma_2\cdots\sigma_{n-3}\sigma_{n-1}}{t_0\sigma_1\sigma_3\cdots\sigma_{n-4}\sigma_{n-2}}, \text{ if } n \text{ is odd.} \tag{6.113}$$

If $p = 2q + 1$ is odd then

$$\sigma_{2q+1+j} = \sigma_j \text{ for } j = 0, 1, \ldots, 2q. \tag{6.114}$$

Further, $t_{2p} = t_{4q+2}$ has even index so by (6.112)

$$
\begin{aligned}
t_{2p} &= \frac{t_0\sigma_1\sigma_3\cdots\sigma_{2q-3}\sigma_{2q-1}\sigma_{2q+1}\sigma_{2q+3}\cdots\sigma_{4q-1}\sigma_{4q+1}}{\sigma_0\sigma_2\cdots\sigma_{2q-2}\sigma_{2q}\sigma_{2q+2}\sigma_{2q+4}\cdots\sigma_{4q-2}\sigma_{4q}} \\
&= t_0\left(\frac{\sigma_1\sigma_3\cdots\sigma_{2q-3}\sigma_{2q-1}}{\sigma_{2q+2}\sigma_{2q+4}\cdots\sigma_{4q-2}\sigma_{4q}}\right)\left(\frac{\sigma_{2q+1}\sigma_{2q+3}\cdots\sigma_{4q-1}\sigma_{4q+1}}{\sigma_0\sigma_2\cdots\sigma_{2q-2}\sigma_{2q}}\right) \tag{6.115} \\
&= t_0
\end{aligned}
$$

where the last equality holds because each of the two ratios in (6.115) equals 1 by (6.114). It follows that $\{t_n\}$ has period $2p$. This may not be a prime period; for example, if $\sigma_j = \sigma$ for $j = 0, 1, \ldots, p - 2$ where $\sigma > 0$ and $\sigma \neq 1$ and if also $\sigma_{p-1} = t_0 = 1$ then by (6.113)

$$t_p = \frac{\sigma_0\sigma_2\cdots\sigma_{p-3}\sigma_{p-1}}{t_0\sigma_1\sigma_3\cdots\sigma_{p-4}\sigma_{p-2}} = \frac{\sigma^{(p-1)/2}}{\sigma^{(p-1)/2}} = 1 = t_0$$

i.e., $\{t_n\}$ has period p.
(b) Suppose that p is even. Then again by (6.112)

$$t_p = t_0\frac{\sigma_1\sigma_3\cdots\sigma_{p-3}\sigma_{p-1}}{\sigma_0\sigma_2\cdots\sigma_{p-4}\sigma_{p-2}} = t_0$$

where the last equality is true by (6.111).
(c) If p is even then as in Part (b)

$$t_p = t_0\frac{\sigma_1\sigma_3\cdots\sigma_{p-3}\sigma_{p-1}}{\sigma_0\sigma_2\cdots\sigma_{p-4}\sigma_{p-2}}. \tag{6.116}$$

Since σ has even period p, by (6.112)

$$t_{2p} = t_0 \left(\frac{\sigma_1 \sigma_3 \cdots \sigma_{p-3} \sigma_{p-1}}{\sigma_0 \sigma_2 \cdots \sigma_{p-4} \sigma_{p-2}} \right) \left(\frac{\sigma_{p+1} \sigma_{p+3} \cdots \sigma_{2p-3} \sigma_{2p-1}}{\sigma_p \sigma_{p+2} \cdots \sigma_{2p-4} \sigma_{2p-2}} \right)$$

$$= t_0 \left(\frac{\sigma_1 \sigma_3 \cdots \sigma_{p-3} \sigma_{p-1}}{\sigma_0 \sigma_2 \cdots \sigma_{p-4} \sigma_{p-2}} \right)^2 .$$

Inductively, we see that

$$t_{mp} = t_0 \left(\frac{\sigma_1 \sigma_3 \cdots \sigma_{p-3} \sigma_{p-1}}{\sigma_0 \sigma_2 \cdots \sigma_{p-4} \sigma_{p-2}} \right)^m . \tag{6.117}$$

If (6.111) does not hold then the ratio inside the parentheses in (6.117) is either greater than one or less than one. If greater than one then $t_{mp} \to \infty$ monotonically as $m \to \infty$ and

$$t_{mp+1} = \frac{\sigma_{mp}}{t_{mp}} = \frac{\sigma_0}{t_{mp}} \to 0.$$

Similarly, if the ratio in (6.117) is less than one then $t_{mp} \to 0$ and $t_{mp+1} \to \infty$ monotonically as $m \to \infty$. Thus the proof of (c) is complete. ∎

Example 6.10
Consider the expow equation

$$x_{n+1} = x_{n-1} e^{\alpha_n - b x_n - b x_{n-1}}, \quad x_0, x_{-1}, b > 0 \tag{6.118}$$

where α_n is a periodic sequence of real numbers with minimal or prime period $p \geq 1$. By Corollary 6.6 with $k = m = 1$ this equation reduces in order to

$$y_{n+1} = t_{n+1} y_n e^{-b y_n}, \quad \text{where } t_{n+1} = \frac{e^{\alpha_n}}{t_n}.$$

If p is either odd or it is even and satisfies (6.111) then by Lemma 6.2 the solution $\{t_n\}$ of the factor equation above is periodic with period p or $2p$. A full cycle of $\{t_n\}$ consists of q **distinct** points where $q \leq 2p$. It follows that each orbit of (6.118) is confined to q distinct curves of type

$$\xi_i(u) = t_i u e^{-bu}$$

in the plane where the t_i are the distinct values of t_n. For sufficiently large values of α_n some of the mappings ξ_i are chaotic (due to their having a period 3 solution or a snap-back repeller. Thus the collection of points (y_{n-1}, y_n) on each curve ξ_i tends to be densely distributed on a segment of ξ_i. These dense patches are visibly highlighted in a numerically generated plot of the orbit of (6.118) in its state-space (or "phase plane"). Figure 6.5 depicts this situation for (6.118) with $b = 1$ and

$$\alpha_n = 5 + 0.6 \sin \frac{\pi n}{3}$$

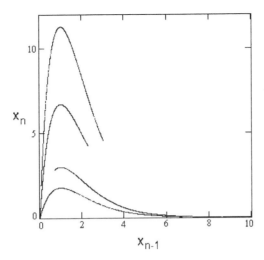

FIGURE 6.5
   A nonperiodic phase plane orbit of Eq. (6.118) with periodic α_n.

which has period 6. Thus e^{α_n} also has period 6 and further, it satisfies (6.111) with

$$e^{\alpha_1} e^{\alpha_3} e^{\alpha_5} = e^{\alpha_2} e^{\alpha_4} e^{\alpha_6} = 3269017.37.$$

The sequence $\{t_n\}$ has only 4 distinct points per cycle:

n	1	2	3	4	5	6
t_n	8.16	30.6	8.16	18.2	4.85	18.2

so that we observe patches on only four distinct curves in Figure 6.5. We note that these patches are not continuous curves but made of 20000 tightly packed points. Clearly this is not a periodic orbit for Eq. (6.118). For smaller values of α_n for which the mappings ξ_i have periodic points, numerical simulations show that the solutions of (6.118) are also periodic; these periods must be integer multiples of p by the preceding argument.

   If p is even but (6.111) is not satisfied then the unbounded subsequences of $\{t_n\}$ cause a spread of points in the orbit of (6.118) since now there are an infinite number of curves like the ξ_i above. Figure 6.6 shows a numerically generated orbit of this type where the sequence

$$\alpha_n = \ln\left(120 + 20\sin\frac{\pi n}{4}\right)$$

has period 8 with

$$e^{\alpha_1} e^{\alpha_3} e^{\alpha_5} e^{\alpha_7} = 201640000, \quad e^{\alpha_2} e^{\alpha_4} e^{\alpha_6} e^{\alpha_8} = 201600000.$$

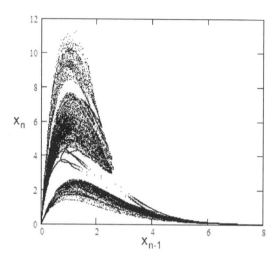

FIGURE 6.6

A phase plane orbit of Eq. (6.118) with α_n still periodic.

The planar orbit in Figure 6.6 is a plot of 60000 points ($b = 1$). As more points are generated numerically and plotted, the peak of the cone will rise without bound since the sequence $\{t_n\}$ is unbounded. ▯

Our study of expow equations above points to a large variety of equations with semiconjugate factorizations and thus, reducible in order. We have also shown that such equations are capable of generating rich dynamic behaviors. Further, the behaviors of positive solutions in the last two examples above might be difficult to explain without the reduction of order that results from the semiconjugate factorization. On the other hand, the interested reader will have noticed that many types of expow equations that are amenable to analysis using Lemma 6.2 and Theorem 6.3 have not been considered in this first study. Some of those equations and the behaviors of their solutions on various subgroups of $\mathbb{C}_0$ (e.g., $(0, \infty)$ or the circle group $\mathbb{T}$) can provide significant challenges and rewards in the future studies of expow equations.

Finally, many expow equations do not satisfy identity (6.52) and therefore, the above discussion is not applicable to such equations. Whether these latter types of expow equations possess form symmetries and semiconjugate factorizations of a different kind remains an open question.

6.4   Notes

Much of the material in this section is taken from Sedaghat (2008, 2009d, 2009e, 2010c). The biological model in Example 6.8 is from Franke et al. (1999). Separable rational difference equations and their applications are discussed in Philos et al. (1994), DeVault et al. (1998), El-Metwally et al. (2001), Grove and Ladas (2005), and Chan et al. (2005). These references establish important facts about convergence and periodic character of solutions of separable rational equations of order greater than one. Example 6.4 confirms that this class of equations contains members with complicated aperiodic and chaotic solutions.

The reader may have noticed the similarity between several results appearing in Sections 5.6 and 6.3. Specifically, Theorem 5.6 together with Corollary 5.8, Lemma 5.6 and the Proposition following it in Section 5.6 may be compared to Theorem 6.3 in Section 6.3. Further, Corollaries 5.9 and 5.10 in Section 5.6 may be compared to Corollaries 6.3 and 6.7 in Section 6.3. This similarity is striking, considering the different classes to which the equations involved belong and the different methods by which each set of results are derived. This suggests a possibly deep connection between the two sets of results and therefore, also between the two different factorization types. Exploring this issue is left to future work on this topic.

6.5   Problems

6.1 This problem generalizes Corollary 6.1(b). Let $\mathcal{F}$ be a nontrivial field and consider the difference equation

$$x_{n+1} = \alpha_n + ax_n + b\psi(x_{n-i}) + \psi(x_{n-k}) \qquad (6.119)$$

where $0 \leq i < k$, $a, b, \alpha_n \in \mathcal{F}$ for all $n \geq 0$ and $\psi : \mathcal{F} \to \mathcal{F}$ is a given function.
(a) Show that Eq. (6.119) has the separable form symmetry (6.8) if

$$a^{k-i}b + 1 = 0, \quad \text{or} \quad b = -a^{-k+i}, \text{ for } i \geq 0. \qquad (6.120)$$

Use Theorem 6.1 as in Example 6.2.
(b) If (6.120) holds then show that the SC factorization of (6.119) is

$$t_{n+1} = \alpha_n + at_n, \quad t_0 = x_0 + \sum_{j=1}^{k} a^{-k+j-1}\psi(x_{-j})$$

$$x_{n+1} = t_{n+1} - \sum_{j=1}^{k} a^{-k+j-1}\psi(x_{n-j+1}).$$

Note that the SC factorization is the same for all $i < k$.

6.2 Consider the following special case of Eq. (6.119):

$$x_{n+1} = x_n - \psi(x_{n-i}) + \psi(x_{n-k}). \tag{6.121}$$

(a) Verify that Eq. (6.121) has a separable form symmetry and find its SC factorization.

(b) Show that for each set of initial values $x_0, x_{-1}, \ldots, x_{-k}$ the corresponding solution of (6.121) is the same as the solution of the cofactor equation

$$x_{n+1} = t_0 - \sum_{j=1}^{k} \psi(x_{n-j+1})$$

with initial values $x_0, x_{-1}, \ldots, x_{-k+1}$ and

$$t_0 = x_0 + \sum_{j=1}^{k} \psi(x_{-j+1}).$$

(c) Let $\beta \neq 0$ in $\mathcal{F} = \mathbb{R}$ and define

$$\psi(u) = -\frac{\beta}{u}.$$

If $k = 1$ then determine the asymptotic behavior of solutions of

$$x_{n+1} = x_n + \frac{\beta}{x_n} - \frac{\beta}{x_{n-1}}$$

using the corresponding solutions of the first-order equation

$$x_{n+1} = t_0 + \frac{\beta}{x_n}.$$

(d) If $\beta > 0$ and $t_0 = x_0 - \sum_{j=1}^{k}(\beta/x_{-j}) > 0$ and $x_{-j} > 0$ for $j = 1, \ldots, k$ then find the asymptotic behavior of solutions of

$$x_{n+1} = x_n + \frac{\beta}{x_{n-i}} - \frac{\beta}{x_{n-k}}$$

for arbitrary $i < k$ using the asymptotic behaviors of the corresponding solutions of

$$x_{n+1} = t_0 + \sum_{j=1}^{k} \frac{\beta}{x_{n-j+1}}.$$

It can be shown (Philos et al. 1994) that every positive solution of the above equation converges to its unique positive fixed point

$$L = \frac{t_0 + \sqrt{t_0^2 + 4k\beta}}{2}.$$

6.3 (a) Show that the following special case of Eq. (6.119)

$$x_{n+1} = \alpha_n + 2x_n + \psi(x_{n-k+1}) + \psi(x_{n-k}) \tag{6.122}$$

has a separable form symmetry in the finite field $\mathcal{F} = \mathbb{Z}_3$ and find its SC factorization.

(b) Find a value of the time index i so that

$$x_{n+1} = \alpha_n + 2x_n + \psi(x_{n-i}) + \psi(x_{n-k})$$

has a SC factorization in $\mathbb{Z}_5$.

6.4 Give the details of the proof of Corollary 6.2.

6.5 (a) Find the SC factorizations of the following equations over $\mathbb{C}\backslash\{0\}$

$$\text{(i) } x_{n+1} = \frac{\alpha_n x_n}{x_{n-1}}, \quad \text{(ii) } x_{n+1} = \frac{\alpha_n x_{n-1}}{x_n}.$$

(b) Let $\alpha_n = \alpha$ be a nonzero constant. Show that every solution of Equation (i) in (a) is periodic with period 6 in $\mathbb{C}\backslash\{0\}$ either by using Equation (i) directly or by using its SC factorization. What can be said about Equation (ii) with constant α_n?

6.6 (a) Prove that the group

$$G_a = \{a^n : n \in \mathbb{Z}, \; a \neq 0, 1\}$$

is invariant under the difference equation

$$x_{n+1} = \alpha x_n^{m_0} x_{n-1}^{m_1} \cdots x_{n-k}^{m_k} \quad \text{where:}$$
$$m_j \in \mathbb{Z} \text{ for } j = 0, \ldots, k, \text{ and } \alpha, x_0, \ldots, x_{-k} \in G_a.$$

(b) Study the behavior of solutions of $x_{n+1} = x_n x_{n-1}$ in G_a for (i) $a = 2$, (ii) $a = i$ and (iii) $a = e^i$.

6.7 Prove that the group

$$\mathbb{C}_r = \{z : \operatorname{Re} z, \operatorname{Im} z \in \mathbb{Q}, \; z \neq 0\}$$

is invariant under rational equations with coefficients and initial values in $\mathbb{C}_r$ (as long as the denominator is not equal to zero).

6.8 Find the SC factorizations of equations (6.65) and (6.66).

6.9 Let ψ be as defined in Example 6.6. For every positive integer k show that the difference equation

$$x_{n+1} = \alpha_n \psi(x_n) \psi(x_{n-2k+1}) x_{n-2k+1}$$

has a SC factorization and determine this factorization. Note that this equation generalizes Eq. (6.64).

6.10 Prove Lemma 6.1.

6.11 Give the details of the proof of Theorem 6.3(a).

6.12 Consider the second-order difference equation

$$x_{n+1} = \frac{a_n x_{n-1}}{(ax_n + b)(cx_{n-1} + d)} \tag{6.123}$$

where $a, b, c, d \in \mathbb{C}\backslash\{0\}$ and a_n is a sequence of nonzero complex numbers.
(a) If $ad = bc$ then show that (6.123) has the following SC factorization

$$t_{n+1} = \frac{a_n}{t_n},$$

$$x_{n+1} = \frac{t_{n+1} x_n}{cx_n + d}.$$

(b) Let $a, b, c, d > 0$ with $ad = bc$ and assume that a_n is a periodic sequence of positive real numbers with period $p \geq 1$. Use the SC factorization in (a) to determine possible types of solutions for (6.123).

6.13 Consider the third-order difference equation

$$x_{n+1} = \frac{a_n x_{n-1}(ax_n + b)}{cx_{n-2} + d} \tag{6.124}$$

where we assume that $a, b, c, d \in \mathbb{C}\backslash\{0\}$ and a_n is a sequence of nonzero complex numbers. If $ad = bc$ then show that Eq. (6.124) has a SC factorization and determine the factor and cofactor equations in this case.

6.14 Supply the details involving the polynomial Q_1 in the proof of Corollary 6.3.

6.15 Complete the proof of Corollary 6.4.

6.16 Consider the third-order expow equation

$$x_{n+1} = x_n^{1-a} x_{n-1}^a e^{a_n - b_0 x_n - b_1 x_{n-1} - b_2 x_{n-2}}, \quad a \geq 0, \ b_0, b_1, b_2, a_n \in \mathbb{R}. \tag{6.125}$$

(a) Let $a = 0$. Show that Eq. (6.125) has an SC factorization over the multiplicative group $(0, \infty)$ for some values of b_0, b_1, b_2. Determine this SC factorization and the corresponding constraint on b_0, b_1, b_2.
(b) Let $a > 0$. Show that Eq. (6.125) has two SC factorizations over the multiplicative group $(0, \infty)$ depending on the possible values of b_0, b_1, b_2. Determine each of these factorizations and the corresponding constraints on b_0, b_1, b_2. Further, show that each of the two cofactor equations has an SC factorization and determine constraints on b_0, b_1, b_2 that correspond to each factorization. For these values of b_0, b_1, b_2 find the complete SC factorization of (6.125) into a triangular system of first-order equations.

(c) Use the results in Part (b) to obtain a complete SC factorization of

$$x_{n+1} = \sqrt{x_n x_{n-1}} e^{-x_n + 0.5x_{n-1} + 0.5x_{n-2}} \tag{6.126}$$

into a triangular system of first-order equations. Use this system to discuss the global behavior of the positive solutions of Eq. (6.126).

(d) Repeat Part (c) for the equation

$$x_{n+1} = \sqrt{x_n x_{n-1}} e^{x_n - 0.5x_{n-1} - 0.5x_{n-2}}.$$

6.17 Show that the third-order expow equation

$$x_{n+1} = x_{n-2} e^{\alpha_n - bx_n - bx_{n-1} - bx_{n-2}}, \quad b, \alpha_n \in \mathbb{R}, \ b \neq 0.$$

has an SC factorization of the type mentioned in Corollary 6.3 over the complex numbers $\mathbb{C}$ but not over $\mathbb{R}$. Determine this SC factorization.

6.18 Consider the following delay expow equation

$$x_{n+1} = x_{n-2} e^{\alpha_n - b_0 x_n - b_k x_{n-k}}, \quad k \geq 1 \tag{6.127}$$

which may be compared with Eq. (6.82) in Corollary 6.4.

(a) Show that if $b_0 + b_k = 0$ then Eq. (6.127) has a SC factorization

$$t_{n+1} = e^{\alpha_n} t_n,$$
$$x_{n+1} = t_{n+1} x_n^{-1} x_{n-1}^{-1} e^{-b_0 x_n - b_0 x_{n-1} - \cdots - b_0 x_{n-k+1}}. \tag{6.128}$$

(b) If $k = 3m$ with $m \geq 1$ then show that the cofactor equation (6.128) has a SC factorization over $\mathbb{C}$ as

$$r_{n+1} = t_{n+1} r_n^c, \quad c = \frac{-1 + i\sqrt{3}}{2},$$
$$x_{n+1} = r_{n+1} x_n^{\bar{c}} e^{-b_0 (x_n - \bar{c} x_{n-1} + x_{n-3} - \bar{c} x_{n-4} + \cdots + x_{n-k+3} - \bar{c} x_{n-k+2})}.$$

where $\bar{c} = (-1 - i\sqrt{3})/2$ is the complex conjugate of c.

(c) Suggest a generalization of the above process for the expow equation

$$x_{n+1} = x_{n-j} e^{\alpha_n - b_0 x_n - b_k x_{n-k}}, \quad j, k \geq 1.$$

6.19 Consider the powers-only expow equation

$$x_{n+1} = x_n^\alpha x_{n-1}^\beta, \quad \alpha, \beta \in \mathbb{R}, \ \beta \neq 0. \tag{6.129}$$

(a) Show that Eq. (6.129) has a semiconjugate factorization

$$t_{n+1} = t_n^p, \quad t_0 = x_0 x_{-1}^q$$
$$x_{n+1} = t_{n+1} x_n^{-q}$$

for suitable values of the parameters p and q (see Section 6.3.4).

(b) Verify that $p, q \in \mathbb{R}$ if and only if $\alpha^2 + 4\beta \geq 0$. In this case the solutions of the SC factorization in Part (a) have single-valued moduli.

(c) If $\alpha^2 + 4\beta \geq 0$ then find an explicit formula for the solutions of (6.129) by solving the first order equations in the SC factorization in (a).

7

Time-Dependent Form Symmetries

To simplify our initial exposition to semiconjugate factorization, it was assumed in Chapter 3 that the form symmetry H was independent of n. Now that the basic ideas and results are in place, we are ready to extend the concept of semiconjugate factorization to all nonautonomous difference equations by considering form symmetries that are functions of the "independent variable" n (viewed intuitively as a discrete "time" variable). Much of the development in this chapter consists of direct extensions of the basic results in Chapters 3 and 5. In particular, we extend the SC factorization Theorem 3.1 and the useful invertible-map criterion Theorem 5.1 in Section 5.1. As might be expected, some new facts and features emerge in this more general setting. For instance, these extensions generalize the results in Chapter 5 on type-$(k, 1)$ order reductions to include SC factorizations of linear equations with variable coefficients. In the case of linear difference equations, our results uncover new, previously unknown features that make up the structure and define the essential properties of these difference equations.

7.1 The semiconjugate relation and factorization

Let G be a nontrivial group and let F_n be the unfolding on G^{k+1} of the function f_n in the difference equation

$$x_{n+1} = f_n(x_n, \ldots, x_{n-k}) \tag{7.1}$$

for each n. Then (7.1) is equivalent to

$$X_{n+1} = F_n(X_n), \quad X_n = (x_n, \ldots, x_{n-k}). \tag{7.2}$$

As in Chapter 3, we are interested in a lower dimensional factor equation

$$Y_{n+1} = \Phi_n(Y_n), \quad Y_n = (y_n, \ldots, y_{n-m+1}), \ m \le k \tag{7.3}$$

corresponding to (7.2). If there exists a sequence of maps $H_n : G^{k+1} \to G^m$ such that for every solution $\{X_n\}$ of (7.2) the sequence

$$Y_n = H_n(X_n), \quad n = 0, 1, 2, \ldots \tag{7.4}$$

is a solution of (7.3) then

$$\Phi_n(H_n(X_n)) = \Phi_n(Y_n) = Y_{n+1} = H_{n+1}(X_{n+1}) = H_{n+1}(F_n(X_n)).$$

Therefore, (7.4) is satisfied for all solutions of (7.2) and (7.3) if and only if the sequence $\{H_n\}$ of maps satisfies the following equality for all n

$$H_{n+1} \circ F_n = \Phi_n \circ H_n. \tag{7.5}$$

This semiconjugate relation is illustrated by the diagram

$$
\begin{array}{ccc}
G^{k+1} & \xrightarrow{\;F_n\;} & F_n(G^{k+1}) \\
\downarrow{\scriptstyle H_n} & & \downarrow{\scriptstyle H_{n+1}} \\
H_n(G^{k+1}) = G^m & \xrightarrow{\;\Phi_n\;} & \Phi_n(H_n(G^{k+1})) = H_{n+1}(F_n(G^{k+1}))
\end{array}
$$

If the mappings H_n are independent of n, i.e., $H_n = H$ for all n then Eq. (7.5) reduces to Eq. (3.5) in Section 3.1.5. This leads to the following definition in which it is convenient to think of n as a discrete "time" index.

DEFINITION 7.1   *Let $k \geq 1$, $1 \leq m \leq k$ and G be a nontrivial group. If there is a sequence of surjective maps $H_n : G^{k+1} \to G^m$ such that Eq. (7.5) is satisfied for a given pair of function sequences $\{F_n\}$ and $\{\Phi_n\}$ then we say that F_n is **semiconjugate** to Φ_n for each n and refer to the sequence $\{H_n\}$ as a **time-dependent form symmetry** for Eq. (7.2) or equivalently, for Eq. (7.1). Since $m < k + 1$, the form symmetry $\{H_n\}$ is **order-reducing**.*

Technically, a time-dependent form symmetry can also be defined as a *single map*

$$H : \mathbb{N} \times G^{k+1} \to G^m, \quad H(n; u_0, \dots, u_k) = H_n(u_0, \dots, u_k).$$

We choose the sequence definition due to its more intuitive content and in keeping with our practice in this book where difference equations are defined in terms of map sequences on groups, as in (7.1).

Considerations similar to those in Section 3.1.5 lead to the following theorem that generalizes Theorem 3.1 and makes precise the concept of semiconjugate factorization of difference equations. For the notation that is common to the next theorem and Theorem 3.1, see Section 3.1.5.

THEOREM 7.1
Let $k \geq 1$, $1 \leq m \leq k$, let $h_n : G^{k-m+1} \to G$ for $n \geq -m + 1$ be a sequence of functions on a given nontrivial group G and define the functions $H_n : G^{k+1} \to G^m$ by

$$H_n(u_0, \dots, u_k) = [u_0 * h_n(u_1, \dots, u_{k+1-m}), \dots, u_{m-1} * h_{n-m+1}(u_m, \dots, u_k)]. \tag{7.6}$$

Then the following statements are true:

(a) *For each $n \geq 0$, the function H_n defined by (7.6) is surjective.*

(b) *If $\{H_n\}$ is an order-reducing form symmetry then the difference equation (7.1) has an equivalent SC factorization consisting of the system of factor-cofactor equations*

$$t_{n+1} = g_n(t_n, \ldots, t_{n-m+1}), \tag{7.7}$$

$$x_{n+1} = t_{n+1} * h_{n+1}(x_n, \ldots, x_{n-k+m})^{-1} \tag{7.8}$$

whose orders m and $k + 1 - m$, respectively, add up to the order of (7.1).

(c) *The SC factor map $\Phi_n : G^m \to G^m$ in (7.5) is the unfolding of Eq. (7.7) for each $n \geq 0$. In particular, each Φ_n is of scalar type.*

PROOF   (a) Let n be a fixed nonnegative integer and for $j = 0, \ldots, m-1$ denote the j-th coordinate function of H_n by

$$\eta_{j+1}(u_0, \ldots, u_k) = u_j * h_{n-j}(u_{j+1}, \ldots, u_{j+k+1-m}). \tag{7.9}$$

Now choose an arbitrary point $(v_1, \ldots, v_m) \in G^m$ and define

$$u_{m-1} = v_m * h_{n-m+1}(u_m, u_{m+1} \ldots, u_k)^{-1}, \tag{7.10}$$

$$u_m = u_{m+1} = \ldots u_k = \bar{u}$$

where $\bar{u}$ is a fixed element of G, e.g., the identity. Then

$$
\begin{aligned}
v_m &= u_{m-1} * h_{n-m+1}(\bar{u}, \bar{u} \ldots, \bar{u}) \\
&= u_{m-1} * h_{n-m+1}(u_m, u_{m+1} \ldots, u_k) \\
&= \eta_m(u_0, \ldots, u_k) \\
&= \eta_m(u_0, \ldots, u_{m-2}, \underbrace{v_m * h_{n-m+1}(\bar{u}, \bar{u} \ldots, \bar{u})^{-1}}_{u_{m-1}}, \bar{u} \ldots, \bar{u})
\end{aligned}
$$

for any selection of elements $u_0, \ldots, u_{m-2} \in G$. Using the same idea, define

$$u_{m-2} = v_{m-1} * h_{n-m+2}(u_{m-1}, \bar{u} \ldots, \bar{u})^{-1}$$

with u_{m-1} defined by (7.10) so as to get

$$
\begin{aligned}
v_{m-1} &= u_{m-2} * h_{n-m+2}(u_{m-1}, \bar{u} \ldots, \bar{u}) \\
&= u_{m-2} * h_{n-m+2}(u_{m-1}, u_m \ldots, u_{k-1}) \\
&= \eta_{m-1}(u_0, \ldots, u_k) \\
&= \eta_{m-1}(u_0, \ldots, u_{m-3}, \underbrace{v_{m-1} * h_{n-m+2}(u_{m-1}, \bar{u} \ldots, \bar{u})^{-1}}_{u_{m-2}}, u_{m-1}, \bar{u} \ldots, \bar{u})
\end{aligned}
$$

for any choice of $u_0, \ldots, u_{m-3} \in G$. Continuing in this way, by induction we obtain elements $u_{m-1}, \ldots, u_0 \in G$ such that

$$v_i = \eta_i(u_0, \ldots, u_{m-1}, \bar{u} \ldots, \bar{u}), \quad i = 1, \ldots, m.$$

Therefore, $H_n(u_0, \ldots, u_{m-1}, \bar{u} \ldots, \bar{u}) = (v_1, \ldots, v_m)$ and it follows that H_n is onto G^m.

(b) To show that the SC factorization system consisting of equations (7.7) and (7.8) is equivalent to Eq. (7.1) we show that (i) each solution $\{x_n\}$ of (7.1) uniquely generates a solution of the system (7.7) and (7.8) and, conversely, (ii) each solution $\{(t_n, y_n)\}$ of the system (7.7) and (7.8) correseponds uniquely to a solution $\{x_n\}$ of (7.1). To establish (i) let $\{x_n\}$ be the unique solution of (7.1) corresponding to a given set of initial values $x_0, \ldots x_{-k} \in G$. Define the sequence

$$t_n = x_n * h_n(x_{n-1}, \ldots, x_{n-k+m-1}) \tag{7.11}$$

for $n \geq -m + 1$. Then for each $n \geq 0$ if H_n is defined by (7.6) it follows from the semiconjugate relation (7.5) that

$$\begin{aligned}
x_{n+1} &= f_n(x_n, \ldots, x_{n-k}) \\
&= g_n(x_n * h_n(x_{n-1}, \ldots, x_{n-k+m-1}), \ldots, \\
&\quad x_{n-m+1} * h_{n-m+1}(x_{n-m}, \ldots, x_{n-k})) * [h_{n+1}(x_n, \ldots, x_{n-k+m})]^{-1} \\
&= g_n(t_n, \ldots, t_{n-m+1}) * [h_{n+1}(x_n, \ldots, x_{n-k+m})]^{-1}.
\end{aligned}$$

Therefore, $g_n(t_n, \ldots, t_{n-m+1}) = x_{n+1} * h_{n+1}(x_n, \ldots, x_{n-k+m}) = t_{n+1}$ so that $\{t_n\}$ is the unique solution of the factor equation (7.7) with initial values

$$t_{-j} = x_{-j} * h_{-j}(x_{-j-1}, \ldots, x_{-j-k+m-1}), \quad j = 0, \ldots, m-1.$$

Further, since $x_{n+1} = t_{n+1} * [h_{n+1}(x_n, \ldots, x_{n-k+m})]^{-1}$ for $n \geq 0$ by (7.11), $\{x_n\}$ is the unique solution of the cofactor equation (7.8) with initial values $y_{-i} = x_{-i}$ for $i = 0, 1, \ldots, k - m$ and with the values t_n obtained above.

To establish (ii) let $\{(t_n, y_n)\}$ be a solution of the factor-cofactor system with initial values

$$t_0, \ldots, t_{-m+1}, y_{-m}, \ldots y_{-k} \in G.$$

Note that these numbers determine $y_{-m+1}, \ldots, y_0$ through the cofactor equation

$$y_{-j} = t_{-j} * [h_{-j}(y_{-j-1}, \ldots, y_{-j-1-k+m})]^{-1}, \quad j = 0, \ldots, m-1. \tag{7.12}$$

Now for $n \geq 0$ we obtain

$$\begin{aligned}
y_{n+1} &= t_{n+1} * [h_{n+1}(y_n, \ldots, y_{n-k+m})]^{-1} \\
&= g_n(t_n, \ldots, t_{n-m+1}) * [h_{n+1}(y_n, \ldots, y_{n-k+m})]^{-1} \\
&= g_n(y_n * h_n(y_{n-1}, \ldots, y_{n-k+m-1}), \ldots, \\
&\quad y_{n-m+1} * h_{n-m+1}(y_{n-m}, \ldots, y_{n-k})) * h_{n+1}(y_n, \ldots, y_{n-k+m})^{-1} \\
&= f_n(y_n, \ldots, y_{n-k}).
\end{aligned}$$

Thus $\{y_n\}$ is the unique solution of Eq. (7.1) that is generated by the initial values (3.22) and $y_{-m}, \ldots y_{-k}$. This completes the proof of (b).

(c) We show that each coordinate function $\phi_{j,n}$ is the projection into coordinate $j-1$ for $j > 1$. From the definition of H_n in (7.6) and the semiconjugate relation (7.5) we infer that

$$
\begin{aligned}
H_{n+1}(F_n(u_0, \ldots, u_k)) &= H_{n+1}(f_n(u_0, \ldots, u_k), u_0, \ldots, u_{k-1}) \\
&= (f_n(u_0, \ldots, u_k) * h_{n+1}(u_0, \ldots, u_{k-m}), \\
&\quad u_0 * h_n(u_1, \ldots, u_{k-m+1}), \ldots, \\
&\quad u_{m-2} * h_{n-m+2}(u_{m-1}, \ldots, u_{k-1})).
\end{aligned}
$$

Matching the corresponding component functions in the above equality for $j \geq 2$ yields

$$
\phi_{j,n}(u_0 * h_n(u_1, \ldots, u_{k+1-m}), \ldots, u_{m-1} * h_{n-m+1}(u_m, \ldots, u_k)) = \\
u_{j-2} * h(u_{j-1}, u_j \ldots, u_{j+k-m-1})
$$

which shows that $\phi_{j,n}$ maps its j-th coordinate to its $(j-1)$-st. Therefore, for each n and every $(t_1, \ldots, t_m) \in H_n(G^{k+1})$ we have

$$
\Phi_n(t_1, \ldots, t_m) = [g_n(t_1, \ldots, t_m), t_1, \ldots, t_{m-1}]
$$

i.e., $\Phi_n|_{H_n(G^{k+1})}$ is of scalar type. Since by Part (a) $H_n(G^{k+1}) = G^m$ for every n, it follows that Φ_n is of scalar type. ∎

As with Theorem 3.1, the SC factorization in the above theorem does not require the determination of component functions $\phi_{j,n}$ for $j \geq 2$.

Clearly, concepts of order reduction (types, factor and cofactor chains, etc) also extend to time-dependent form symmetries with essentially the same definitions as before and there is no need for repeating them in this chapter.

7.2 Invertible-map criterion revisited

In Section 5.1 we obtained a useful criterion by which we could determine whether a given difference equation of type (7.1) had form symmetries leading to type-$(k,1)$ order reductions. We now show that the same useful idea extends to the time-dependent context.

Consider the following special case of (7.6) with $m = k$

$$
H_n(u_0, u_1, \ldots, u_k) = [u_0 * h_n(u_1), u_1 * h_{n-1}(u_2), \ldots, u_{k-1} * h_{n-k+1}(u_k)] \quad (7.13)
$$

with $h_n : G \to G$ being a sequence of surjective self-maps of the underlying group G. If the form symmetry (7.13) exists then it admits a type-$(k,1)$

reduction and its SC factorization is

$$t_{n+1} = \phi_n(t_n, \ldots, t_{n-k+1}), \tag{7.14}$$

$$x_{n+1} = t_{n+1} * h_{n+1}(x_n)^{-1}. \tag{7.15}$$

The initial values of the factor equation (7.14) are given in terms of the initial values of (7.1) as

$$t_{-j} = x_{-j} * h_{-j}(x_{-j+1}), \quad j = 0, 1, \ldots, k - 1.$$

The next result is a direct extension of Theorem 5.1.

THEOREM 7.2

(Time-dependent invertible map criterion) Assume that $h_n : G \to G$ is a sequence of bijections of G. For arbitrary $u_0, v_1, \ldots, v_k \in G$ and every n define $\zeta_{0,n}(u_0) \equiv u_0$ and for $j = 1, \ldots, k$,

$$\zeta_{j,n}(u_0, v_1, \ldots, v_j) = h_{n-j+1}^{-1}(\zeta_{j-1,n}(u_0, v_1, \ldots, v_{j-1})^{-1} * v_j) \tag{7.16}$$

with the usual distinction observed between map inversion and group inversion. Then Eq. (7.1) has the time-dependent form symmetry $\{H_n\}$ defined by (7.13) if and only if the quantity

$$f_n(\zeta_{0,n}, \zeta_{1,n}(u_0, v_1), \ldots, \zeta_{k,n}(u_0, v_1, \ldots, v_k)) * h_{n+1}(u_0) \tag{7.17}$$

is independent of u_0 for every n. In this case Eq. (7.1) has a SC factorization whose factor functions in (7.14) are given by

$$\phi_n(v_1, \ldots, v_k) = f_n(\zeta_{0,n}, \zeta_{1,n}(u_0, v_1), \ldots, \zeta_{k,n}(u_0, v_1, \ldots, v_k)) * h_{n+1}(u_0). \tag{7.18}$$

PROOF Assume first that (7.17) is independent of u_0 for all $v_1, \ldots, v_k$ so that the functions

$$\phi_n(v_1, \ldots, v_k) = f_n(\zeta_{0,n}, \zeta_{1,n}, \ldots, \zeta_{k,n}) * h_{n+1}(u_0) \tag{7.19}$$

are well defined. Next, if H_n is given by (7.13) then for all $u_0, u_1, \ldots, u_k$

$$\phi_n(H_n(u_0, u_1, \ldots, u_k)) = \phi_n(u_0 * h_n(u_1), u_1 * h_{n-1}(u_2), \ldots, u_{k-1} * h_{n-k+1}(u_k)).$$

Now, by (7.16) for each n and all u_0, u_1

$$\zeta_{1,n}(u_0, u_0 * h_n(u_1)) = h_n^{-1}(u_0^{-1} * u_0 * h_n(u_1)) = u_1.$$

Similarly, for each n and all u_0, u_1, u_2

$$\zeta_{2,n}(u_0, u_0 * h_n(u_1), u_1 * h_{n-1}(u_2)) = h_{n-1}^{-1}(\zeta_{1,n}(u_0, u_0 * h_n(u_1))^{-1} *$$
$$u_1 * h_{n-1}(u_2))$$
$$= u_2.$$

Suppose by way of induction that

$$\zeta_{l,n}(u_0 * h_n(u_1), \ldots, u_{l-1} * h_{n-l+1}(u_k)) = u_l$$

for $1 \leq l < j$. Then

$$\zeta_{j,n}(u_0 * h_n(u_1), \ldots, u_{j-1} * h_{n-j+1}(u_j)) = h_{n-j+1}^{-1}(u_{j-1}^{-1} * u_{j-1} * h_{n-j+1}(u_j)) = u_j.$$

Thus by (7.19)

$$\phi_n(H_n(u_0, u_1, \ldots, u_k)) = f_n(u_0, \ldots, u_k) * h_{n+1}(u_0).$$

Now if F_n and Φ_n are the unfoldings of f_n and ϕ_n, respectively, then

$$
\begin{aligned}
H_{n+1}(F_n(u_0, \ldots, u_k)) &= [f_n(u_0, \ldots, u_k) * h_{n+1}(u_0), u_0 * h_n(u_1), \\
&\qquad \ldots, u_{k-2} * h_{n-k+2}(u_{k-1})] \\
&= [\phi_n(H_n(u_0, u_1, \ldots, u_k)), u_0 * h_n(u_1), \\
&\qquad \ldots, u_{k-2} * h_{n-k+2}(u_{k-1})] \\
&= \Phi_n(H_n(u_0, \ldots, u_k))
\end{aligned}
$$

and it follows that $\{H_n\}$ is a semiconjugate form symmetry for Eq. (7.1). The existence of a SC factorization with factor functions defined by (7.18) now follows from Theorem 7.1.

Conversely, if $\{H_n\}$ as given by (7.13) is a time-dependent form symmetry of Eq. (7.1) then the semiconjugate relation implies that for arbitrary $u_0, \ldots, u_k$ in G there are functions ϕ_n such that

$$f_n(u_0, \ldots, u_k) * h_{n+1}(u_0) = \phi_n(u_0 * h_n(u_1), \ldots, u_{k-1} * h_{n-k+1}(u_k)). \quad (7.20)$$

For every $u_0, v_1, \ldots, v_k$ in G and with functions $\zeta_{j,n}$ as defined above, note that

$$\zeta_{j-1,n}(u_0, v_1, \ldots, v_{j-1}) * h_{n-j+1}(\zeta_{j,n}(u_0, v_1, \ldots, v_j)) = v_j, \quad j = 1, 2, \ldots, k.$$

Therefore, abbreviating $\zeta_{j,n}(u_0, v_1, \ldots, v_j)$ by $\zeta_{j,n}$ we have

$$
\begin{aligned}
f_n(\zeta_{0,n}, \zeta_{1,n}, \ldots, \zeta_{k,n}) * h_{n+1}(u_0) &= \phi_n(\zeta_{0,n} * h_n(\zeta_{1,n}), \zeta_{1,n} * h_{n-1}(\zeta_{2,n}), \\
&\qquad \ldots, \zeta_{k-1,n} * h_{n-k+1}(\zeta_{k,n})) \\
&= \phi_n(v_1, \ldots, v_k)
\end{aligned}
$$

which is independent of u_0. ∎

7.3 Time-dependent linear form symmetry

Among the various form symmetries discussed in Chapter 5, the linear form symmetry was one of the most interesting and also one that naturally allows dependence on n. The time-dependent version of this form symmetry is defined next.

DEFINITION 7.2   *Let $\mathcal{F}$ be a nontrivial field and $\{\alpha_n\}$ a sequence of elements of $\mathcal{F}$ such that $\alpha_n \in \mathcal{F}\backslash\{0\}$ for all $n \geq -k + 1$. A (time-dependent)* **linear form symmetry** *is defined as the following special case of (7.13) with $h_n(u) = -\alpha_{n-1}u$*

$$[u_0 - \alpha_{n-1}u_1, u_1 - \alpha_{n-2}u_2, \ldots, u_{k-1} - \alpha_{n-k}u_k]. \qquad (7.21)$$

We call the sequence $\{\alpha_n\}$ of nonzero elements in $\mathcal{F}$ the **eigensequence** *of the linear form symmetry. If Eq. (7.1) has a linear form symmetry then $\{\alpha_n\}$ may be called an eigensequence of (7.1).*

Use of the term "eigen" from the linear theory is apt here for two reasons. Note that the sequence $\{\alpha_n\}$ characterizes the linear form symmetry (7.21) completely. Further, we find below that linear difference equations indeed have linear form symmetries.

The existence of a linear form symmetry implies a type-$(k, 1)$ order reduction for Eq. (7.1) and a SC factorization where the cofactor equation (7.15) is determined more specifically as

$$x_{n+1} = t_{n+1} + \alpha_n x_n. \qquad (7.22)$$

The following *necessary and sufficient* condition for the existence of a time-dependent linear form symmetry is an application of Theorem 7.2 and a generalization of Corollary 5.5.

COROLLARY 7.1

Equation (7.1) has a time-dependent linear form symmetry of type (7.21) on a nontrivial field $\mathcal{F}$ if and only if there is an eigensequence $\{\alpha_n\}$ in $\mathcal{F}\backslash\{0\}$ that makes the following quantity

$$f_n(u_0, \zeta_{1,n}(u_0, v_1), \ldots, \zeta_{k,n}(u_0, v_1, \ldots, v_k)) - \alpha_n u_0 \qquad (7.23)$$

independent of u_0, where for all $n \geq 0$ and for $j = 1, \ldots, k$,

$$\zeta_{j,n}(u_0, v_1, \ldots, v_j) = \frac{\zeta_{j-1,n}(u_0, v_1, \ldots, v_{j-1}) - v_j}{\alpha_{n-j}}$$

$$= \frac{1}{\prod_{i=1}^{j}\alpha_{n-i}}\left(u_0 - \sum_{i=1}^{j}v_i\prod_{p=1}^{i}\alpha_{n-p}\right).$$

PROOF   The conclusions follow immediately from Theorem 7.2 using $h_n(u) = -\alpha_{n-1}u$. The last equality above is established from the equality preceding it by routine calculation.   ∎

REMARK 7.1 If Eq. (7.1) has a linear form symmetry then by Corollary 7.1 an eigensequence of (2.3) can be defined equivalently as a sequence $\{\alpha_n\}$ in $\mathcal{F}\backslash\{0\}$ for which the quantity in (7.23) is independent of u_0 for every $n \geq 0$.

∎

Let us illustrate the preceding corollary with an example.

Example 7.1
Consider the third-order difference equation

$$x_{n+1} = a_n x_n + b_n x_{n-1} + g_n(x_n + c x_{n-1} + d x_{n-2}) \qquad (7.24)$$

where $c, d \in \mathbb{R}$ with $d \neq 0$ and for all n, $a_n, b_n \in \mathbb{R}$ and $g_n : \mathbb{R} \rightarrow \mathbb{R}$. By Corollary 7.1 a linear form symmetry for (7.24) exists if and only if the quantity in (7.17), i.e.,

$$a_n u_0 + b_n \zeta_{1,n} + g_n(u_0 + c\zeta_{1,n} + d\zeta_{2,n}) - \alpha_n u_0 \qquad (7.25)$$

is independent of u_0 for all n. Substituting

$$\zeta_{1,n} = \frac{u_0 - v_1}{\alpha_{n-1}}, \quad \zeta_{2,n} = \frac{\zeta_{1,n} - v_2}{\alpha_{n-2}} = \frac{u_0 - v_1 - \alpha_{n-1}v_2}{\alpha_{n-1}\alpha_{n-2}}$$

in (7.25) and rearranging terms gives

$$\left(a_n + \frac{b_n}{\alpha_{n-1}} - \alpha_n\right)u_0 - \frac{b_n}{\alpha_{n-1}}v_1 +$$
$$g_n\left(\left(1 + \frac{c}{\alpha_{n-1}} + \frac{d}{\alpha_{n-1}\alpha_{n-2}}\right)u_0 - \frac{c}{\alpha_{n-1}}v_1 - \frac{d}{\alpha_{n-2}}v_2\right)$$

which is independent of u_0 for all n if the coefficients of the u_0 terms are zeros, i.e., for all n,

$$\alpha_n = a_n + \frac{b_n}{\alpha_{n-1}}, \qquad (7.26)$$

$$\alpha_{n-1} = -c - \frac{d}{\alpha_{n-2}}. \qquad (7.27)$$

A single sequence $\{\alpha_n\}$ that is never zero for all $n \geq 0$ must exist that satisfies both (7.26) and (7.27). If such a sequence exists then the SC factorization of (7.24) is obtained via Theorem 7.2. We first calculate the factor functions using (7.18) as

$$\phi_n(v_1, v_2) = -\frac{b_n}{\alpha_{n-1}}v_1 + g_n\left(-\frac{c}{\alpha_{n-1}}v_1 - \frac{d}{\alpha_{n-2}}v_2\right)$$
$$= (a_n - \alpha_n)v_1 + g_n\left(-\frac{c}{\alpha_{n-1}}v_1 + (c + \alpha_{n-1})v_2\right)$$

Therefore, the SC factorization is

$$t_{n+1} = (a_n - \alpha_n)t_n + g_n\left(-\frac{c}{\alpha_{n-1}}t_n + (c + \alpha_{n-1})t_{n-1}\right),$$

$$x_{n+1} = \alpha_n x_n + t_{n+1}.$$

▯

If $a_n = a$ and $b_n = b$ are constants for all n then the pair of equations (7.26) and (7.27) share a constant solution, i.e., a common fixed point, if the pair of quadratic equations

$$\lambda^2 - a\lambda - b = 0, \quad \lambda^2 + c\lambda + d = 0$$

have a common nonzero root. This is similar to what we observed in Section 5.6 on the joint eigenvalues of difference equations with linear arguments and is not surprising. If a_n, b_n are constants then Eq. (7.24) is indeed of the type studied in Section 5.6.

In general, Eq. (7.26) need not have constant solutions; however, *common solutions* to equations (7.26) and (7.27) may still exist, as the next example illustrates.

Example 7.2
Consider the following special case of Eq. (7.24)

$$x_{n+1} = (-1)^{n+1}x_n - 2x_{n-1} + g_n(x_n + x_{n-2}). \tag{7.28}$$

In this case, (7.26) and (7.27) take the following forms

$$\alpha_n = (-1)^{n+1} - \frac{2}{\alpha_{n-1}} \tag{7.29}$$

$$\alpha_{n-1} = -\frac{1}{\alpha_{n-2}}. \tag{7.30}$$

Every solution of Eq. (7.30) is a sequence of period 2

$$\left\{q, -\frac{1}{q}, q, -\frac{1}{q}, \ldots\right\} \tag{7.31}$$

where $q = \alpha_{-2} \in \mathbb{R}$. Now (7.30) yields $\alpha_{-1} = -1/q$, which we substitute as an initial value in Eq. (7.29) to get

$$\alpha_0 = -1 + 2q.$$

Now to obtain the period-two sequence in (7.31) as a solution of (7.29), we require the above value of α_0 to be equal to q; thus

$$\alpha_0 = q \Rightarrow 2q - 1 = q \Rightarrow q = 1.$$

We check that if $\alpha_0 = q = 1$ in (7.29) then

$$\alpha_1 = 1 - \frac{2}{q} = -1 = -\frac{1}{q}, \ \alpha_2 = -1 - \frac{2}{-1} = 1 = q, \ \text{etc}$$

so that both (7.29) and (7.30) generate the same sequence $\{\alpha_n\}$ where $\alpha_n = (-1)^n$ for $n \geq -2$. Now from Example 7.1 we obtain the following SC factorization of (7.28):

$$t_{n+1} = 2(-1)^{n+1}t_n + g_n\left((-1)^{n+1}t_{n-1}\right),$$
$$x_{n+1} = (-1)^n x_n + t_{n+1}.$$

▯

7.4 SC factorization of linear equations

It is worth noting at this stage that (7.26) and (7.27) above are examples of Riccati equations; see Example 4.10. In this section, we discover that Riccati equations occur naturally in the context of reductions of order for nonautonomous *linear* equations in association with the linear rather than the inversion form symmetry. The idea in Example 4.10 is also applied later in this section to obtain a criterion for the existence of the linear form symmetry that does not explicitly involve the Riccati equation; instead, the criterion is based on the existence of a nonzero solution of the homogeneous part, which is an HD1 equation. This interplay between the linear form symmetry in a field $\mathcal{F}$ and the inversion form symmetry in the multiplicative group of units $\mathcal{F}\backslash\{0\}$ is a feature that is unique to linear equations.

7.4.1 Factorization theorem for linear equations

The following application of Corollary 7.1 and Theorem 7.2 gives a type-$(k,1)$ semiconjugate factorization for nonautonomous and nonhomogeneous linear difference equations.

COROLLARY 7.2

(The general linear equation) Let $\{a_{i,n}\}$, $i = 1,\ldots,k$ and $\{b_n\}$ be given sequences in a nontrivial field $\mathcal{F}$ such that $a_{k,n} \neq 0$ for all n. The nonhomogeneous linear equation of order $k+1$

$$x_{n+1} = a_{0,n}x_n + a_{1,n}x_{n-1} + \cdots + a_{k,n}x_{n-k} + b_n \tag{7.32}$$

has the linear form symmetry with eigensequences $\{\alpha_n\}$ *in* $\mathcal{F}\backslash\{0\}$ *that satisfy the following Riccati difference equation of order* k

$$\alpha_n = a_{0,n} + \frac{a_{1,n}}{\alpha_{n-1}} + \frac{a_{2,n}}{\alpha_{n-1}\alpha_{n-2}} + \cdots + \frac{a_{k,n}}{\alpha_{n-1}\cdots\alpha_{n-k}}. \tag{7.33}$$

The corresponding SC factorization of (7.32) is

$$t_{n+1} = b_n - \sum_{i=1}^{k}\sum_{j=i}^{k} \frac{a_{j,n}}{\alpha_{n-i}\cdots\alpha_{n-j}} t_{n-i+1} \tag{7.34}$$

$$x_{n+1} = \alpha_n x_n + t_{n+1}. \tag{7.35}$$

PROOF   We determine a sequence $\{\alpha_n\}$ of nonzero elements of $\mathcal{F}$ such that for each n the quantity (7.23) is independent of u_0 for the following function

$$f_n(u_0,\ldots,u_k) = a_{1,n}u_0 + a_{2,n}u_1 + \cdots + a_{k,n}u_k + b_n.$$

For arbitrary $u_0, v_1, \ldots, v_k \in \mathcal{F}$ and $j = 0, 1, \ldots, k$ define $\zeta_{j,n}(u_0, v_1, \ldots, v_j)$ as in Corollary 7.1. Then (7.23) is

$$-\alpha_n u_0 + b_n + a_{1,n}u_0 + a_{2,n}\zeta_{1,n}(u_0, v_1) + \cdots + a_{k,n}\zeta_{k,n}(u_0, v_1, \ldots, v_k) =$$

$$b_n + \left[\sum_{j=1}^{k} \frac{a_{j,n}}{\prod_{i=1}^{j}\alpha_{n-i}} - \alpha_n\right] u_0 - \sum_{j=1}^{k} a_{j,n}\sum_{i=1}^{j} \frac{v_i}{\prod_{p=i}^{j}\alpha_{n-p}}.$$

The above quantity is independent of u_0 if and only if the coefficient of u_0 is zero for all n; i.e., if $\{\alpha_n\}$ is a solution of the difference equation

$$\alpha_n = \sum_{j=1}^{k} \frac{a_{j,n}}{\prod_{i=1}^{j}\alpha_{n-i}}$$

which is Eq. (7.33). Thus by Corollary 7.1 Eq. (7.32) has the linear form symmetry (7.21). Its SC factor equation is obtained using (7.18) in Theorem 7.2 and the above calculations

$$t_{n+1} = b_n - \sum_{j=1}^{k} a_{j,n}\sum_{i=1}^{j} \frac{t_{n-i+1}}{\prod_{p=i}^{j}\alpha_{n-p}}$$

$$= b_n - \sum_{i=1}^{k}\sum_{j=i}^{k} \frac{a_{j,n}}{\alpha_{n-i}\cdots\alpha_{n-j}} t_{n-i+1}.$$

Finally, the cofactor equation is derived immediately from the linear form symmetry and the proof is complete.   ∎

The Riccati equation (7.33) may also be viewed as a *constraint relation* that can be written more symmetrically as

$$\frac{a_{0,n}}{\alpha_n} + \frac{a_{1,n}}{\alpha_n \alpha_{n-1}} + \frac{a_{2,n}}{\alpha_n \alpha_{n-1} \alpha_{n-2}} + \cdots + \frac{a_{k,n}}{\alpha_n \alpha_{n-1} \cdots \alpha_{n-k}} = 1 \quad (7.36)$$

involving the variable coefficients of (7.32) and the sequence $\{\alpha_n\}$ that satisfies it.

We emphasize that in Corollary 7.2 the only requirement is to find *any* one solution of (7.33), or equivalently, of (7.36) that is well defined; i.e., it is not necessary to gather substantial information about the solutions of Riccati equation (or relation). The next example illustrates Corollary 7.2.

Example 7.3

Consider the second-order difference equation

$$x_{n+1} = (-1)^{n+1} x_n + x_{n-1} + b_n \quad (7.37)$$

where b_n, x_0, x_{-1} are in a nontrivial field $\mathcal{F}$. The associated Riccati equation of (7.37) is

$$r_n = (-1)^{n+1} + \frac{1}{r_{n-1}}. \quad (7.38)$$

First, let $\mathcal{F}$ be a field of characteristic zero such as $\mathbb{Q}$, $\mathbb{R}$ or $\mathbb{C}$. In such fields, straightforward calculations show that if $r_0 \neq 0, -1$ then

$$r_1 = \frac{r_0 + 1}{r_0}, \quad r_2 = -\frac{1}{r_0 + 1}, \quad r_3 = -r_0,$$

$$r_4 = -\frac{r_0 + 1}{r_0}, \quad r_5 = \frac{1}{r_0 + 1}, \quad r_6 = r_0.$$

It follows that all solutions of the Riccati equation (7.38) that start outside the singularity set $\{0, -1\}$ are eigensequences with period 6 (which is minimal in $\mathbb{Q}$ or $\mathbb{R}$ but need not be so in $\mathbb{C}$). The SC factorization of the linear equation (7.37) is obtained by Corollary 7.2 as

$$t_{n+1} = -\frac{1}{r_{n-1}} t_n + b_n,$$

$$x_{n+1} = r_n x_n + t_{n+1}.$$

For the sake of comparison, consider next a finite field, $\mathcal{F} = \mathbb{Z}_3 = \{0, 1, 2\}$ with addition and multiplication modulo 3, in which the group of units $\{1, 2\}$ is nontrivial. Then r_0 has only two possible nonzero values 1,2. It is readily seen that $r_0 = 1$ gives a solution $\{1, 2, 1, 2, \ldots\}$ of period two for (7.38) and since $2 = -1$ in $\mathbb{Z}_3$, we may indicate this solution as $r_n = (-1)^{n+1}$. Now Corollary 7.2 yields the following SC factorization of the linear equation (7.37) on $\mathbb{Z}_3$

$$t_{n+1} = (-1)^{n+1} t_n + b_n,$$

$$x_{n+1} = (-1)^{n+1} x_n + t_{n+1}.$$

We also note that $r_0 = 2$ gives $r_1 = 1+2 = 0$ in $\mathbb{Z}_3$ so 2 is in the singularity set of (7.38). In fact, this singularity set is $\{0,2\} = \{0,-1\}$. □

Recall from Example 4.10 that the Riccati equation (7.33) is itself the SC factor equation of the *homogeneous part* of (7.32) relative to the *inversion* form symmetry. We discussed the latter form symmetry in Section (5.3) in relation to HD1 equations; even with variable coefficients a homogeneous linear difference equation is HD1 relative to the multiplicative group of units in a field. This observation leads to the following version of Corollary 7.2 that does not involve the Riccati equation explicitly.

COROLLARY 7.3

Let $\{a_{i,n}\}$, $i = 1,\ldots,k$ and $\{b_n\}$ be given sequences in a nontrivial field $\mathcal{F}$ such that $a_{k,n} \neq 0$ for all n. Assume that the homogeneous part of (7.32) has a solution $\{y_n\}$ in $\mathcal{F}$ such that $y_n \neq 0$ for all n. Then $\{y_n/y_{n-1}\}$ is an eigensequence of (7.32) whose SC factorization is given by the pair of equations (7.34) and (7.35).

The next result gives a special case of Corollary 7.3 that is more definitive.

COROLLARY 7.4

In Eq. (7.32) let $\{a_{i,n}\}$, $i = 1,\ldots,k$ and $\{b_n\}$ be sequences of real numbers with $a_{i,n} \geq 0$ for all i,n and $a_{k,n} > 0$ for all n. Then (7.32) has an eigensequence $\{y_{n+1}/y_n\}$ and a SC factorization in $\mathbb{R}$ given by the pair of equations (7.34) and (7.35).

PROOF If we choose $y_{-j} = 1$ for $j = 0,\ldots,k$ then the corresponding solution $\{y_n\}$ of the homogeneous part of (7.32) is a sequence of positive real numbers. Now an application of Corollary 7.3 completes the proof. ∎

7.4.2 Constant coefficients without field extensions

The occurrence of a Riccati equation in Corollary 7.2 raises an interesting question. How does Eq. (7.33) fit into the framework defined previously in Section 5.6.4 for equations with *constant* coefficients? Since the factor equation (7.34) is again the same type of equation as (7.32), Corollary 7.2 can be applied repeatedly as long as each of the associated Riccati equations possesses at least one solution. In the case of constant coefficients where $a_{j,n} \equiv a_j$ for all n, that solution may be taken to be the constant solution represented by a fixed point of the autonomous Riccati equation, (7.33) in the field $\mathcal{F}$. Our earlier work on difference equations with linear arguments led to exactly such constant solutions since *the fixed points of Eq. (7.33) with constant coefficients are none other than the eigenvalues of the homogeneous* part of (7.32), also with constant coefficients. If $\mathcal{F}$ is an algebraically closed field such as $\mathbb{C}$ then the eigenvalues (and thus the fixed points) can always be found when the coefficients are constant. If $\mathcal{F}$ is not algebraically closed (like $\mathbb{R}$ or $\mathbb{Q}$) then the eigenvalues may exist only in an extension field of $\mathcal{F}$. In this case, the SC factorization discussed in Section 5.6.4 cannot be defined in $\mathcal{F}$.

We notice that in Corollary 7.2 with variable coefficients, polynomial roots are not relevant. Hence, the issue of field extensions does not arise and we deal instead with the associated Riccati equation and its solutions (eigensequences) in the given underlying field $\mathcal{F}$. This idea remains valid even if the Riccati equation (and thus the original linear equation) has constant coefficients. Therefore, *in the case of constant coefficients, if the Riccati equation is used instead of the characteristic polynomial then extensions of $\mathcal{F}$ are not required for defining the factor or cofactor equations.* It is only necessary to demonstrate the existence of at least one valid solution of the Riccati equation; or equivalently, show that the singularity set of the Riccati equation is not all of $\mathcal{F}$.

The following result on nonhomogeneous linear equations with constant coefficients is an immediate consequence of Corollary 7.2. It improves Corollary 5.11 in Section 5.6.4 by allowing eigensequences when eigenvalues do not exist in the given field $\mathcal{F}$.

COROLLARY 7.5

Let $\{b_n\}$ be a given sequence in a nontrivial field $\mathcal{F}$ and let $\{a_i\}$, $i = 1, \ldots, k$ be constants in $\mathcal{F}$ such that $a_k \neq 0$.
(a) The nonhomogeneous linear equation of order $k + 1$ with constant coefficients

$$x_{n+1} = a_0 x_n + a_1 x_{n-1} + \cdots + a_k x_{n-k} + b_n \qquad (7.39)$$

has a linear form symmetry with eigensequence $\{\alpha_n\}$ for every solution $\{\alpha_n\}$ in $\mathcal{F}$ of the following autonomous Riccati equation of order k

$$\alpha_n = a_0 + \frac{a_1}{\alpha_{n-1}} + \frac{a_2}{\alpha_{n-1}\alpha_{n-2}} + \cdots + \frac{a_k}{\alpha_{n-1} \cdots \alpha_{n-k}}. \qquad (7.40)$$

(b) Every fixed point of (7.40) in $\mathcal{F}$ is a nonzero root of the characteristic polynomial of (7.39), i.e., a solution of

$$\lambda^{k+1} - a_0 \lambda^k - a_1 \lambda^{k-1} - \cdots - a_{k-1}\lambda - a_k \qquad (7.41)$$

or an eigenvalue of the homogeneous part of (7.39) in $\mathcal{F}$. As constant solutions of (7.40) such eigenvalues are constant eigensequences of (7.39).

Example 7.4

Consider the nonhomogeneous linear equation

$$x_{n+1} = x_n + x_{n-1} + c_n \qquad (7.42)$$

where the constants c_n are rationals, i.e., $\mathcal{F} = \mathbb{Q}$. Then the associated Riccati equation

$$r_n = 1 + \frac{1}{r_{n-1}} \tag{7.43}$$

has two irrational fixed points $\pm\gamma$ where γ is the golden ratio

$$\gamma = \frac{1 + \sqrt{5}}{2}.$$

These constant solutions $\gamma^{\pm}$ yield the SC factorization

$$t_{n+1} = -\frac{1}{\gamma^+} t_n + c_n = \gamma^- t_n + c_n,$$

$$x_{n+1} = \gamma^+ x_n + t_{n+1}$$

over $\mathbb{R}$. This factorization is not valid over $\mathbb{Q}$; however, (7.43) is a rational function and any rational initial value $r_0 > 0$ yields a valid solution of (7.43). In particular, the Fibonacci sequence $\{\varphi_n\}$

$$\{1, 1, 2, 3, 5, 8, \ldots\}$$

is a solution of the homogeneous part of (7.42) and ratios $\alpha_n = \varphi_{n+1}/\varphi_n$ of its consecutive terms give the solution of (7.43) with $r_0 = 1$. By Corollary 7.3 such a solution of (7.43) yields the following SC factorization of Eq. (7.45) that is valid in $\mathbb{Q}$:

$$t_{n+1} = c_n - \frac{\varphi_{n-1}}{\varphi_n} t_n$$

$$x_{n+1} = \frac{\varphi_{n+1}}{\varphi_n} x_n + t_{n+1}.$$

Note that $\lim_{n\to\infty} \varphi_{n+1}/\varphi_n = \gamma^+$ in the above factorization; in this way the factorization over $\mathbb{Q}$ is related in a simple way to the factorization over $\mathbb{R}$.
□

The next example illustrates factorization over the real numbers when the eigenvalues are complex.

Example 7.5

For a given sequence d_n of real numbers the homogeneous part of the equation

$$x_{n+1} = x_n - x_{n-1} + d_n \tag{7.44}$$

has two complex eigenvalues

$$\alpha_{\pm} = \frac{1 \pm i\sqrt{3}}{2}$$

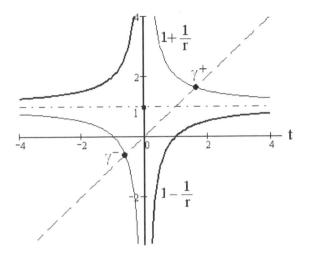

FIGURE 7.1
Graphs of Riccati equations, one having two fixed points $\gamma^{\pm}$, the other having none.

that are roots of $\lambda^2 - \lambda + 1$. Thus, (7.44) has no constant eigensequences in $\mathbb{R}$ but it does have nonconstant real eigensequences since the Riccati equation

$$\alpha_n = 1 - \frac{1}{\alpha_{n-1}}$$

with the initial value $\alpha_0 = 2$ has a solution

$$\left\{ 2, \frac{1}{2}, -1, 2, \frac{1}{2}, -1, \ldots \right\}$$

of period three in $\mathbb{R}$. The corresponding real SC factorization is

$$t_{n+1} = -\frac{1}{\alpha_{n-1}} t_n, \quad x_{n+1} = \alpha_n x_n + t_{n+1}.$$

In contrast to the factorization of Eq. (7.42) there is no obvious relationship between the factorization of (7.44) over the real numbers and its factorization with constant eigensequences over the complex numbers. ▯

Figure 7.1 illustrates preceding two examples. In one case there are two real fixed point, namely, $\gamma^{\pm}$ while in the other case there are no real fixed points.

Example 7.6
Consider the non-homogeneous linear equation with constant coefficients

$$x_{n+1} = ax_n + bx_{n-1} + \gamma_n, \quad b \neq 0 \tag{7.45}$$

where the fixed coefficients a, b and the sequence of constants γ_n are real, i.e., $\mathcal{F} = \mathbb{R}$. Let S be the singularity set of the Riccati equation

$$r_n = a + \frac{b}{r_{n-1}}. \tag{7.46}$$

It can be shown that S is a proper subset of $\mathbb{R}$ (see Grove, et al (2001) or Kulenovic and Ladas (2002), Section 1.6). Thus, we may choose real $r_0 \notin S$. Then the SC factorization

$$t_{n+1} = \gamma_n - \frac{b}{r_{n-1}} t_n = \gamma_n - (r_n - a) t_n$$
$$x_{n+1} = r_n x_n + t_{n+1}$$

of Eq. (7.45) is defined on $\mathbb{R}$ for all a, b even if the eigenvalues of the homogeneous part of (7.45) are complex (i.e., if $a^2 + 4b < 0$). $\quad\square$

In the next example we consider a finite field where a complete analysis of all parameter values in (7.45) is possible.

Example 7.7

Consider Eq. (7.45) again but now on the field $\mathbb{Z}_3 = \{0, 1, 2\}$ with addition and multiplication defined modulo 3. There are a limited number of cases to consider here. Since $b \neq 0$, there are six possible cases for the two parameters: $a = 0, 1, 2$ and $b = 1, 2$. Similarly, for the associated Riccati equation (7.46) a nonzero initial value r_0 can have only the two possible values $1, 2$. Examining these cases one at a time (details left to the reader as an exercise) we arrive at the following conclusions:

If $b = 2$ and $a \neq 0$ then (7.46) has one constant solution in $\mathbb{Z}_3$ in each case ($\rho = 2$ and $\rho = 1$, respectively). Hence, SC factorizations exist by Corollary 7.2 based on these solutions.

If $b = 2$ and $a = 0$ then (7.46) has no constant solutions in $\mathbb{Z}_3$. However, each of its solutions are periodic with period 2 so a SC factorization over $\mathbb{Z}_3$ exists by Corollary 7.2 based on this solution.

If $b = 1$ and $a = 0$ then all solutions of (7.46) in $\mathbb{Z}_3$ are constant sequences.

If $b = 1$ and $a \neq 0$ then Eq. (7.46) has no valid solutions (constant or not) in $\mathbb{Z}_3$. Therefore, Corollary 7.2 does not apply to Eq. (7.45) on $\mathbb{Z}_3$ in these cases. $\quad\square$

The existence of SC factorizations of linear equations over arbitrary finite fields or over fields of positive characteristic appears to be a challenging problem that may involve significant technical issues in algebra and number theory.

7.4.3 Implications of the existence of constant solutions of the Riccati equation

When they exist in a given field $\mathcal{F}$, constant solutions have the practical advantage of simplifying calculations and derivations of results considerably. The appearance of a Riccati equation in Corollary 7.2 thus raises a question: *What restrictions are imposed on the variable coefficients in Eq. (7.32) by the existence of constant solutions of (7.33), if any?* The next result states essentially that the more constant solutions there are, the more related, and restricted, are the coefficients of the Riccati equation.

PROPOSITION 7.1

Assume that the Riccati equation (7.33) has m distinct constant solutions $\rho_1, \ldots, \rho_m$ in a nontrivial field $\mathcal{F}$, $1 \leq m \leq k$. Then the variable coefficients of the linear non-homogeneous equation (7.32) satisfy

$$a_{j,n} = c_{j,1}a_{m,n} + \cdots + c_{j,k-m+1}a_{k,n} + c_{j,k-m}, \quad 0 \leq j \leq m-1 \qquad (7.47)$$

where the quantities $c_{j,1}, \ldots, c_{j,k-m}$ are rational functions of $\rho_1, \ldots, \rho_m$. Therefore, m of the variable coefficients in (7.32) are determined by the remaining $k+1-m$ coefficients. In particular, if Eq. (7.33) has k distinct constant solutions in $\mathcal{F}$ then the coefficients of Eq. (7.32) have only one arbitrary variable sequence; i.e., the variable coefficients are all of type

$$a_{j,n} = a_j + b_j\sigma_n, \quad a_j, b_j, \sigma_n \in \mathcal{F}, \ j = 0, 1, \ldots, k, \ \sigma_n, n \geq 0 \qquad (7.48)$$

that was encountered in Section 5.6.4 on difference equations with linear arguments.

PROOF Substituting each of the constant solutions $\rho_1, \ldots, \rho_m$ in (7.33) gives one of the equations in the following system of m equations

$$a_{0,n} + \frac{1}{\rho_j}a_{1,n} + \frac{1}{\rho_j^2}a_{2,n} + \cdots + \frac{1}{\rho_j^k}a_{k,n} = \rho_j, \quad 1 \leq j \leq m.$$

Each equation above is linear in $a_{0,n}, \ldots, a_{k,n}$ so the first m of these, i.e., $a_{0,n}, \ldots, a_{m-1,n}$ can be eliminated as functions of the remaining parameters and rational functions of $\rho_1, \ldots, \rho_m$ as specified in (7.47).

Next, if (7.33) has k distinct constant solutions $\rho_1, \ldots, \rho_k$ then setting $m = k$, each of the coefficients $a_{j,n}$ can be expressed in terms of $a_{k,n}$ alone as

$$a_{j,n} = c_{j,1}a_{k,n} + c_{j,0}, \quad 0 \leq j \leq k-1.$$

Defining $\sigma_n = a_{k,n}$ for all n and $a_j = c_{j,0}$, $b_j = c_{j,1}$ for $0 \leq j \leq k-1$ and $a_k = 0$, $b_k = 1$ we obtain (7.48). ∎

7.5 Notes

The material in this section appears in part in Sedaghat (2010a, 2010b).
The results presented here are the most general in this book for recursive
difference equations. In particular, these results extend those in Chapter
5 and they include semiconjugate factorizations of general (nonautonomous
and nonhomogeneous) linear equations over arbitrary algebraic fields. Clearly,
other types of SC factorizations, e.g., type-$(1, k)$ are also possible with time-
dependent form symmetries which are left to future investigations in this
interesting and challenging area.

7.6 Problems

7.1 (a) Using the ideas in Examples 7.1 and 7.2 show that the difference
equation

$$x_{n+1} = a_n x_n + b_n x_{n-1} + g_n(x_n - c x_{n-2}),$$
$$a_n, b_n \in \mathbb{R} \text{ for all } n, \ c \in \mathbb{R}, \ c \neq 0$$

has the linear form symmetry if $b_0 \neq c$, there is $q \neq 0$ such that

$$a_0 = q\left(1 - \frac{b_0}{c}\right)$$

and for all $n \geq 1$

$$a_n = \frac{c - b_n}{s_n}, \quad s_n = \begin{cases} q, & \text{if } n \text{ is even} \\ c/q, & \text{if } n \text{ is odd} \end{cases}.$$

(b) Use the results in (a) with $q = 1$ to find a SC factorization for the
following difference equation

$$x_{n+1} = (-1)^n(1 + b_n)x_n + b_n x_{n-1} + g_n(x_n + x_{n-2})$$

where b_n is a sequence of real numbers such that $b_0 \neq -1$.

(c) Use the results in (b) to find a SC factorization for the difference equation

$$x_{n+1} = \frac{(-1)^n}{n+1}x_n - \frac{n}{n+1}x_{n-1} + \frac{\beta_n}{x_n + x_{n-2} + \gamma_n}$$

where $\beta_n, \gamma_n \in \mathbb{R}$ for all n.

7.2 Provide the details in Example 7.7.

Time-Dependent Form Symmetries

7.3 Let $\{b_n\}$ be any sequence of real numbers. Referring to Example 7.7, show that both of following difference equations have SC factorizations over $\mathbb{R}$ and determine these factorizations using the constant solutions of the associated Riccati equations.

$$x_{n+1} = x_n + x_{n-1} + b_n$$
$$x_{n+1} = 2x_n + x_{n-1} + b_n.$$

7.4 (a) Verify that if $\{x_n\}$ is the solution of the linear equation

$$x_{n+1} = x_n + x_{n-1}$$

in the finite field $\mathbb{Z}_{11}$ with initial values $x_{-1} = 1$ and $x_0 = 4$ then $x_n \neq 0$ for all $n \geq -1$.

(b) Let $\{b_n\}$ be an arbitrary sequence in $\mathbb{Z}_{11}$. Conclude from (a) and Corollary 7.3 that the linear nonhomogeneous equation

$$x_{n+1} = x_n + x_{n-1} + b_n$$

has a SC factorization over $\mathbb{Z}_{11}$. Determine this factorization.

We note that $\mathbb{Z}_{11}$ is not algebraically closed and does not contain the irrational eigenvalues of the Fibonacci equation. Hence the theory in Section 5.6.4 does not apply to this problem.

7.5 If p is a prime number and $p \geq 5$ then prove that the difference equation

$$x_{n+1} = 3x_n + (p-2)x_{n-1} + b_n$$

has a SC factorization in the field $\mathbb{Z}_p$ for any sequence $\{b_n\}$ in $\mathbb{Z}_p$.

Hint. Note that $p - 2 = -2$ in $\mathbb{Z}_p$ and look for a constant solution of the homogeneous part.

7.6 (a) In Proposition 7.1 let $k = 2$. For $m = 1, 2$ determine the coefficients $c_{j,i}$ explicitly.

(b) Repeat Part (a) for $k = 3$ and $m = 1, 2, 3$.

8

Nonrecursive Difference Equations

Let S be a nonempty set and $\{f_n\}, \{g_n\}$ sequences of functions $f_n, g_n : S^{k+1} \to S$ where k is a positive integer. We call the equation

$$f_n(x_n, x_{n-1}, \ldots, x_{n-k}) = g_n(x_n, x_{n-1}, \ldots, x_{n-k}) \tag{8.1}$$

a *"scalar" difference equation of order k* in the set S. Unlike the difference equations studied in previous chapters, Eq. (8.1) is not recursive and generally it does not unfold to a (vector) self-map of S^{k+1}. Thus semiconjugacy does not apply to equations of type (8.1). On the other hand, (8.1) clearly generalizes the types of equations studied in previous chapters since the recursive equation of order k,

$$x_n = \psi_n(x_{n-1}, \ldots, x_{n-k}) \tag{8.2}$$

is a special case of (8.1) where

$$f_n(u_0, u_1, \ldots, u_k) = u_0$$
$$g_n(u_0, u_1, \ldots, u_k) = \psi_n(u_1, \ldots, u_k)$$

for all n and all $u_0, u_1, \ldots, u_k \in S$.

Consideration of nonrecursive (or implicit) difference equations such as (8.1) may occur naturally. For example, the following equation on the set of real numbers $\mathbb{R}$:

$$|x_n| = a|x_{n-1} - x_{n-2}|, \quad 0 < a < 1 \tag{8.3}$$

states that the magnitude of a quantity x_n at time n is a fraction of the difference between its values in the two immediately preceding times; however, (8.3) leaves the sign of x_n undetermined. For instance, if we set $x_0 = -1$ and $x_1 = 1$ then we obtain $|x_2| = 2a$ but the actual value of x_2 can be either $2a$ or $-2a$. As a possible physical interpretation of (8.3) we imagine a node in a circuit that in every second n fires a pulse x_n that may go either to the right (if $x_n > 0$) or to the left (if $x_n < 0$) but the amplitude $|x_n|$ of the pulse obeys Eq. (8.3). What sorts of signal patterns satisfy Eq. (8.3)? It happens that many different types of patterns are possible, most of which do not arise if the absolute value is absent from the left-hand side of (8.3); see the next section of this chapter for additional details pertaining to (8.3) and similar equations.

The question arises as to whether the notions of form symmetry and reduction of order can be still defined in the nonrecursive context, which is more general. In this chapter, we show that for Eq. (8.1) the basic concepts of order reduction such as form symmetry and factorization into factor and cofactor pairs of equations can still be defined as before, even without the standard semiconjugate relation. A concept that is similar to the semiconjugate relation but does not require the unfolding map is sufficient for defining form symmetries and deriving the lower order factor and cofactor equations.

Eq. (8.1) represents a generalization of the recursive equation (8.2) in a different direction than the customary one; namely, to vector maps (recall that the unfolding of (8.2) is a special vector map of the k-dimensional state-space). In another departure from the recursive format of (8.2), uniqueness of solutions is lost as there may be an infinite number of solutions of (8.1) that pass through any prescribed point in the state-space (as usual, a *solution* of Eq. (8.1) in S is defined as any sequence $\{x_n\}$ in S that satisfies the equality (8.1) for every n). Therefore, we no longer think of solutions as orbits generated through map iteration starting from some initial value. However, preserving the pattern of previous chapters in keeping with forward solutions (n a positive integer) it may be helpful at times to interpret or plot solutions of (8.1) as orbits $\{(x_n, x_{n-1}, \dots, x_{n-k})\}$ in the state-space S^{k+1} that contain, or pass through a given "initial point" $(x_0, x_{-1}, \dots, x_{-k+1})$ corresponding to $n = 0$.

Analyzing the solutions of a nonrecursive difference equation such as (8.1) is generally more difficult than analyzing the solutions of recursive equations. In addition to the loss of uniqueness, even the existence of solutions for (8.1) in a particular set is not guaranteed. However, studying the form symmetries and reduction of order in nonrecursive equations is worth the effort. The greater generality of these equations not only leads to the resolution of a wider class of problems, but it also provides for increased flexibility in handling *recursive* equations. As illustrated in the next section (among other things) some recursive equations may have *nonrecursive* form symmetries that the more specialized theory based on semiconjugacy cannot determine.

8.1 Examples and discussion

We begin our discussion with a few examples to illustrate the existence of nonrecursive form symmetries and how they can reduce a given equation to a pair of lower order equations resembling a factor/cofactor pair.

Example 8.1

A recursive equation with nonrecursive form symmetry. Consider the recursive difference equation

$$y_n = \sqrt{ay_{n-1}^2 + by_{n-1} - aby_{n-2} + c} \qquad (8.4)$$

where $a, b, c \in \mathbb{R}$. The occurrence of a square root limits the occurrence of real solutions for which the quantity under the root must be nonnegative for all n. What conditions on parameters a, b, c or on initial values y_0, y_{-1} imply that (8.4) has real solutions? We now show that an answer to this question can be obtained by first answering the question: Does (8.4) have a factorization into lower order difference equations?

An SC factorization would be nice to have but it is not obviously indicated. We try another approach. By squaring both sides of (8.4) and re-labeling the variable as x_n we obtain

$$x_n^2 - ax_{n-1}^2 - bx_{n-1} + abx_{n-2} = c. \tag{8.5}$$

Any real solution $\{y_n\}$ of (8.4) is clearly also a (nonnegative) solution of the nonrecursive equation (8.5). Of course, (8.5) is a less restricted equation as it may have solutions in $\mathbb{R}$ that cannot be generated by (8.4); e.g., solutions in which $x_n < 0$ for infinitely many values of n. In fact, if $\{\beta_n\}$ is a fixed but arbitrarily selected binary sequence taking values in $\{-1, 1\}$ then every real solution of the equation

$$s_n = \beta_n \sqrt{as_{n-1}^2 + bs_{n-1} - abs_{n-2} + c} \tag{8.6}$$

is also a solution of (8.5) since after squaring, we see that s_n satisfies Eq. (8.5).

Splitting into a pair of lower order equations. Now, rearranging terms in (8.5) yields

$$x_n^2 - bx_{n-1} = a(x_{n-1}^2 - bx_{n-2}) + c$$

which reveals an order-reducing substitution $x_n^2 - bx_{n-1} = t_n$ that gives the system of first-order equations

$$t_n = at_{n-1} + c \tag{8.7}$$
$$x_n^2 - bx_{n-1} = t_n. \tag{8.8}$$

This is entirely similar to a SC factorization with a factor (8.7) that in this case is nonhomogeneous linear. The quadratic equation (8.8) is the analog of the cofactor equation, although it is a nonrecursive first-order equation. Each solution of the nonrecursive equation (8.5) may be found by finding a solution of the first order equation (8.7) and then using this solution for t_n in (8.8). In the next section we make these concepts precise.

It is an important fact that neither of the two recursive forms $uh(v)$ or $u + h(v)$ can express the quadratic form $u^2 - bv$. Therefore, $x_n^2 - bx_{n-1}$ is not a form symmetry of the type that was discussed in previous chapters. Nevertheless, solving (8.8) for x_n and choosing the nonnegative square root gives a "factorization"

$$t_n = at_{n-1} + c,$$
$$x_n = \sqrt{bx_{n-1} + t_n}$$

for Eq. (8.4) over the real numbers that is not a SC factorization of the type discussed in previous chapters.

A discussion of solutions. We now use the above factorization to determine conditions on parameters a, b, c and initial values y_0, y_{-1} that imply (8.4) has real solutions. For illustration, we consider an interesting special case here with parameter values

$$a = 1, c = 0, \ b > 0.$$

Then (8.7) is the trivial equation $t_n = t_{n-1}$ and each solution of the recursive equation

$$y_n = \sqrt{y_{n-1}^2 + b(y_{n-1} - y_{n-2})} \tag{8.9}$$

is also a solution of the first-order, recursive equation

$$x_n = \sqrt{bx_{n-1} + t_0}, \quad t_0 = x_0^2 - bx_{-1} \tag{8.10}$$

where $x_0 = y_0$ and $x_{-1} = y_{-1}$. This equation is easier to understand than the second-order (8.9). For each point (x_0, x_{-1}) and the corresponding t_0, the solution of (8.10) is an orbit of the one-dimensional map

$$g(u, t_0) = \sqrt{bu + t_0}$$

which is a member of a family of maps (a different map for each distinct value of t_0). A routine examination of $g(u, t_0)$ leads to the following conclusions (details left to the reader as an exercise):

1. If $t_0 > 0$ (i.e., $x_0^2 > bx_{-1}$) then the orbit with initial value x_0 converges to the unique, attracting fixed point $\bar{x}(t_0)$, i.e.,

$$\lim_{n \to \infty} x_n = \bar{x}(t_0)$$

where

$$\bar{x}(t_0) = \frac{1}{2}\left(b + \sqrt{b^2 + 4t_0}\right).$$

Further, to ensure that a real solution is obtained, it is necessary that

$$bx_0 + t_1 = bx_0 + t_0 \geq 0$$

i.e.,

$$x_{-1} \leq x_0 + \frac{1}{b}x_0^2, \quad x_0 \in (-\infty, \infty). \tag{8.11}$$

See Figure 8.1.

2. If $t_0 \leq 0$ then fixed points exist if $b^2 + 4t_0 \geq 0$, i.e., if

$$x_{-1} \leq \frac{b}{4} + \frac{1}{b}x_0^2, \quad x_0 \in (-\infty, \infty). \tag{8.12}$$

In this case, there are two (nonnegative) fixed points:

$$x^{\pm}(t_0) = \frac{1}{2}\left(b \pm \sqrt{b^2 + 4t_0}\right)$$

Of these, $x^-(t_0)$ is repelling and $x^+(t_0)$ is attracting. Thus, each orbit of $g(u, t_0)$ with initial value

$$x_0 > x^-(t_0) \tag{8.13}$$

converges to $x^+(t_0)$ if, to satisfy (8.11) and (8.12),

$$x_{-1} \leq \min\left\{x_0, \frac{b}{4}\right\} + \frac{1}{b}x_0^2, \quad x_0 \in (-\infty, \infty). \tag{8.14}$$

If (8.14) holds then inequality (8.13) holds trivially if $x_0 > b/2$. Otherwise, it is readily verified that (8.13) holds if

$$x_{-1} < x_0 \leq \frac{b}{2}.$$

3. If $x_0 = x_{-1}$ then $t_0 = x_0^2 - bx_0$ and each (positive) solution of (8.10) is constant. Thus on the $45°$ line $v = u$ in the state-space $\mathbb{R}^2$ every real solution of (8.9) is constant.

4. From Items 1-3 above, we conclude that the real solution of Eq. (8.9) converges to $\bar{x}(t_0)$ (or $x^+(t_0)$) if the point (x_0, x_{-1}) is in the following region of the state-space

$$\left\{(x_0, x_{-1}) : x_{-1} \leq \min\left\{x_0, \frac{b}{4}\right\} + \frac{1}{b}x_0^2\right\} \cap \left[0, \frac{b}{2}\right] \times (-\infty, x_0). \tag{8.15}$$

See Figure 8.2. Points outside of this region either stay fixed at $x^-(t_0)$ or do not result in real solutions. In particular, it follows that every real solution of (8.9) that is not constant converges to $\bar{x}(t_0)$ (or $x^+(t_0)$).

□

Figure 8.1 illustrates Item 1 in the preceding example with

$$b = 2; \quad x_{-1} = 1, \ x_0 = 1.8;$$
$$t_0 = 1.24, \ \bar{x}(t_0) = x^+(t_0) = 2.497.$$

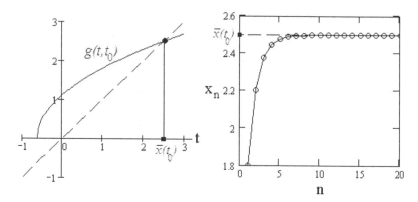

FIGURE 8.1
    Illustrating a case in Example 8.1 (see the text for details).

The graph of $g(u, t_0)$ is shown in the left panel of Figure 8.1 from which it is clear that $\bar{x}(t_0) = x^+(t_0)$ is an attracting fixed point for all (x_0, x_{-1}) in the region (8.15) that are on the parabola

$$x_{-1} = \frac{1}{b}x_0^2 - \frac{t_0}{b}.$$

The corresponding solution $\{x_n\}$ that converges to $\bar{x}(t_0)$ appears in the right panel. Region (8.15) is shown in Figure 8.2 as the shaded part with the point $(x_0, x_{-1}) = (1.8, 1)$ identified as a black dot inside it. The unshaded or clear region thus represents the singularity set of Eq. (8.9).

The asymptotic behaviors of solutions discussed in the preceding example is not as easily inferred from a *direct* investigation of (8.9). The bifurcation in state-space that occurs when t_0 changes sign to negative is not readily apparent from the second-order equation (8.9). Further, the reduction of order to the recursive Eq. (8.10) is derived *indirectly* via the nonrecursive equation (8.5) and its factorization rather than a semiconjugate relation.

In the next example we discuss the possibility of repeated reductions of order and some other issues.

Example 8.2
A third-order equation with absolute values. Let $\{a_n\}$ be a given sequence of nonzero real numbers and consider the following difference equation on $\mathbb{R}$:

$$|x_n + x_{n-1}| = a_n|x_{n-1} - x_{n-3}|. \tag{8.16}$$

Splitting into a pair of lower order equations. By adding and subtracting x_{n-2} inside the absolute value on the right-hand side of (8.16) we find that

$$|x_n + x_{n-1}| = a_n|x_{n-1} + x_{n-2} - (x_{n-2} + x_{n-3})|. \tag{8.17}$$

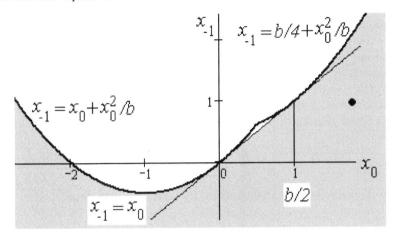

FIGURE 8.2
  The shaded region shows the part of the state-space where real solutions
  exist; therefore, the clear region is the singularity set of Eq. (8.9).

The substitution (analogous to a linear form symmetry)

$$t_n = x_n + x_{n-1} \tag{8.18}$$

in (8.17) results in the second-order absolute value difference equation

$$|t_n| = a_n|t_{n-1} - t_{n-2}| \tag{8.19}$$

that is related to (8.16) via (8.18). Eq. (8.19) is analogous to a factor equation
for (8.16) while (8.18), written as

$$x_n = t_n - x_{n-1} \tag{8.20}$$

is analogous to a (recursive) cofactor equation.
  The factor equation (8.19) is of course not recursive; it is a generalization
of (8.3) in the chapter introduction above. Further, if $\{\beta_n\}$ is any fixed but
arbitrarily chosen binary sequence taking values in $\{-1, 1\}$ then every real
solution $\{s_n\}$ of the recursive equation

$$s_n = \beta_n a_n|s_{n-1} - s_{n-2}| \tag{8.21}$$

is also a solution of (8.19) since upon taking the absolute value we see that
$\{s_n\}$ satisfies Eq. (8.19). This observation shows that the single nonrecursive
equation (8.19) has as many solutions as can be generated by the entire,
uncountably infinite class of equations (8.21). Indeed, a large number of
solutions!
  We also point out that from each initial point (x_0, x_{-1}) the recursive equation (8.21) has a unique real solution $\{s_n\}$. It follows that for each one of

the uncountably many sequences β_n there is a distinct orbit $\{(x_n, x_{n-1})\}$ of (8.19) that passes through a given point (x_0, x_{-1}). Thus there is a striking lack of uniqueness for solutions passing through any prescribed point in the state-space.

A further splitting into lower-order equations. Eq. (8.19), which has order two, also admits a reduction of order as follows. If $\{t_n\}$ is a solution of (8.19) that is never zero for all n then we may divide both sides of (8.19) by $|t_{n-1}|$ to get

$$\left| \frac{t_n}{t_{n-1}} \right| = a_n \left| 1 - \frac{t_{n-2}}{t_{n-1}} \right|$$

where the substitution (analogous to the inversion form symmetry)

$$r_n = \frac{t_n}{t_{n-1}}$$

yields the first-order difference equation

$$|r_n| = a_n \left| 1 - \frac{1}{r_{n-1}} \right| \tag{8.22}$$

that is related to (8.19) via the (recursive) equation

$$t_n = r_n t_{n-1}. \tag{8.23}$$

A discussion of solutions. For illustration, we discuss the solutions of Eq. (8.16) when

$$a_n = a \text{ is constant and } 0 < a < 1 \tag{8.24}$$

i.e., equation (8.3) in this chapter's Introduction. We choose Eq. (8.19) for this discussion rather than (8.22) even though the latter equation has order one, because (8.22) has a nonempty singularity set among all real numbers and excludes solutions of (8.19) that can equal to zero for some values of n.

First, we consider the nonnegative solutions of (8.19) under conditions (8.24). If $t_n \geq 0$ for all n then such a sequence $\{t_n\}$ satisfies the recursive equation

$$y_n = a|y_{n-1} - y_{n-2}|. \tag{8.25}$$

Claim: Under conditions (8.24) every solution of (8.25) converges to zero. To prove this claim, it is no loss of generality to assume that $y_0, y_1 \geq 0$ and define

$$\mu = \max\{y_0, y_1\}.$$

We now observe that

$$y_2 = a|y_1 - y_0| \leq a \max\{y_0, y_1\} \leq a\mu,$$
$$y_3 = a|y_2 - y_1| \leq a \max\{y_2, y_1\} \leq a\mu.$$

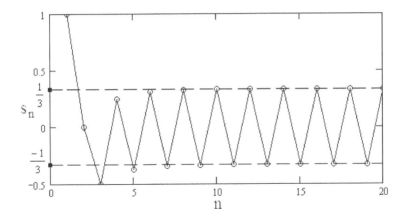

FIGURE 8.3

A solution of Eq. (8.3) that converges to a 2-cycle.

Further,

$$y_4 = a|y_3 - y_2| \le a \max\{y_3, y_2\} \le a^2\mu,$$
$$y_5 = a|y_4 - y_3| \le a \max\{y_4, y_3\} \le a^2\mu.$$

This reasoning by induction yields

$$y_{2n}, y_{2n+1} \le a^n \mu, \quad \text{for all } n$$

and under conditions (8.24) it follows that

$$\lim_{n \to \infty} y_n = 0.$$

This proves the above claim. From this observation and Eq. (8.20) it follows that if $\{t_n\}$ is a nonnegative solution of Eq. (8.3) then the corresponding solution $\{x_n\}$ of (8.16) under conditions (8.24) converges to a 2-cycle as $n \to \infty$.

Do all solutions of (8.3) converge to zero? The answer is quickly found to be negative; indeed, for some values of $a \in (0, 1)$ there are binary sequences $\{\beta_n\}$ in $\{-1, 1\}$ for which the following special case of equation (8.21) has nonconverging solutions:

$$s_n = \beta_n a|s_{n-1} - s_{n-2}|. \tag{8.26}$$

In Figure 8.3 the parameter values

$$a = \frac{1}{2}, \quad \beta_n = (-1)^{n+1}, \quad s_{-1} = s_0 = 1 \tag{8.27}$$

generate solutions of (8.26) that converge to the 2-cycle $\{-1/3, 1/3\}$.

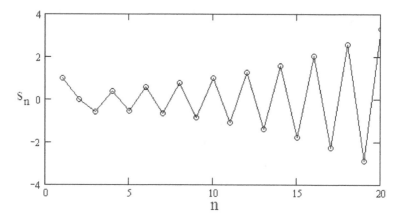

FIGURE 8.4

An unbounded solution of Eq. (8.3).

If we set $a = 0.6$ in (8.27) with other parameter values unchanged then the corresponding solution of (8.26) is unbounded as seen in Figure 8.4 Both of these solutions are also solutions of Eq. (8.3).

If the sequence $\{\beta_n\}$ is aperiodic then there are solutions of (8.3) with more complicated behavior. Figures 8.5 and 8.6 illustrate a solution of (8.26) with $a = 0.8$, initial values $s_{-1} = s_0 = 1$ and the sequence

$$\beta_n = \begin{cases} -1 & \text{if } r_n < 0.45 \\ 1 & \text{if } r_n \geq 0.45 \end{cases} \tag{8.28}$$

where the sequence $\{r_n\}$ is the solution of the first-order logistic equation below with the given initial value:

$$r_n = 3.75 r_{n-1}(1 - r_{n-1}), \quad r_0 = 0.4.$$

Figure 8.6 shows the state-space orbit of the solution in Figure 8.5. We emphasize that *this is also a solution of Eq. (8.3)!*

Solutions of (8.26) with the "chaotic" binary sequence $\{\beta_n\}$ defined by (8.28) depend sensitively on the threshold value in (8.28). For instance, if this value is changed to 0.4 then some solutions may converge to zero while if the threshold value is changed to 0.5 some solutions appear to be unbounded.

The preceding observations also indicate that the class of solutions of the third-order equation (8.16) contains a large variety of complicated solutions, *even with a constant value $a \in (0, 1)$.* ▯

We note that equations studied in Example 8.2, namely, (8.16) and (8.19) are of the same *quadratic* type as Eq. (8.5) in Example 8.1 because (8.16) and (8.19) are equivalent, respectively, to

$$(x_n + x_{n-1})^2 = a_n^2 (x_{n-1} - x_{n-3})^2$$

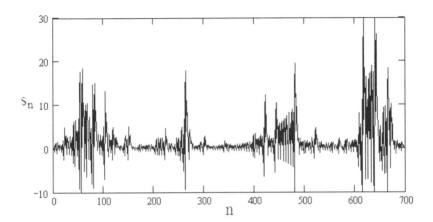

FIGURE 8.5

A complicated, nonperiodic solution of Eq. (8.3).

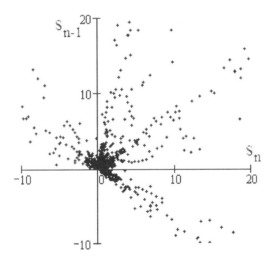

FIGURE 8.6

The state-space orbit of the solution in Figure 8.5.

and

$$t_n^2 = a_n^2 (t_{n-1} - t_{n-2})^2.$$

However, there is also an important difference; each of the equations in Example 8.2 possess real solutions through every point of its state-space but Eq. (8.5) may not have solutions through some points in its state-space, i.e, the real plane. Since taking of a general square-root is involved in Example 8.1 the existence of a real solution is not always guaranteed. A more detailed discussion of quadratic difference equations appears in a later section of this chapter.

The next example illustrates a completely different type of nonrecursive form than the preceding two examples despite its use of a quadratic difference equation.

Example 8.3

A nonrecursive equation on the set of positive integers. Consider the second-order difference equation

$$x_n x_{n-1} = x_{n-1} x_{n-2} + 2n \qquad (8.29)$$

over the set $\mathbb{N}_0$ of all nonnegative integers. Although the operations of addition and multiplication used in (8.29) are inherited from an ambient field such as $\mathbb{Q}$ or $\mathbb{R}$, note that it is not possible to write (8.29) in a recursive form (8.2) in $\mathbb{N}_0$. Even collecting terms on one side is not permitted in $\mathbb{N}_0$ since that requires subtraction and additive inverses.

Splitting into a pair of lower order equations. The substitution $t_n = x_n x_{n-1}$ in (8.29) yields the system of first-order equations

$$t_n = t_{n-1} + 2n,$$
$$x_n x_{n-1} = t_n.$$

Both of these equations are well defined on $\mathbb{N}_0$ and the first equation above is easily solved by iteration

$$t_n = n(n+1) + t_0, \quad t_0 = x_0 x_{-1}.$$

A discussion of solutions. Every solution of (8.29) is a solution of

$$x_n x_{n-1} = n(n+1) + x_0 x_{-1} \qquad (8.30)$$

which is easier to work with than (8.29). In particular, if $\{x_n\}$ is a solution of (8.30) then since $x_n x_{n-1} \to \infty$ as $n \to \infty$ there is a subsequence of $\{x_n\}$ that converges to ∞. Let $\{x_{n_j}\}$ be such a subsequence. Since for every j the following numbers are integers

$$x_{n_j - 1} = \frac{n_j(n_j + 1)}{x_{n_j}} + \frac{x_0 x_{-1}}{x_{n_j}}$$

it follows that integer solutions of (8.30) are possible if $x_0 x_{-1} = 0$. The equality

$$x_1 x_0 = x_0 x_{-1} + 2$$

that is implied by either (8.29) or (8.30) shows that $x_0 \neq 0$. If $x_{-1} = 0$ and $x_0 > 0$ then the same equality implies that $x_0 = 1$ or $x_0 = 2$ for solutions in $\mathbb{N}_0$. Each of these two values of x_0 yields a valid solution of (8.30) in $\mathbb{N}_0$ that we can find by simple iteration

$$x_0 = 1 \Rightarrow x_n = n + 1 \text{ for all } n \geq -1,$$

$$x_0 = 2 \Rightarrow x_n = \begin{cases} 2(n+1), & \text{if } n \text{ is even}, \\ (n+1)/2, & \text{if } n \text{ is odd}. \end{cases}$$

Of the two solutions listed above, the first one can be expressed in $\mathbb{N}_0$. Since the *formula* for the second solution involves division by 2 it is meaningful in an ambient field like $\mathbb{R}$ rather than $\mathbb{N}_0$, a situation that is reminiscent of what occurs in linear equations having real coefficients and complex eigenvalues.
□

The theory discussed in the next section applies to arbitrary sets regardless of any algebraic structure that may be used to define formulas.

8.2   Form symmetries, factors, and cofactors

In this section we define the concepts of order-reducing form symmetry and the associated factorization for Eq. (8.1) on an arbitrary set S. In analogy to semiconjugate factorizations, we seek a decomposition of Eq. (8.1) into a pair of difference equations of lower orders. A factor equation of type

$$\phi_n(t_n, t_{n-1}, \ldots, t_{n-m}) = \psi_n(t_n, t_{n-1}, \ldots, t_{n-m}), \quad 1 \leq m \leq k - 1 \quad (8.31)$$

may be derived from (8.1) where $\phi_n, \psi_n : S^{m+1} \to S$ for all n if there is a sequence of mappings $H_n : S^{k+1} \to S^{m+1}$ such that

$$f_n = \phi_n \circ H_n \text{ and } g_n = \psi_n \circ H_n \quad (8.32)$$

for all $n \geq 1$. If we denote

$$H_n(u_0, \ldots, u_k) = [h_{0,n}(u_0, \ldots, u_k), h_{1,n}(u_0, \ldots, u_k), \ldots, h_{m,n}(u_0, \ldots, u_k)]$$

then for each solution $\{x_n\}$ of Eq. (8.1)

$$\phi_n(h_{0,n}(x_n, \ldots, x_{n-k}), h_{1,n}(x_n, \ldots, x_{n-k}), \ldots, h_{m,n}(x_n, \ldots, x_{n-k})) =$$
$$\psi_n(h_{0,n}(x_n, \ldots, x_{n-k}), h_{1,n}(x_n, \ldots, x_{n-k}), \ldots, h_{m,n}(x_n, \ldots, x_{n-k}))$$

In order for a sequence $\{t_n\}$ in S defined by the substitution

$$t_n = h_{0,n}(x_n, \ldots, x_{n-k})$$

to be a solution of (8.31), the functions H_n must have a special form that is defined next. A review of the discussion in Section 3.2 may be helpful in understanding the motivation for the next definition.

DEFINITION 8.1   *A sequence of functions $\{H_n\}$ is an **order-reducing form symmetry** of Eq. (8.1) on a nonempty set S if there is an integer m, $1 \leq m < k$, and sequences of functions $\phi_n, \psi_n : S^{m+1} \to S$ and $h_n : S^{k-m+1} \to S$ such that*

$$H_n(u_0, \ldots, u_k) = [h_n(u_0, \ldots, u_{k-m}), h_{n-1}(u_1, \ldots, u_{k-m+1}), \quad (8.33)$$

$$\ldots, h_{n-m}(u_m, \ldots, u_k)] \quad (8.34)$$

and the sequences $\{\phi_n\}$, $\{\psi_n\}$, $\{f_n\}$, $\{g_n\}$ and $\{H_n\}$ satisfy the relations (8.32) for all $n \geq 1$.

REMARK 8.1   The reader may ask why in (8.32) we used the same H_n for both f_n and g_n. Well, if we define

$$g_n = \psi_n \circ H_n', \quad H_n' : S^{k+1} \to S^{l+1}, \quad \psi_n : S^{l+1} \to S$$

where the integer l may be different from m then the factor equation is not properly defined. For a solution $\{x_n\}$ of (8.1), we obtain

$$\phi_n(h_{0,n}(x_n, \ldots, x_{n-k}), h_{1,n}(x_n, \ldots, x_{n-k}), \ldots, h_{m,n}(x_n, \ldots, x_{n-k})) =$$
$$\psi_n(h_{0,n}'(x_n, \ldots, x_{n-k}), h_{1,n}'(x_n, \ldots, x_{n-k}), \ldots, h_{l,n}'(x_n, \ldots, x_{n-k}))$$

which after the substitutions

$$t_n = h_n(x_n, \ldots, x_{n-k+m}), \quad s_n = h_n'(x_n, \ldots, x_{n-k+l})$$

in the format defined by (8.33) yields

$$\phi_n(t_n, t_{n-1}, \ldots, t_{n-m}) = \psi_n(s_n, s_{n-1}, \ldots, s_{n-l}) \quad (8.35)$$

as the potential factor equation. This is clearly problematic since the above equation does not define a difference equation like (8.1) unless one of t_n or s_n can be expressed as a function of the other, i.e., there is a common, lower-dimensional form symmetry that allows us to reduce (8.35) to a single variable t_n, s_n or third variable that has even lower dimension. This further substitution reduces H_n' and/or H_n to a single, lower-dimensional form symmetry as in (8.32). ∎

If $S = (G, *)$ is a group, then the notion of form symmetry in Definition 8.1 generalizes the recursive form symmetry in Definition 7.1 where the components of H_n are defined as

$$h_{n-j}(u_j, \ldots, u_{k-m+j}) = u_j * \widetilde{h}_{n-j}(u_{j+1}, \ldots, u_{k-m+j}), \quad j = 0, 1, \ldots, m.$$

We have the following basic factorization theorem for nonrecursive difference equations.

THEOREM 8.1
Assume that Eq. (8.1) has an order-reducing form symmetry $\{H_n\}$ defined by (8.33) on a nonempty set S. Then the difference equation (8.1) has a factorization into an equivalent system of factor and cofactor equations

$$\phi_n(t_n, \ldots, t_{n-m}) = \psi_n(t_n, \ldots, t_{n-m}) \tag{8.36}$$

$$h_n(x_n, \ldots, x_{n-k+m}) = t_n \tag{8.37}$$

whose orders m and $k - m$ respectively, add up to the order of (8.1).

PROOF To show the equivalence, we show that for each solution $\{x_n\}$ of (8.1) there is a solution $\{(t_n, y_n)\}$ of the system of equations (8.36), (8.37) such that $y_n = x_n$ for all $n \geq 1$ and conversely, for each solution $\{(t_n, y_n)\}$ of the system of equations (8.36) and (8.37) the sequence $\{y_n\}$ is a solution of (8.1).

First assume that $\{x_n\}$ is a solution of Eq. (8.1) through a given initial point $(x_0, x_{-1}, \ldots, x_{-k+1}) \in G^{k+1}$. Define the sequence $\{t_n\}$ in S as in (8.37) for $n \geq -m + 1$ so that by (8.32)

$$\begin{aligned}
\phi_n(t_n, \ldots, t_{n-m}) &= \phi_n(h_n(x_n, \ldots, x_{n-k+m}), \ldots, h_{n-m}(x_{n-m}, \ldots, x_{n-k})) \\
&= \phi_n(H_n(x_n, \ldots, x_{n-k})) \\
&= f_n(x_n, \ldots, x_{n-k}) \\
&= g_n(x_n, \ldots, x_{n-k}) \\
&= \psi_n(H_n(x_n, \ldots, x_{n-k})) \\
&= \psi_n(t_n, \ldots, t_{n-m}).
\end{aligned}$$

It follows that $\{t_n\}$ is a solution of (8.36). Further, if $y_n = x_n$ for $n \geq -k+m$ then by the definition of t_n, $\{y_n\}$ is a solution of (8.37).

Conversely, let $\{(t_n, y_n)\}$ be a solution of the factor-cofactor system (8.36), (8.37) with initial values

$$t_0, \ldots, t_{-m+1}, y_{-m}, \ldots y_{-k+1} \in G.$$

We note that $y_0, y_{-1}, \ldots, y_{-m+1}$ satisfy the equations

$$h_j(y_j, \ldots, y_{j-k+m}) = t_j, \quad j = 0, -1, \ldots, -m + 1.$$

Now for $n \geq 1$, (8.32) implies

$$
\begin{aligned}
f_n(y_n, \ldots, y_{n-k}) &= \phi_n(H_n(y_n, \ldots, y_{n-k})) \\
&= \phi_n(h_n(y_n, \ldots, y_{n-k+m}), \ldots, h_{n-m}(y_{n-m}, \ldots, y_{n-k})) \\
&= \phi_n(t_n, \ldots, t_{n-m}) \\
&= \psi_n(t_n, \ldots, t_{n-m}) \\
&= \psi_n(H_n(x_n, \ldots, x_{n-k})) \\
&= g_n(x_n, \ldots, x_{n-k}).
\end{aligned}
$$

Therefore, $\{y_n\}$ is a solution of (8.1).

∎

Order-reduction types for nonrecursive equations can be defined as before.

DEFINITION 8.2   *If Eq. (8.40) has the factorization consisting of equations (8.36) and (8.37) then it is said to have a* **type-$(m, k - m)$ reduction** *in order.*

Example 8.4

Let S be the set $(0, \infty)$ which under ordinary addition is a semigroup. For each $a \in S$ and $m \in \{1, 2, 3, \ldots\}$ we may define $ma = a + \cdots + a$ (adding m times) without introducing the field operations. Thus if m_0, m_1, m_2 are given positive integers and $\{c_n\}$ is any sequence of positive real numbers then the third-order difference equation

$$
m_0 x_n + m_1 x_{n-1} + m_2 x_{n-2} + m_3 x_{n-3} = c_n \tag{8.38}
$$

is well defined in the additive semigroup $(0, \infty)$. To obtain a type-$(2, 1)$ reduction of order for (8.38) consider a form symmetry

$$
H(u_0, u_1, u_2, u_3) = [h(u_0, u_1), h(u_1, u_2), h(u_2, u_3)]
$$

where

$$
h(u, v) = u + \mu v, \quad \mu \in \{1, 2, 3, \ldots\}.
$$

If

$$
\begin{aligned}
f_n(u_0, u_1, u_2, u_3) &= m_0 u_0 + m_1 u_1 + m_2 u_2 + m_3 u_3, \\
g_n(u_0, u_1, u_2, u_3) &= c_n
\end{aligned}
$$

then based on the way these functions defined, we consider factor-function sequences $\phi_n, \psi_n : S^3 \to S$

$$
\begin{aligned}
\phi_n(v_1, v_2, v_3) &= l_1 v_1 + l_2 v_2 + l_3 v_3, \quad l_1, l_2, l_3 \in \{1, 2, 3, \ldots\}, \\
\psi_n(v_1, v_2, v_3) &= c_n.
\end{aligned}
$$

To satisfy the relations in (8.32) it is necessary that

$$l_1(u_0 + \mu u_1) + l_2(u_1 + \mu u_2) + l_3(u_2 + \mu u_3) = m_0 u_0 + m_1 u_1 + m_2 u_2 + m_3 u_3$$

for all $u_0, u_1, u_2, u_3 > 0$. By matching the coefficients of u_0, u_1, u_2, u_3 on both sides, this condition implies the following statements:

$$l_1 = m_0, \quad l_1\mu + l_2 = m_1, \quad l_2\mu + l_3 = m_2, \quad l_3\mu = m_3. \tag{8.39}$$

If there are positive integers μ, l_1, l_2, l_3 satisfying conditions (8.39) then a factorization of (8.38) exists within the semigroup $(0, \infty)$ as follows:

$$l_1 t_n + l_2 t_{n-1} + l_3 t_{n-2} = c_n,$$
$$x_n + \mu x_{n-1} = t_n.$$

Equations listed in (8.39) may be solved in the field of real numbers using the usual algebraic operations. Of course, we seek solutions that are positive integers. From (8.39) we find that if the cubic polynomial equation

$$m_0\mu^3 - m_1\mu^2 + m_2\mu - m_3 = 0.$$

has a positive integer solution μ then

$$l_1 = m_0, \quad l_2 = m_1 - m_0\mu, \quad l_3 = m_2 - \mu l_2 = \frac{m_3}{\mu}.$$

If furthermore $m_1 > m_0\mu$ and $m_2 > \mu(m_1 - m_0\mu)$ then l_1, l_2, l_3 are positive integers and we have a valid solution of (8.39) in $(0, \infty)$. For instance, if $m_0 = 1$, $m_1 = 3$, $m_2 = 3$ and $m_3 = 1$ then $\mu = 1$ is a root of the cubic equation above and we calculate $l_1 = 1$, $l_2 = 2$ and $l_3 = 1$. It is worth noting that in this case the factor equation

$$t_n + 2t_{n-1} + t_{n-2} = c_n$$

once again has the form symmetry $h(u_0, u_1) = u_0 + u_1$ (i.e., $\mu = 1$) and a factorization into two first-order equations

$$r_n + r_{n-1} = c_n,$$
$$t_n + t_{n-1} = r_n.$$

□

8.3   Semi-invertible map criterion

A group structure is necessary for obtaining certain results such as an extension of the useful invertible map criterion to nonrecursive equations. In this section and the rest of this chapter, we assume that $S = (G, *)$ is a nontrivial group unless otherwise stated. Denoting the identity of G by ι, the difference equation (8.1) can be written in the equivalent form

$$E_n(x_n, x_{n-1}, \ldots, x_{n-k}) = \iota, \quad k \geq 2, \ n = 1, 2, 3, \ldots \qquad (8.40)$$

where $E_n = f_n * [g_n]^{-1}$ (the brackets indicate group inversion is involved, rather than map inversion). Our aim in this section is to obtain an extension of the useful invertible map criterion of Sections 5.1 and 7.2 to the nonrecursive equation (8.40). Using this extension we obtain various form symmetries that lead to type-$(k-1,1)$ order reductions, analogously to the recursive cases in Chapter 5.

A type-$(k-1,1)$ reduction of Eq. (8.40) is characterized by the following factorization

$$\phi_n(t_n, \ldots, t_{n-k+1}) = \iota \qquad (8.41)$$
$$h_n(x_n, x_{n-1}) = t_n \qquad (8.42)$$

in which the cofactor equation has order one. This system occurs if the following is a form symmetry of (8.40):

$$H_n(u_0, \ldots, u_k) = [h_n(u_0, u_1), h_{n-1}(u_1, u_2), \ldots, h_{n-k+1}(u_{k-1}, u_k)]. \quad (8.43)$$

A specific example of the above factorization is the system of equations (8.7) and (8.8) in Section 8.1 that factor Eq. (8.5). More examples are discussed below.

The following type of coordinate function h_n is of particular interest in this section.

DEFINITION 8.3   *A coordinate function $h : G^2 \to G$ on a nontrivial group G is* **separable** *if*

$$h(u, v) = \mu(u) * \theta(v)$$

for given self-maps μ, θ of G into itself. A separable h is **right semi-invertible** *if θ is a bijection and* **left semi-invertible** *if μ is a bijection. If both θ and μ are bijections then h is* **semi-invertible**. *A form symmetry $\{H_n\}$ is (right, left) semi-invertible if the coordinate function h_n is (right, left) semi-invertible for every n.*

Note that a semi-invertible h is *not* a bijection in general, as the next example shows.

Example 8.5
(a) Consider the linear map $h(u, v) = u - v$ on $\mathbb{R}$ which is separable relative to ordinary addition with $\mu(u) = u$ and $\theta(v) = -v$. Since both μ and θ are

bijections of $\mathbb{R}$, the function h is semi-invertible. However, h is not a bijection since $h(a, a) = 0$ for all a.

(b) The function $h(u, v) = u^2 - v$ is right semi-invertible but not left semi-invertible on $\mathbb{R}$ because the function $\mu(u) = u^2$ is not a bijection of $\mathbb{R}$. ▯

Clearly, functions of type $u * \widetilde{h}(v)$ where $\widetilde{h}$ is a bijection are semi-invertible functions. Therefore, semi-invertible functions generalize the types of maps discussed previously in Sections 5.1 and 7.2. The invertible map criterion discussed in those sections also extends to all right semi-invertible form symmetries; this is established in the next theorem which is extends Theorem 7.2 to nonrecursive equations.

THEOREM 8.2

(Semi-invertible map criterion) Assume that $h_n(u, v) = \mu_n(u) * \theta_n(v)$ is a sequence of right semi-invertible functions with bijections θ_n of a group G. For arbitrary $u_0, v_1, \ldots, v_k \in G$ define $\zeta_{0,n} \equiv u_0$ and for $j = 1, \ldots, k$

$$\zeta_{j,n}(u_0, v_1, \ldots, v_j) = \theta_{n-j+1}^{-1}([\mu_{n-j+1}(\zeta_{j-1,n}(u_0, v_1, \ldots, v_{j-1}))]^{-1} * v_j) \quad (8.44)$$

with the usual distinction observed between map inversion and group inversion. Then Eq. (8.40) has the form symmetry (8.43) and the associated factorization into equations (8.41) and (8.42) if and only if the following quantity

$$E_n(u_0, \zeta_{1,n}(u_0, v_1), \ldots, \zeta_{k,n}(u_0, v_1, \ldots, v_k)) \quad (8.45)$$

is independent of u_0 for all n. In this case, the factor functions ϕ_n are given by

$$\phi_n(v_1, \ldots, v_k) = E_n(u_0, \zeta_{1,n}(u_0, v_1), \ldots, \zeta_{k,n}(u_0, v_1, \ldots, v_k)). \quad (8.46)$$

PROOF Assume that the quantity in (8.45) is independent of u_0 for every n. Then the functions ϕ_n in (8.46) are well defined and if H_n is given by (8.43) and $v_{j+1} = h_{n-j}(u_j, u_{j+1})$ for $j = 0, \ldots, k-1$ in (8.44) then

$$\phi_n(H_n(u_0, \ldots, u_k)) = \phi_n(h_n(u_0, u_1), h_{n-1}(u_1, u_2), \ldots, h_{n-k+1}(u_{k-1}, u_k))$$
$$= E_n(u_0, \zeta_{1,n}(u_0, h_n(u_0, u_1)), \ldots,$$
$$\zeta_{k,n}(u_0, h_n(u_0, u_1), \ldots, h_{n-k+1}(u_{k-1}, u_k)).$$

Now, observe that

$$\zeta_{1,n}(u_0, h_n(u_0, u_1)) = \theta_n^{-1}([\mu_n(u_0)]^{-1} * \mu_n(u_0) * \theta_n(u_1)) = u_1.$$

By way of induction, assume that for $j < k$

$$\zeta_{j,n}(u_0, h_n(u_0, u_1), \ldots, h_{n-j+1}(u_{j-1}, u_j)) = u_j \quad (8.47)$$

and note that

$$\zeta_{j+1,n}(u_0,\ldots,h_{n-j}(u_j,u_{j+1})) = \theta_{n-j}^{-1}([\mu_{n-j}(\zeta_{j,n}(u_0,\ldots,h_{n-j+1}(u_{j-1},u_j)))]^{-1}$$
$$*\mu_{n-j}(u_j) * \theta_{n-j}(u_{j+1}))$$
$$= \theta_{n-j}^{-1}([\mu_{n-j}(u_j)]^{-1} * \mu_{n-j}(u_j) * \theta_{n-j}(u_{j+1}))$$
$$= u_{j+1}.$$

It follows that (8.47) is true for all $j = 0, 1, \ldots, k$ and thus

$$\phi_n(H_n(u_0,\ldots,u_k)) = E_n(u_0,\ldots,u_k)$$

i.e., $\{H_n\}$ as defined by (8.43) is a form symmetry of Eq. (8.40) and therefore Theorem 8.1 implies the existence of the associated factorization into equations (8.41) and (8.42).

Conversely, suppose that $\{H_n\}$ as defined by (8.43) is a form symmetry of Eq. (8.40). Then there are functions ϕ_n such that for all $u_0, v_1, \ldots, v_k \in G$

$$E_n(u_0,\zeta_{1,n},\ldots,\zeta_{k,n}) = \phi_n(H_n(u_0,\zeta_{1,n},\ldots,\zeta_{k,n}))$$
$$= \phi_n(h_n(u_0,\zeta_{1,n}),\ldots,h_{n-k+1}(\zeta_{k-1,n},\zeta_{k,n}))$$

where $\zeta_{j,n} = \zeta_{j,n}(u_0,v_1,\ldots,v_j)$ for $j = 1,\ldots,k$. Since

$$h_{n-j+1}(\zeta_{j-1,n},\zeta_{j,n}) = \mu_{n-j+1}(\zeta_{j-1,n}) * \theta_{n-j+1}(\zeta_{j,n})$$
$$= \mu_{n-j+1}(\zeta_{j-1,n}) * \theta_{n-j+1}(\theta_{n-j+1}^{-1}([\mu_{n-j+1}(\zeta_{j-1,n})]^{-1} * v_j))$$
$$= \mu_{n-j+1}(\zeta_{j-1,n}) * [\mu_{n-j+1}(\zeta_{j-1,n})]^{-1} * v_j$$
$$= v_j$$

it follows that $E_n(u_0,\zeta_{1,n},\ldots,\zeta_{k,n})$ is independent of u_0 for all n, as stated.

∎

If $\{H_n\}$ is a semi-invertible (right and left) form symmetry of Eq. (8.40) then the following result states that the cofactor equation (8.42) can be expressed in recursive form.

COROLLARY 8.1
Assume that the functions h_n in Theorem 8.2 are semi-invertible so that both μ_n and θ_n are bijections. Then Eq. (8.40) has the following factorization

$$\phi_n(t_n,\ldots,t_{n-k+1}) = \iota$$
$$x_n = \mu_n^{-1}(t_n * [\theta_n(x_{n-1})]^{-1})$$

in which the cofactor equation is recursive.

Special choices of μ_n and θ_n give analogs of the identity, inversion and linear form symmetry that we discussed in Chapter 5. In particular, if θ_n is linear for all n then the selection

$$\mu_n(u) = u^2, \quad \theta_n(v) = -bv$$

gives the cofactor equation (8.8) in Section 8.1. More generally, if $\{a_n\}, \{b_n\}$ and $\{c_n\}$ are sequences in a field $\mathcal{F}$ then the difference equation

$$x_n^2 - a_n x_{n-1}^2 - b_n x_{n-1} + a_n b_{n-1} x_{n-2} = c_n \qquad (8.48)$$

can be factored as

$$t_n = a_n t_{n-1} + c_n$$
$$x_n^2 - b_n x_{n-1} = t_n.$$

Difference equations such as (8.48) which consist only of polynomial terms of degree one and two make up a significant class of difference equations that lie just beyond the linear equations. We discuss these "quadratic equations" and the challenges that their factorizations pose more methodically in the next section.

8.4 Quadratic difference equations

In Section 7.4 we showed that a linear difference equation

$$x_n + a_{1,n} x_{n-1} + \cdots + a_{k,n} x_{n-k} = b_n$$

has a SC factorization over a field $\mathcal{F}$ if a solution of the homogeneous part existed that was never zero (equivalently, if the associated Riccati equation of lower order had a solution).

A natural generalization of the linear equation is the *quadratic difference equation* over a field $\mathcal{F}$

$$\sum_{i=0}^{k} \sum_{j=i}^{k} a_{i,j,n} x_{n-i} x_{n-j} + \sum_{j=0}^{k} b_{j,n} x_{n-j} = c_n \qquad (8.49)$$

that is defined by the quadratic expression

$$E_n(u_0, u_1, \ldots, u_k) = \sum_{i=0}^{k} \sum_{j=i}^{k} a_{i,j,n} u_i u_j + \sum_{j=0}^{k} b_{j,n} u_j - c_n.$$

Linear equations may be viewed as degenerate special cases of quadratic ones where $a_{i,j,n} = 0$ for all i, j, n. When discussing Eq. (8.49) we implicitly assume that

There is $j \in \{0, 1, \ldots, k\}$ such that $a_{0,j,n} \neq 0$ for all n \qquad (8.50)

although this restriction is not the weakest possible for preserving quadratic terms (see Example 8.9 below for an equation that violates the above condition but satisfies a weaker one). Condition (8.50) leads to the following terminology in classifying quadratic difference equations.

DEFINITION 8.4   *The quadratic difference equation (8.49) is* **improper** *or* **degenerate** *if its coefficients do not satisfy Condition (8.50).*

Equations of type (8.49) under Condition (8.50) also include the familiar *rational recursive equations* of type

$$x_n = \frac{-\sum_{i=1}^{k}\sum_{j=i}^{k} a_{i,\,j,n}x_{n-i}x_{n-j} - \sum_{j=1}^{k} b_{j,n}x_{n-j} + c_n}{\sum_{j=1}^{k} a_{0,\,j,n}x_{n-j} + b_{0,n}} \tag{8.51}$$

as a special case with

$$a_{0,0,n} = 0 \text{ for all } n. \tag{8.52}$$

These rational equations have been studied extensively; see the Notes section of this chapter. As seen in Section 5.6.5, their solutions in some cases exhibit a rich variety of qualitatively different behaviors. We note that all solutions of (8.51) are also solutions of the quadratic (8.49) when condition (8.52) is satisfied.

In particular, Ladas rational difference equations are among extensively studied special cases of (8.51) in which no quadratic terms are present in the numerator, i.e., when $a_{i,\,j,n} = 0$ for all i, j, n.

Other special cases of (8.51) include recursive polynomial equations that occur when $a_{0,\,j,n} = 0$ for all $j = 1, 2, \ldots, k$ and all $n \geq 1$ but $b_{0,n}$ are nonzero. Well known examples include the Henon difference equation

$$x_n = a + bx_{n-2} - x_{n-1}^2$$

and the logistic equation with delay

$$x_n = x_{n-1}(a - bx_{n-2} - x_{n-1})$$

that generalizes the familiar one-dimensional logistic map.

Unlike the linear case, the existence of a factorization for Eq. (8.49) is not clear and in general, finding any factorization into lower order equations is a challenging problem. On the other hand, the examples discussed in Section 8.1 had factorizations that could be found easily; therefore, special cases of (8.49) may be amenable to analysis; see, e.g., Corollary 8.6.

REMARK 8.2   The nature of the field $\mathcal{F}$ is important to the factorization problem. For instance, if $\mathcal{F} = \mathbb{Z}_2$ then all quadratic expressions without mixed terms reduce to linear ones since for all $u \in \mathbb{Z}_2$, $u = 0$ or $u = 1$ so that

$u^2 = u$. In particular, on the field $\mathbb{Z}_2$ Eq. (8.5) in Section 8.1 reduces to the nonhomogeneous linear equation

$$x_n = (a + b)x_{n-1} + abx_{n-2} + c$$

whose factorization has already been discussed.

∎

8.4.1   Existence and variety of real solutions

Before proceeding further with our main discussion, i.e., factorizations and reductions of orders, it is helpful to consider the existence of solutions for the quadratic difference equation (8.49). In this section, we consider this issue for the case $\mathcal{F} = \mathbb{R}$, i.e., when all coefficients and parameters are real numbers. Real solutions are important in both applied modeling and in numerical simulations and offer substantial insights into the nature of the problem without technical or structural issues that may arise with regard to the existence of square-roots in other fields.

As noted above, if $a_{0,0,n} = 0$ for all but finitely many n then under Condition (8.50) there is a unique recursive equation, namely, (8.51) that is equivalent to (8.49). The existence of solutions in this case depends on the singularity set of (8.51), i.e., the set of initial values that result in a zero denominator after a finite number of iterations; see, e.g., Section 5.4 on the Riccati difference equation.

In this section, we assume that

$$a_{0,0,n} \neq 0 \text{ for all } n \tag{8.53}$$

as this represents a new and substantially different case and also ensures that the quadratic difference equation is not improper. Although Condition (8.50) is satisfied in this case, the quadratic difference equation does not have a unique recursive form and direct iteration is not possible. But an idea discussed in Examples 8.1 and 8.2 can be used to study the existence of solutions for nonrecursive quadratic equations satisfying (8.53). In this section we obtain conditions implying the existence of real solutions for (8.49) when all coefficients and parameters are real and (8.53) holds.

Before considering the general case, it is instructive to discuss the following special case.

LEMMA 8.1

Consider the second-order, quadratic difference equation

$$x_n^2 = Q_n(x_{n-1}, \ldots, x_{n-k}) \tag{8.54}$$

where $\{Q_n\}$ is a sequence of quadratic functions

$$Q_n(u_1, \ldots, u_k) = \sum_{i=1}^{k} \sum_{j=i}^{k} a_{i,j,n} u_i u_j + \sum_{j=1}^{k} b_{j,n} u_j + c_n$$

in which all the coefficients $a_{i,j,n}, b_{j,n}, c_n$ are real numbers, not all zeros. Then $\{x_n\}$ is a real solution of (8.54) if and only if $\{x_n\}$ is a real solution of some member of the family of recursive equations

$$s_n = \beta_n \sqrt{Q_n(s_{n-1}, \ldots, s_{n-k})} \tag{8.55}$$

where $\{\beta_n\}$ is a binary sequence with values in the set $\{-1, 1\}$.

PROOF   Let $\{x_n\}$ be a real solution of the recursive equation (8.55) with a given binary sequence $\{\beta_n\}$. Then

$$x_n = \beta_n \sqrt{Q_n(x_{n-1}, \ldots, x_{n-k})} \tag{8.56}$$

for all n and squaring both sides of the above equation shows that $\{x_n\}$ satisfies (8.54).

Conversely, let $\{x_n\}$ be a real solution of (8.54). For $n = 1$

$$x_1^2 = Q_1(x_0, \ldots, x_{-k+1})$$

Since x_1 is real, the square-root of the right-hand side of above equality is real. Therefore, for $\beta_1 \in \{-1, 1\}$ (depending on the sign of x_1)

$$x_1 = \beta_1 \sqrt{Q_1(x_0, \ldots, x_{-k+1})}.$$

Continuing in this way for $n = 2, 3, \ldots$ a binary sequence $\{\beta_n\}$ is defined for which (8.56) holds. For this binary sequence $\{\beta_n\}$, $\{x_n\}$ is evidently a solution of (8.55).   ∎

REMARK 8.3   (Recursive class) The set of equations (8.55) may be called the **recursive class** of the nonrecursive equation (8.54). We use the term recursive class intuitively in this book rather than give a formal definition. The recursive class in Lemma 8.1 is an uncountable set and further, every solution of each equation in the recursive class is also a valid solution of (8.54). The "if and only if" statement in Lemma 8.1 implies that the recursive class of equations (8.55) for all $\{\beta_n\}$ is equivalent to the nonrecursive equation (8.54) in the sense that every solution of (8.54) is also a solution of some equation in its recursive class.   ∎

Lemma 8.1 applies to (8.5) and (8.19) as has already been pointed out in Examples 8.1 and 8.2. We now give another example.

Example 8.6

Let us find the real solutions of the following quadratic difference equation without mixed-product terms

$$x_n^2 + x_{n-1}^2 + x_{n-2}^2 = 1. \tag{8.57}$$

To assure the existence of real solutions of Eq. (8.57), let $x_0, x_{-1} \in [-1, 1]$ such that

$$x_0^2 + x_{-1}^2 \leq 1.$$

This is equivalent to choosing an initial point in the unit disk in $\mathbb{R}^2$ and examining the orbits containing this point. To simplify notation, set

$$x_0 = a, \ x_{-1} = b \quad \text{with } a^2 + b^2 \leq 1.$$

Writing (8.57) in the form (8.54), consider the recursive difference equation

$$s_n = \beta_n \sqrt{1 - s_{n-1}^2 - s_{n-2}^2}, \quad s_0 = x_0, \ s_{-1} = x_{-1} \tag{8.58}$$

where $\{\beta_n\}$ is a given though arbitrarily chosen binary sequence in $\{-1, 1\}$. Note that

$$s_1 = \beta_1 \sqrt{1 - (a^2 + b^2)} \in [-1, 1]$$

is real; similarly

$$s_2 = \beta_2 \sqrt{1 - s_1^2 - s_0^2} = \beta_2 \sqrt{1 - 1 + a^2 + b^2 - a^2} = \beta_2 |b| \in [-1, 1]$$

$$s_3 = \beta_3 \sqrt{1 - s_2^2 - s_1^2} = \beta_3 \sqrt{1 - b^2 - 1 + a^2 + b^2} = \beta_3 |a| \in [-1, 1].$$

Further iteration generates a sequence $\{s_n\}$ with a general form

$$\left\{ b, a, \beta_1 \sqrt{1 - (a^2 + b^2)}, \ldots, \beta_{3n-1}|b|, \beta_{3n}|a|, \beta_{3n+1} \sqrt{1 - (a^2 + b^2)}, \ldots \right\} \tag{8.59}$$

with $n \geq 1$ that is a solution of (8.58) and its orbit is in the closed unit disk in $\mathbb{R}^2$.

The set of all equations of type (8.58) for every possible binary sequence $\{\beta_n\}$ is the recursive class of (8.57). The sequence (8.59) provides significant insight into the nature of the real solutions of (8.57) becuase it shows that the sequence $\{\beta_n\}$ essentially determines the behavior of the real solutions of (8.57). For example, if $\{\beta_n\}$ is periodic then so is (8.59). In particular, for the case $\beta_n = 1$ for all n (i.e., the positive square-root case) (8.59) is periodic with period three except in the case described next.

A special class of solutions of (8.57) are obtained from its constant solutions. Setting all variables in (8.57) equal to a common value u and solving we get

$$3u^2 = 1 \Rightarrow u = \pm \frac{1}{\sqrt{3}}.$$

By direct substitution is it easy to see that the sequence

$$\left\{ \frac{\beta_n}{\sqrt{3}} \right\}_{n=1}^{\infty} \tag{8.60}$$

is a solution of (8.57) for every binary sequence $\{\beta_n\}$ with values in the set $\{-1, 1\}$. Also note that (8.60) may also be obtained from (8.59) with $a = b = 1/\sqrt{3}$. It is evident from (8.60) that (8.57) has real periodic solutions of all possible periods as well as bounded, nonperiodic solutions!   ☐

It is worth noting that Eq. (8.57) has a factorization that is easy to find. The substitution

$$y_n = x_n^2 \tag{8.61}$$

transforms (8.57) into the nonhomogeneous linear equation

$$y_n + y_{n-1} + y_{n-2} = 1.$$

After factoring the linear equation as discussed in previous chapters, we may substitute x_n back using (8.61) to obtain a factorization of (8.57) into two first-order equations. A generalization of this idea, namely, using an *order-preserving* form symmetry such as (8.61) to transform a nonlinear equation into a linear one, is discussed later in this chapter; see Remark 8.5. In Example 8.14 we obtain another formula for the solutions of (8.57) using the classical theory of linear equations. This formula is equivalent to (8.59) but not as clearly descriptive of the nature of solutions.

In the next example we determine the recursive class and the nature of solutions of a quadratic equation containing mixed terms. In this case we cannot rely on linear equations to simplify our task.

Example 8.7

Let a, b, c be real numbers such that

$$a \neq 0 \text{ and } b, c > 0. \tag{8.62}$$

Let us find the recursive class of the quadratic equation

$$x_n^2 = a x_n x_{n-1} + b x_{n-2}^2 + c. \tag{8.63}$$

This equation also belongs to the class of quadratic difference equations discussed in Corollary 8.3 below. We solve for the term x_n by completing the squares:

$$x_n^2 - a x_n x_{n-1} + \frac{a^2}{4} x_{n-1}^2 = \frac{a^2}{4} x_{n-1}^2 + b x_{n-2}^2 + c$$

$$\left(x_n - \frac{a}{2} x_{n-1} \right)^2 = \frac{a^2}{4} x_{n-1}^2 + b x_{n-2}^2 + c.$$

Now as in Example 8.6 we may take the square root using binary sequences for completeness

$$x_n - \frac{a}{2}x_{n-1} = \beta_n \sqrt{\frac{a^2}{4}x_{n-1}^2 + bx_{n-2}^2 + c}$$

$$x_n = \frac{a}{2}x_{n-1} + \beta_n \sqrt{\frac{a^2}{4}x_{n-1}^2 + bx_{n-2}^2 + c} \qquad (8.64)$$

where $\{\beta_n\}$ is any given sequence with $\beta_n \in \{-1, 1\}$ chosen arbitrarily for every n. The recursive class of (8.63) is the set of all equations of type (8.64) for all possible binary sequences $\{\beta_n\}$.

Under conditions (8.62), for each fixed sequence $\{\beta_n\}$ every solution of the recursive equation (8.64) with real initial values is real because the quantity under the square root is always nonnegative. Furthermore, since

$$\sqrt{\frac{a^2}{4}x_{n-1}^2 + bx_{n-2}^2 + c} > \left|\frac{a}{2}x_{n-1}\right|$$

it follows that for each n,

$$x_n > 0 \quad \text{if} \quad \beta_n = 1,$$
$$x_n < 0 \quad \text{if} \quad \beta_n = -1.$$

This sign-switching implies that a significant variety of oscillating solutions are possible for Eq. (8.63) under conditions (8.62). Indeed, since β_n is chosen arbitrarily, for every sequence of positive integers

$$\{m_1, m_2, m_3, \ldots\}$$

there is a solution of (8.63) that starts with positive values of x_n for m_1 terms by setting $\beta_n = 1$ for $1 \leq n \leq m_1$. Then $x_n < 0$ for the next m_2 terms with $\beta_n = -1$ for n in the range

$$m_1 + 1 \leq n \leq m_1 + m_2$$

and so on with the sign of x_n switching according to the sequence $\{m_n\}$. Figure 8.7 illustrates this situation with parameter values

$$a = 0.6, \quad b = 0.5, \quad c = 1, \quad x_{-1} = x_0 = 1$$
$$\beta_n = \begin{cases} 1, & \text{if } \sin(n/3) + \cos(n\sqrt{2}/3) > 0 \\ -1, & \text{otherwise.} \end{cases}$$

☐

The method of completing the square discussed in Example 8.7 can be used to determine the recursive class of *every* quadratic difference equation.

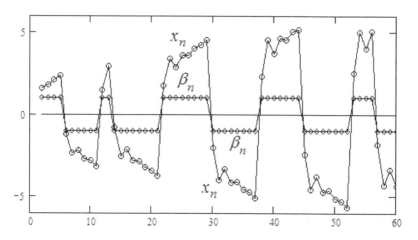

FIGURE 8.7

Oscillation of a solution x_n is determined by the oscillation pattern of the binary sequence β_n.

This useful feature enables the calculation of solutions of quadratic equations through iteration, a feature that is not shared by nonrecursive difference equations in general. The next result extends Lemma 8.1 to the general quadratic case.

LEMMA 8.2

Consider the quadratic difference equation

$$\sum_{i=0}^{k}\sum_{j=i}^{k} a_{i,j,n} x_{n-i} x_{n-j} + \sum_{j=0}^{k} b_{j,n} x_{n-j} + c_n = 0 \qquad (8.65)$$

in which all the coefficients $a_{i,j,n}, b_{j,n}, c_n$ are real numbers with $a_{0,0,n} \neq 0$ for all n. Then $\{x_n\}$ is a real solution of (8.65) if and only if $\{x_n\}$ is a real solution of some member of the family of recursive equations, namely, the recursive class of (8.65)

$$s_n = L_n(s_{n-1}, \ldots, s_{n-k}) + \beta_n \sqrt{L_n^2(s_{n-1}, \ldots, s_{n-k}) + Q_n(s_{n-1}, \ldots, s_{n-k})} \qquad (8.66)$$

where $\{\beta_n\}$ is a fixed but arbitrarily chosen binary sequence with values in the *set $\{-1, 1\}$ and for each n,*

$$L_n(u_1, \ldots, u_k) = \frac{-1}{2a_{0,0,n}} \left(b_{0,n} + \sum_{j=1}^{k} a_{0,j,n} u_j \right), \tag{8.67}$$

$$Q_n(u_1, \ldots, u_k) = \frac{-1}{a_{0,0,n}} \left(\sum_{i=1}^{k} \sum_{j=i}^{k} a_{i,j,n} u_i u_j + \sum_{j=1}^{k} b_{j,n} u_j + c_n \right). \tag{8.68}$$

PROOF If

$$E_n(u_0, u_1, \ldots, u_k) = \sum_{i=0}^{k} \sum_{j=i}^{k} a_{i,j,n} u_i u_j + \sum_{j=1}^{k} b_{j,n} u_j + c_n$$

then using definitions (8.67) and (8.68) we may write

$$E_n(u_0, u_1, \ldots, u_k) = a_{0,0,n}[x_n^2 - 2L_n(u_1, \ldots, u_k)x_n - Q_n(u_1, \ldots, u_k)] \tag{8.69}$$

Since $a_{0,0,n} \neq 0$ for all n, the solution set of (8.65) is identical to the solution set of

$$x_n^2 - 2L_n(u_1, \ldots, u_k)x_n - Q_n(u_1, \ldots, u_k) = 0. \tag{8.70}$$

Completing the square in (8.70) gives

$$[x_n - L_n(u_1, \ldots, u_k)]^2 = L_n^2(u_1, \ldots, u_k) + Q_n(u_1, \ldots, u_k).$$

Now, arguing as in Lemma 8.1, this equation is equivalent to (8.66). ∎

Example 8.7 provides a quick illustration of the preceding result with

$$L(u_1, u_2) = -\frac{a}{2} u_1, \quad Q(u_1, u_2) = bu_2^2 + c$$

where $L_n = L$ and $Q_n = Q$ are independent of n. Similarly, in Example 8.6 we have

$$L(u_1, u_2) = 0, \quad Q(u_1, u_2) = 1 - u_1^2 - u_2^2.$$

We are now ready to present the existence theorem for real solutions of (8.65). Let $\{\beta_n\}$ be a fixed but arbitrarily chosen binary sequence in $\{-1, 1\}$ and define the functions L_n and Q_n as in Lemma 8.2. Further, let the functions on the right-hand side of (8.66) be labeled f_n, i.e.,

$$f_n(u_1, \ldots, u_k) = L_n(u_1, \ldots, u_k) + \beta_n \sqrt{L_n^2(u_1, \ldots, u_k) + Q_n(u_1, \ldots, u_k)}. \tag{8.71}$$

These functions on $\mathbb{R}^k$ unfold to the self-maps

$$F_n(u_1, \ldots, u_k) = (f_n(u_1, \ldots, u_k), u_1, \ldots, u_{k-1}).$$

We emphasize that each function sequence $\{F_n\}$ is determined by a given or *fixed* binary sequence $\{\beta_n\}$ as well as the function sequences $\{L_n\}$ and $\{Q_n\}$ that are given by (8.65). Clearly the functions f_n are real-valued at a point $(u_1, \ldots, u_k) \in \mathbb{R}^k$ if and only if

$$L_n^2(u_1, \ldots, u_k) + Q_n(u_1, \ldots, u_k) \geq 0. \qquad (8.72)$$

In what follows, we denote the set of all points in $\mathbb{R}^k$ for which inequality (8.72) holds by $\mathcal{S}$; i.e.,

$$\mathcal{S} = \bigcap_{n=0}^{\infty} \left\{ (u_1, \ldots, u_k) \in \mathbb{R}^k : L_n^2(u_1, \ldots, u_k) + Q_n(u_1, \ldots, u_k) \geq 0 \right\}. \qquad (8.73)$$

THEOREM 8.3

Assume that the set $\mathcal{S}$ defined by (8.73) is nonempty.

(a) The quadratic difference equation (8.65) has a real solution $\{x_n\}_{n=-k+1}^{\infty}$ if and only if the point $P_0 = (x_0, x_{-1}, \ldots, x_{-k+1})$ is in $\mathcal{S}$ and there is a binary sequence $\{\beta_n\}$ in $\{-1, 1\}$ such that the forward orbit of P_0 under the associated maps $\{F_n\}$ is contained in $\mathcal{S}$; i.e.,

$$\{F_n \circ F_{n-1} \circ \cdots \circ F_1(P_0)\}_{n=1}^{\infty} \subset \mathcal{S}.$$

(b) If the maps $\{F_n\}$ have a nonempty invariant set $M \subset \mathcal{S}$ for all n, i.e.,

$$F_n(M) \subset M \subset \mathcal{S}$$

then the quadratic difference equation (8.65) has real solutions.

PROOF (a) By Lemma 8.2, $\{x_n\}_{n=-k+1}^{\infty}$ is a real solution of (8.65) if and only if there is a binary sequence $\{\beta_n\}$ in $\{-1, 1\}$ such that $\{x_n\}_{n=-k+1}^{\infty}$ is a real solution of the recursive equation (8.66). Using the notation in (8.71), Eq. (8.66) can be written as

$$s_n = f_n(s_{n-1}, \ldots, s_{n-k}). \qquad (8.74)$$

Now the forward orbit of the solution $\{x_n\}_{n=-k+1}^{\infty}$ of (8.74) in the state-space is the sequence

$$\mathcal{O} = \{F_n \circ F_{n-1} \circ \cdots \circ F_1(P_0)\}_{n=1}^{\infty} = \{(x_{n-1}, \ldots, x_{n-k})\}_{n=1}^{\infty}$$

in $\mathbb{R}^k$ that starts from $P_0 \in \mathcal{S}$. It is clear from the definition of $\mathcal{S}$ that each x_n is real if and only if $\mathcal{O} \subset \mathcal{S}$. This observation completes the proof of (a).

(b) Let $P_0 \in M$. Then $\{F_n \circ F_{n-1} \circ \cdots \circ F_1(P_0)\}_{n=1}^{\infty} \subset M \subset \mathcal{S}$ so by (a)

$$\{(x_{n-1}, \ldots, x_{n-k})\}_{n=1}^{\infty} \subset \mathcal{S}.$$

It follows that each x_n is real and thus, $\{x_n\}_{n=-k+1}^{\infty}$ is a solution of (8.74). An application of Lemma 8.2 completes the proof. ∎

The next definition is motivated by the above result. We note that if $\mathcal{S}$ has any invariant subset $M = M(\{\beta_n\})$ (relative to some binary sequence $\{\beta_n\}$ in $\{-1,1\}$ or equivalently, some map sequence $\{F_n\}$) then the union of all such invariant sets in $\mathcal{S}$ is again invariant relative to all relevant binary sequences $\{\beta_n\}$ (or map sequences $\{F_n\}$). Invariant sets may exist (i.e., they are nonempty) for some binary sequences $\{\beta_n\}$ in $\{-1,1\}$ and not others. However, the union of all invariant sets,

$$\mathcal{M} = \bigcup \{M(\{\beta_n\}) : \{\beta_n\} \text{ is a binary sequence in } \{-1,1\}\}$$

is the largest or maximal invariant set in $\mathcal{S}$ and as such, $\mathcal{M}$ is unique. In particular, if $\mathcal{S}$ is invariant relative to some binary sequence then $\mathcal{M} = \mathcal{S}$.

DEFINITION 8.5   *The largest invariant subset $\mathcal{M}$ of $\mathcal{S}$ is the **state-space of real solutions** of the quadratic equation (8.65).*

REMARK 8.4   The ambient set $\mathcal{S}$ defined by (8.73) contains the orbits of all real solutions of Eq. (8.65) by Theorem 8.3. Hence, a necessary condition for the existence of real solutions for the quadratic equation is that $\mathcal{S}$ be nonempty. Of course, $\mathcal{S}$ may be empty in some cases, e.g.,

$$x_n^2 + x_{n-1}^2 + x_{n-2}^2 + 1 = 0$$

where

$$\mathcal{S} = \{(u, v) \in \mathbb{R}^2 : -u^2 - v^2 - 1 \geq 0\} = \emptyset.$$

In cases such as the above, Eq. (8.65) obviously has no real solutions. ∎

The existence of a state-space $\mathcal{M}$ may signal the occurrence of a variety of solutions for (8.65). In Example 8.7, where $\mathcal{S} = \mathbb{R}^2$ is trivially invariant (so that $\mathcal{M} = \mathbb{R}^2$) we observed the occurrence of a wide variety of oscillatory behaviors. Generally, when $\mathcal{S} = \mathbb{R}^k$ every solution of (8.65) with its initial point $P_0 \in \mathbb{R}^k$ is a real solution. The next result presents sufficient conditions that imply $\mathcal{S} = \mathbb{R}^k$.

COROLLARY 8.2
The state-space of real solutions of (8.65) is $\mathbb{R}^k$ if the following conditions

hold for all n:

$$A_{j,n} > 0, \ B_{i,j,n} = 0, \tag{8.75}$$

$$\sum_{j=1}^{k} \frac{C_{j,n}^2}{A_{j,n}} \leq b_{0,n}^2 - 4a_{0,0,n}c_n \tag{8.76}$$

where for $k \geq 2$, $j = 1, \ldots, k$, and $n \geq 0$

$$A_{j,n} = a_{0,j,n}^2 - 4a_{0,0,n}a_{j,j,n},$$
$$B_{i,j,n} = a_{0,i,n}a_{0,j,n} - 2a_{0,0,n}a_{i,j,n}, \ i < j,$$
$$C_{j,n} = a_{0,j,n}b_{0,n} - 2a_{0,0,n}b_{j,n}.$$

PROOF By straightforward calculation the inequality $L_n^2 + Q_n \geq 0$, i.e., (8.72) is seen to be equivalent to

$$\sum_{j=1}^{k} A_{j,n}u_j^2 + 2\sum_{i=1}^{k-1}\sum_{j=i+1}^{k} B_{i,j,n}u_iu_j + 2\sum_{j=1}^{k} C_{j,n}u_j + b_{0,n}^2 \geq 4a_{0,0,n}c_n \tag{8.77}$$

for all $(u_1, \ldots, u_k) \in \mathbb{R}^k$ with the coefficients $A_{j,n}$, $B_{i,j,n}$ and $C_{j,n}$ as defined in the statement of the corollary. By conditions (8.76) the double summation term in (8.77) drops out and we may complete the squares in the remaining terms to obtain the inequality

$$\sum_{j=1}^{k} A_{j,n}\left(u_j + \frac{C_{j,n}}{A_{j,n}}\right)^2 \geq \sum_{j=1}^{k} \frac{C_{j,n}^2}{A_{j,n}} - b_{0,n}^2 + 4a_{0,0,n}c_n.$$

By (8.75) the left-hand side of the above inequality is nonnegative while its right-hand side is nonpositive so (8.77) holds under conditions (8.75) and (8.76). The proof is completed by applying Theorem 8.3. ∎

The next result is obtained by an immediate application of the above corollary to the nonhomogeneous quadratic equation of order two with constant coefficients. We previously encountered a special case of the following corollary in Example 8.7.

COROLLARY 8.3
The quadratic difference equation

$$x_n^2 + a_{0,1}x_nx_{n-1} + a_{0,2}x_nx_{n-2} + a_{1,1}x_{n-1}^2 + a_{1,2}x_{n-1}x_{n-2}+$$
$$+a_{2,2}x_{n-2}^2 + b_0x_n + b_1x_{n-1} + b_2x_{n-2} + c_n = 0$$

has $\mathbb{R}^2$ as a state-space of real solutions if the following conditions are satisfied:

$$a_{1,1} < \frac{a_{0,1}^2}{4}, \quad a_{2,2} < \frac{a_{0,2}^2}{4}, \quad a_{1,2} = \frac{1}{2}a_{0,1}a_{0,2} \text{ and}$$

$$c_n \leq \frac{b_0^2}{4} - \frac{(a_{0,1}b_0 - 2b_1)^2}{4(a_{0,1}^2 - 4a_{1,1})} - \frac{(a_{0,2}b_0 - 2b_2)^2}{4(a_{0,2}^2 - 4a_{2,2})} \text{ for all } n.$$

Example 8.8

The state-space of solutions of the difference equation

$$x_n^2 + x_n x_{n-2} - x_{n-1}^2 + 2x_n - x_{n-1} + x_{n-2} = \frac{\sin 2n}{n+1} \tag{8.78}$$

is $\mathbb{R}^2$ because all inequalities in Corollary 8.3 hold; in particular,

$$-\frac{\sin 2n}{n+1} = c_n \leq 1 - 0 - \frac{1}{4} = \frac{3}{4} \quad \text{for } n \geq 1 \tag{8.79}$$

as required. The recursive class of Eq. (8.78) is calculated using Lemma 8.2 as follows:

$$L_n(u, v) = -\frac{1}{2}(2 + v) = -1 - \frac{1}{2}v$$

$$Q_n(u, v) = u^2 + u - v + \frac{\sin 2n}{n+1}$$

so that

$$s_n = -1 - \frac{1}{2}s_{n-2} + \beta_n \sqrt{s_{n-1}^2 + s_{n-1} + \frac{1}{4}s_{n-2}^2 + 1 + \frac{\sin 2n}{n+1}}. \tag{8.80}$$

We note that under the square root above is nonnegative as expected, since

$$s_{n-1}^2 + s_{n-1} + 1 + \frac{\sin 2n}{n+1} = s_{n-1}^2 + s_{n-1} + \frac{1}{4} + \frac{3}{4} + \frac{\sin 2n}{n+1}$$

$$= \left(s_{n-1} + \frac{1}{2}\right)^2 + \frac{3}{4} + \frac{\sin 2n}{n+1}$$

and the last two terms above sum to a nonnegative value for all n by (8.79). Equations (8.80) can be used to generate all the real solutions of (8.78). ∎

In less restricted cases, the set $\mathcal{S}$ may itself be invariant but not equal to $\mathbb{R}^k$. In such cases $\mathcal{M} = \mathcal{S}$ but the state-space of real solutions is a proper subset of $\mathbb{R}^k$. The following result illustrates this case while extending Example 8.6 to more general quadratic equations.

COROLLARY 8.4

Let $a_j, b \in \mathbb{R}$ for $j = 1, \ldots, k$ with $a_k, b \neq 0$ and consider the autonomous quadratic difference equation without mixed product terms

$$x_n^2 + a_1 x_{n-1}^2 + \cdots + a_k x_{n-k}^2 = b. \tag{8.81}$$

(a) If the following inequalities hold

$$a_1 \leq 1, \qquad a_1 a_k \geq 0, \text{ and} \tag{8.82}$$
$$a_j \leq a_1 a_{j-1} \text{ for } j = 2, \ldots, k$$

then the following set, which contains the orbits of all real solutions of (8.81),

$$\mathcal{S} = \left\{ (u_1, \ldots, u_k) : \sum_{j=1}^{k} a_j u_j^2 \leq b \right\}$$

is invariant under each of the map sequences

$$F_n(u_1, \ldots, u_k) = \left(\beta_n \sqrt{b - \sum_{j=1}^{k} a_j u_j^2}, u_1, u_2, \ldots, u_{k-1} \right)$$

that are defined by a binary sequence $\{\beta_n\}$. Thus $\mathcal{S}$ is the state-space of all real solutions of (8.81).

(b) If

$$b > 0, \qquad \sum_{j=1}^{k} a_j > -1 \tag{8.83}$$

(e.g., if $b, a_j > 0$, i.e., $\mathcal{S}$ is elliptic) then (8.81) has a positive fixed point

$$\bar{x} = \sqrt{\frac{b}{1 + \sum_{j=1}^{k} a_j}}$$

and the following are solutions of (8.81)

$$\{\bar{x}\beta_n\}_{n=1}^{\infty}$$

for every fixed but arbitrarily chosen binary sequence $\{\beta_n\}$ in $\{-1, 1\}$. Note that $(\bar{x}, \ldots, \bar{x}) \in \mathcal{S}$ but $\mathcal{S}$ is not necessarily invariant under conditions (8.83).

PROOF   (a) Let $(u_1, \ldots, u_k) \in \mathcal{S}$. Then $F_n(u_1, \ldots, u_k) \in \mathcal{S}$ if and only if

$$a_1 \left(\beta_n \sqrt{b - \sum_{j=1}^{k} a_j u_j^2} \right)^2 + a_2 u_1^2 + \cdots a_k u_{k-1}^2 \leq b \Leftrightarrow$$

$$a_1 \left(b - \sum_{j=1}^{k} a_j u_j^2 \right) + a_2 u_1^2 + \cdots a_k u_{k-1}^2 \leq b \Leftrightarrow$$

$$(a_2 - a_1^2) u_1^2 + \sum_{j=3}^{k-1} (a_j - a_1 a_{j-1}) u_j^2 - a_1 a_k u_k^2 \leq (1 - a_1) b.$$

Now under conditions (8.82) the left-hand side of the last inequality is nonpositive while its right-hand side is nonnegative. It follows that the set $\mathcal{S}$ is invariant under F_n for all n so the last statement of (a) follows from Theorem 8.3.

(b) The number $\bar{x}$ is the positive solution of

$$\bar{x}^2 + a_1\bar{x}^2 + \cdots + a_k\bar{x}^2 = b$$

so $\bar{x}$ is a fixed point of (8.81). Finally, the sequence $\{\bar{x}\beta_n\}_{n=-k+1}^{\infty}$ is easily seen to satisfy (8.81) so the sequence is a solution. Note finally that

$$b = \left(1 + \sum_{j=1}^{k} a_j\right)\bar{x}^2 \geq \sum_{j=1}^{k} a_j\bar{x}^2$$

so that $(\bar{x}, \ldots, \bar{x}) \in \mathcal{S}$ but conditions in (8.82) are not implied. ∎

Similarly to Example 8.6, the substitution $y_n = x_n^2$ transforms Eq. (8.81) into a linear nonhomogeneous equation

$$y_n + a_1 y_{n-1} + \cdots + a_k y_{n-k} = b.$$

However, the inequalities in (8.82) are not readily inferred from the properties of the k-th degree characteristic polynomial of the above nonhomogeneous linear equation.

We close this section with a brief discussion of a case where $a_{0,0,n} \neq 0$ for infinitely many n but not eventually. In this case, the sequence $\{a_{0,0,n}\}$ may play a decisive role in determining the solutions of the difference equation. The next example illustrates this case. This example is also an improper quadratic difference equation; that is, it violates Condition (8.50) (recall Definition 8.4). However, the equation considered in the next example does satisfy the weaker condition

There is $j \in \{0, 1, \ldots, k\}$ such that $a_{0,j,n} \neq 0$ for infinitely many n.

Example 8.9
$(a_{0,0,n} \neq 0$ for infinitely many n but not eventually) Consider the first-order, quadratic difference equation

$$a_n x_n^2 = 4x_{n-1}(1 - x_{n-1}) \tag{8.84}$$

where $\{a_n\}$ is a nonzero binary sequence in $\{0, 1\}$. We first note that if $a_n = 1$ eventually for all large n then the only real solutions of (8.84) are those in the interval $[0, 1]$ since both sides of the equation must be nonnegative. In this case, (8.84) reduces to the recursive equation

$$x_n = 2\sqrt{x_{n-1}(1 - x_{n-1})} \tag{8.85}$$

which has $[0, 1]$ as its invariant interval and indeed, its interval of definition as a real difference equation. The real solutions of Eq. (8.85) in $[0, 1]$ are varied in nature (ranging from periodic to chaotic) but we need not discuss them here; the reader is referred to the standard literature on one-dimensional maps for more details on the solutions of (8.85).

Now, let $m \geq 2$ be a fixed integer and consider the periodic binary sequence

$$a_n = \begin{cases} 0 \text{ if } n = jm \text{ for } j = 0, 1, 2, \ldots \\ 1 \text{ for all other values of } n. \end{cases}$$

What are solutions of (8.84) like for this sequence? We find that there are substantial differences between this case and the preceding one with $a_n = 1$ always. For $n = m$ we obtain

$$0 = a_m x_m^2 = 4x_{m-1}(1 - x_{m-1})$$

so $x_{m-1} = 0$ or 1 while the value of x_m is undetermined by the above equation. Nevertheless, x_m cannot be chosen arbitrarily; to understand the situation better, note that for $n < m$ Eq. (8.84) is identical to (8.85). If we define the map

$$g(u) = 2\sqrt{u(1 - u)}, \quad 0 \leq u \leq 1$$

then $x_{m-1} \in \{0, 1\}$ implies that

$$x_{m-2} \in g^{-1}(\{0, 1\}), \ x_{m-3} \in g^{-2}(\{0, 1\}), \ldots, x_0 \in g^{-m+1}(\{0, 1\}).$$

It follows that the initial value would have to be chosen in the finite set $g^{-m+1}(\{0, 1\})$. Let us start with the set $g^{-1}(\{0, 1\})$. This contains both 0 and 1 since $g(1) = 0 = g(0)$. Further, since $g(1/2) = 1$ and $1/2$ is the only other number with this property. It follows that

$$g^{-1}(\{0, 1\}) = \left\{0, \frac{1}{2}, 1\right\}.$$

Next, the equation $g(u) = 1/2$ has two solutions

$$\frac{2 \pm \sqrt{3}}{4}$$

both in the interval $[0, 1]$ so that

$$g^{-2}(\{0, 1\}) = \left\{0, \frac{2 - \sqrt{3}}{4}, \frac{1}{2}, \frac{2 + \sqrt{3}}{4}, 1\right\}.$$

Given the surjective and unimodal nature of g on $[0, 1]$ for each of the new numbers in $g^{-2}(\{0, 1\})$ we generate two new numbers in $[0, 1]$ so $g^{-3}(\{0, 1\})$ has four new numbers in addition to the five numbers in $g^{-2}(\{0, 1\})$. This process implies that there are $2^{m-1} + 1$ numbers in $g^{-m+1}(\{0, 1\})$ and that

$$g^{-1}(\{0, 1\}) \subset g^{-2}(\{0, 1\}) \subset \cdots \subset g^{-m+1}(\{0, 1\}).$$

In order for Eq. (8.84) to have real solutions the initial value x_0 must be chosen in $g^{-m+1}(\{0,1\})$. For such x_0 the first $m-1$ terms of the solution will end in x_{m-1} which is either 0 or 1. Now, the same thing is true for x_m since $a_{2m} = 0$ and therefore, x_{2m} must be selected in such a way that $x_{2m-1} = 0$ or 1. By this reasoning, we conclude that the values of x_{jm} for every $j = 0, 1, 2, \ldots$ must be chosen from the finite set $g^{-m+1}(\{0,1\})$ and thus, this finite subset of $[0,1]$ contains all solutions of Eq. (8.84). However, the values of x_{jm} can be selected randomly in $g^{-m+1}(\{0,1\})$ for different values of j so the solutions are not all periodic. For instance, if $m = 2$ then the values

$$x_0, x_2, x_4, \ldots \in g^{-1}(\{0,1\}) = \left\{0, \frac{1}{2}, 1\right\}.$$

Suppose that $x_0 = 1/2$. Then $x_1 = 1$ by (8.84) since $a_1 = 1$. Next $a_2 = 0$ so we may choose x_2 to be any one of the three numbers in $g^{-1}(\{0,1\})$. If we again choose $1/2$ and do this for x_n for all even n then a period-2 solution of (8.84) is obtained as

$$\left\{\frac{1}{2}, 1, \frac{1}{2}, 1, \ldots\right\}.$$

Another period-2 solution of (8.84) is obtained by setting $x_0 = 1$ and repeating this choice as x_n for all even n:

$$\{1, 0, 1, 0, \ldots\}$$

On the other hand, we may choose $x_2 = 1$ to obtain $x_3 = 0$ by (8.84) then set $x_4 = 1/2$ and repeat this pattern to obtain a period-4 solution of (8.84) as

$$\left\{\frac{1}{2}, 1, 1, 0, \frac{1}{2}, 1, 1, 0, \ldots\right\}.$$

Other period-4 solutions are possible by different selections of values for even-indexed terms; e.g.,

$$\left\{0, 0, \frac{1}{2}, 1, 0, 0, \frac{1}{2}, 1, , \ldots\right\}$$

and a few other possibilities. In a similar manner we may obtain various solutions for each even period 6, 8, etc. nonperiodic solutions are also possible by choosing values for even-indexed terms in a non-repetitive way. For instance, we may set $x_{2j} = 1/2$ for all odd primes j and $x_{2j} = 1$ for all other values of j. The solution of (8.84) obtained in this way

$$\left\{1, 0, 1, 0, 1, 0, \underset{n=6}{\frac{1}{2}}, 1, 1, 0, \underset{n=10}{\frac{1}{2}}, 1, \ldots, \underset{n=14}{\frac{1}{2}}, \ldots, \underset{n=22}{\frac{1}{2}}, \ldots\right\}$$

is not periodic. $\quad\square$

8.4.2   Quadratic factors and cofactors

We now consider equations of type (8.49) that have a right semi-invertible form symmetry. To simplify the discussion without losing sight of essential ideas, we limit attention to the case $k = 2$, or the second-order equation

$$E_n(x_n, x_{n-1}, x_{n-2}) = 0 \qquad (8.86)$$

on a field $\mathcal{F}$ where

$$\begin{aligned}
E_n(u_0, u_1, u_2) = {} & a_{0,0,n}u_0^2 + a_{0,1,n}u_0u_1 + a_{0,2,n}u_0u_2 + \qquad (8.87) \\
& a_{1,0,n}u_1^2 + a_{1,2,n}u_1u_2 + a_{2,0,n}u_2^2 + \\
& b_{0,n}u_0 + b_{1,n}u_1 + b_{2,n}u_2 + c_n.
\end{aligned}$$

In this expression, to ensure that u_0 (corresponding to the x_n term) does *not* drop out, we assume that for each n,

$$\text{if } a_{0,0,n} = a_{0,1,n} = b_{0,n} = 0 \text{ or } a_{1,2,n} = a_{2,0,n} = b_{2,n} = 0, \qquad (8.88)$$
$$\text{then } a_{0,2,n} \neq 0.$$

If the quadratic expression (8.87) has a form symmetry $\{H_n\}$ where

$$H_n(u_0, u_1, u_2) = [h_n(u_0, u_1), h_{n-1}(u_1, u_2)]$$

then there are sequences $\{\phi_n\}, \{h_n\}$ of functions $\phi_n, h_n : \mathcal{F}^2 \to \mathcal{F}$ that satisfy the relation (8.32), i.e., for all $(u_0, u_1, u_2) \in \mathcal{F}^3$,

$$\phi_n(h_n(u_0, u_1), h_{n-1}(u_1, u_2)) = E_n(u_0, u_1, u_2). \qquad (8.89)$$

Further, the form symmetry H_n is right semi-invertible with respect to $\mathcal{F}$ if there are functions $\mu_n, \theta_n : \mathcal{F} \to \mathcal{F}$ where θ_n is a bijection for all n and

$$h_n(u, v) = \mu_n(u) + \theta_n(v).$$

In general, the functions ϕ_n, h_n satisfying (8.89) need not be of polynomial type even though all E_n are quadratic polynomials. But if we consider only ϕ_n, h_n that are of polynomial type then (8.89) is satisfied if one of ϕ_n or h_n is quadratic and the other linear for each n. In such cases, the factorization of Eq. (8.87) is also of quadratic type since both the factor and cofactor equations are quadratic difference equations. Further, in the right semi-invertible case, θ_n must be linear for the same value of n so a polynomial h_n may be defined as

$$h_n(u, v) = \alpha_n u^2 + \beta_n u + \gamma_n + \delta_n v \qquad (8.90)$$

where $\alpha_n, \beta_n, \gamma_n, \delta_n \in \mathcal{F}$ with $\delta_n \neq 0$ for all n and either $\alpha_n \neq 0$ or $\beta_n \neq 0$ for every n. There are two cases to consider.

Case 1. The factor-function sequence

$$\phi_n(u, v) = a_n u + b_n v + c_n \qquad (8.91)$$

consists of polynomials of degree 1 with variable coefficients for all n. Then according to Eq. (8.89)

$$
\begin{aligned}
E_n(u_0, u_1, u_2) &= a_n(\alpha_n u_0^2 + \beta_n u_0 + \gamma_n + \delta_n u_1) + \\
&\quad b_n(\alpha_{n-1} u_1^2 + \beta_{n-1} u_1 + \gamma_{n-1} + \delta_{n-1} u_2) + c_n \\
&= a_n \alpha_n u_0^2 + b_n \alpha_{n-1} u_1^2 + a_n \beta_n u_0 + \\
&\quad (a_n \delta_n + b_n \beta_{n-1}) u_1 + b_n \delta_{n-1} u_2 + a_n \gamma_n + b_n \gamma_{n-1} + c_n.
\end{aligned}
\tag{8.92}
$$

Conditions (8.88) require that $a_n \neq 0$ for all n. Further, to keep the equation nonlinear of order 2, we assume that $\alpha_n, b_n, \delta_n \neq 0$ for all n. Equation (8.5) in Example 8.1 is a special case of (8.92). Note that *there are no mixed quadratic terms in* (8.92).

As is apparent from the expression (8.92) we may, for economy of notation, normalize h_n by setting $\alpha_n = 1$ and $\gamma_n = 0$ for all n without losing generality.

The next result specifies when a quadratic difference equation of the Case 1 variety has a form symmetry of type (8.90) and a factor-function sequence of type (8.91).

PROPOSITION 8.1
Consider the quadratic difference equation

$$
x_n^2 + A_n x_{n-1}^2 + B_n x_n + C_n x_{n-1} + D_n x_{n-2} + F_n = 0
\tag{8.93}
$$

with A_n, B_n, C_n, D_n, F_n in a nontrivial field $\mathcal{F}$ for all n. If the coefficients satisfy the equality

$$
D_{n+1} + A_{n+1} A_n B_{n-1} = A_{n+1} C_n
\tag{8.94}
$$

for every n then (8.93) has a factorization into first-order equations

$$
t_n + A_n t_{n-1} + F_n = 0,
$$
$$
x_n^2 + A_{n+1} B_n x_n + D_{n+1} x_{n-1} = A_{n+1} t_n.
$$

PROOF Since (8.93) of the same type as (8.92) we match coefficients in the two equations to obtain

$$
a_n = 1, \quad b_n = A_n, \quad \beta_n = B_n, \quad b_n \delta_{n-1} = D_n, \quad c_n = F_n
$$

and

$$
\delta_n + b_n \beta_{n-1} = C_n.
$$

Since $\delta_{n-1} = D_n / b_n = D_n / A_n$ this equation is equivalent to (8.94). Now, the factor-function sequence ϕ_n is really obtained from (8.91) and the cofactor from (8.90). ∎

Note that (8.94) effectively gives the value of D_n in terms of the remaining coefficients. If the coefficients are constants A, B, C, D in the preceding proposition then the following consequence of the above proposition applies to a generalization of Eq. (8.5) we encountered in Example 8.1.

COROLLARY 8.5

The quadratic difference equation

$$x_n^2 + Ax_{n-1}^2 + Bx_n + Cx_{n-1} + A(C - AB)x_{n-2} = F_n$$

has a factorization into first-order equations

$$t_n + At_{n-1} = F_n,$$
$$x_n^2 + ABx_n + A(C - AB)x_{n-1} = At_n.$$

The next case is the reverse of Case 1.

Case 2. $\phi_n(u, v) = a_n u^2 + b_n uv + c_n v^2 + d_n u + e_n v + f_n$ is a polynomial of degree 2 with variable coefficients in a field $\mathcal{F}$ for all n. Then Eq. (8.89) yields a quadratic expression if h_n contains no quadratic terms, i.e., if $\alpha_n = 0$ for all n. Therefore,

$$h_n(u, v) = \beta_n u + \delta_n v + \gamma_n. \tag{8.95}$$

This form symmetry is linear in the variables u, v and is discussed in detail in Section 8.4.3 below. Note that if (8.95) is inserted into the quadratic expression E_n then

$$E_n(u_0, u_1, u_2) = a_n h_n(u_0, u_1)^2 + b_n h_n(u_0, u_1)h_{n-1}(u_1, u_2) + \tag{8.96}$$
$$c_n h_{n-1}(u_1, u_2)^2 + d_n h_n(u_0, u_1) + e_n h_{n-1}(u_1, u_2) + f_n.$$

Conditions (8.88) assure that E_n contains terms u_0 and u_2 as well as quadratic terms for every n. Note that (8.96) subsumes (8.92) and contains mixed quadratic terms. Further, h_n is semi-invertible so the factor equation is recursive.

By the above construction, any difference equation of order two on a field $\mathcal{F}$ that can be written in the form (8.96) has a factorization into a pair of first-order difference equations.

Example 8.10

Consider the quadratic expression (8.96) on $\mathbb{R}$ with the parameter values

$$b_n = c_n = d_n = \gamma_n = 0, \quad a_n = -1, \; \beta_n = 1, \; \delta_n = -\frac{1}{2n}, \; e_n = n.$$

Then the difference equation $E_n(x_n, x_{n-1}, x_{n-2}) = 0$ may be written as

$$4n^2 x_n^2 - 4nx_n x_{n-1} + x_{n-1}^2 - 4n^3 x_{n-1} + 2n^2 x_{n-2} = \sigma_n \tag{8.97}$$

where $\sigma_n = 4n^2 f_n$ is an arbitrary sequence of real numbers. By the preceding discussion, Eq. (8.97) has the factorization

$$t_n^2 = nt_{n-1} + \frac{\sigma_n}{4n^2}$$

$$x_n = \frac{1}{2n}x_{n-1} + t_n.$$

▯

8.4.3 Quadratic equations with a linear form symmetry

The factorization in Example 8.10 can also be obtained from Eq. (8.97) directly by algebraic manipulations followed by the substitution $t_n = x_n - (1/2n)x_{n-1}$. Alternatively, we may obtain the factorization using Theorem 8.2 (see the Problems section below). The next corollary of Theorem 8.2 examines conditions that imply the existence of a linear form symmetry generally for Eq. (8.86). By a *linear form symmetry* in this section we mean the normalized special case of (8.95), i.e.,

$$h_n(u, v) = \alpha_n u + v, \quad \alpha_n \neq 0 \text{ for all } n. \tag{8.98}$$

COROLLARY 8.6
The quadratic difference equation (8.86) has the linear form symmetry with components h_n, h_{n-1} defined by (8.98) if and only if a sequence $\{\alpha_n\}$ exists in the field $\mathcal{F}$ such that $\alpha_n \neq 0$ for all n and all four of the following first-order equations are satisfied:

$$a_{0,0,n} - a_{0,1,n}\alpha_n + a_{1,1,n}\alpha_n^2 + a_{0,2,n}\alpha_n\alpha_{n-1} - a_{1,2,n}\alpha_n^2\alpha_{n-1}$$
$$+ a_{2,2,n}\alpha_n^2\alpha_{n-1}^2 = 0 \tag{8.99}$$
$$a_{0,1,n} - a_{0,2,n}\alpha_{n-1} - 2a_{1,1,n}\alpha_n + 2a_{1,2,n}\alpha_n\alpha_{n-1}$$
$$- 2a_{2,2,n}\alpha_n\alpha_{n-1}^2 = 0 \tag{8.100}$$
$$a_{0,2,n} - a_{1,2,n}\alpha_n + 2a_{2,2,n}\alpha_n\alpha_{n-1} = 0 \tag{8.101}$$
$$b_{0,n} - b_{1,n}\alpha_n + b_{2,n}\alpha_n\alpha_{n-1} = 0. \tag{8.102}$$

In this case, Eq. (8.86) has a factorization with a first-order factor equation

$$(a_{1,1,n} - a_{1,2,n}\alpha_{n-1} + a_{2,2,n}\alpha_{n-1}^2)t_n^2 + a_{2,2,n}t_{n-1}^2 + (a_{1,2,n} - 2a_{2,2,n}\alpha_{n-1})t_nt_{n-1} + (b_{1,n} - b_{2,n}\alpha_{n-1})t_n + b_{2,n}t_{n-1} + c_n = 0.$$

and a cofactor equation $\alpha_n x_n + x_{n-1} = t_n$ also of order one.

PROOF   For any nonzero sequence $\{\alpha_n\}$ in $\mathcal{F}$ the functions h_n defined by (8.98) are semi-invertible with $\mu_n(u) = \alpha_n u$ and $\theta_n(v) = v$ for all n. Note that $\theta_n^{-1} = \theta_n$ for all n and the group structure is the additive group of the field so the quantities $\zeta_{j,n}$ in Theorem 8.2 take the forms

$$\zeta_{1,n} = \zeta_{1,n}(u_0, v_1) = -\alpha_n u_0 + v_1,$$
$$\zeta_{2,n} = \zeta_{2,n}(u_0, v_1, v_2) = \alpha_n \alpha_{n-1} u_0 - \alpha_{n-1} v_1 + v_2.$$

By Theorem 8.2, Eq. (8.86) has the linear form symmetry if and only if the expression $E_n = E_n(u_0, \zeta_{1,n}, \zeta_{2,n})$ is independent of u_0 for all n. Now

$$
\begin{aligned}
E_n = {} & a_{0,0,n} u_0^2 + a_{0,1,n} u_0 (v_1 - \alpha_n u_0) + \\
& a_{0,2,n} u_0 (\alpha_n \alpha_{n-1} u_0 - \alpha_{n-1} v_1 + v_2) + a_{1,1,n} (v_1 - \alpha_n u_0)^2 + \\
& a_{1,2,n} (v_1 - \alpha_n u_0)(\alpha_n \alpha_{n-1} u_0 - \alpha_{n-1} v_1 + v_2) + \\
& a_{2,2,n} (\alpha_n \alpha_{n-1} u_0 - \alpha_{n-1} v_1 + v_2)^2 + b_{0,n} u_0 + \\
& b_{1,n} (v_1 - \alpha_n u_0) + b_{2,n} (\alpha_n \alpha_{n-1} u_0 - \alpha_{n-1} v_1 + v_2) + c_n.
\end{aligned}
$$

Multiplying terms in the above expression gives

$$
\begin{aligned}
E_n = {} & a_{0,0,n} u_0^2 + a_{0,1,n} u_0 v_1 - a_{0,1,n} \alpha_n u_0^2 + a_{0,2,n} \alpha_n \alpha_{n-1} u_0^2 - \\
& a_{0,2,n} \alpha_{n-1} u_0 v_1 + a_{0,2,n} u_0 v_2 + a_{1,1,n} v_1^2 - 2 a_{1,1,n} \alpha_n u_0 v_1 + \\
& a_{1,1,n} \alpha_n^2 u_0^2 + a_{1,2,n} \alpha_n \alpha_{n-1} u_0 v_1 - a_{1,2,n} \alpha_{n-1} v_1^2 + a_{1,2,n} v_1 v_2 - \\
& a_{1,2,n} \alpha_n^2 \alpha_{n-1} u_0^2 + a_{1,2,n} \alpha_n \alpha_{n-1} u_0 v_1 - a_{1,2,n} \alpha_n u_0 v_2 + \\
& a_{2,2,n} \alpha_n^2 \alpha_{n-1}^2 u_0^2 - 2 a_{2,2,n} \alpha_n \alpha_{n-1}^2 u_0 v_1 + 2 a_{2,2,n} \alpha_n \alpha_{n-1} u_0 v_2 + \\
& a_{2,2,n} (\alpha_{n-1} v_1 - v_2)^2 + b_{0,n} u_0 + b_{1,n} v_1 - b_{1,n} \alpha_n u_0 + \\
& b_{2,n} \alpha_n \alpha_{n-1} u_0 - b_{2,n} (\alpha_{n-1} v_1 - v_2) + c_n.
\end{aligned}
$$

Terms containing u_0 or u_0^2 must sum to zeros. Rearranging terms in the preceding expression gives

$$
\begin{aligned}
E_n = {} & (a_{0,0,n} - a_{0,1,n} \alpha_n + a_{0,2,n} \alpha_n \alpha_{n-1} + a_{1,1,n} \alpha_n^2 - a_{1,2,n} \alpha_n^2 \alpha_{n-1} + \\
& a_{2,2,n} \alpha_n^2 \alpha_{n-1}^2) u_0^2 + (a_{0,1,n} - a_{0,2,n} \alpha_{n-1} - 2 a_{1,1,n} \alpha_n + \\
& 2 a_{1,2,n} \alpha_n \alpha_{n-1} - 2 a_{2,2,n} \alpha_n \alpha_{n-1}^2) u_0 v_1 + (a_{0,2,n} - a_{1,2,n} \alpha_n + \\
& 2 a_{2,2,n} \alpha_n \alpha_{n-1}) u_0 v_2 + (b_{0,n} - b_{1,n} \alpha_n + b_{2,n} \alpha_n \alpha_{n-1}) u_0 + \\
& (a_{1,1,n} - a_{1,2,n} \alpha_{n-1} + a_{2,2,n} \alpha_{n-1}^2) v_1^2 + a_{2,2,n} v_2^2 + (a_{1,2,n} - \\
& 2 a_{2,2,n} \alpha_{n-1}) v_1 v_2 + (b_{1,n} - b_{2,n} \alpha_{n-1}) v_1 + b_{2,n} v_2 + c_n.
\end{aligned}
$$

Setting the coefficients of variable terms containing u_0 equal to zeros gives the four first-order equations (8.99)–(8.102). The part of E_n above that does not vanish yields the factor functions

$$
\begin{aligned}
\phi_n(v_1, v_2) = {} & (a_{1,1,n} - a_{1,2,n} \alpha_{n-1} + a_{2,2,n} \alpha_{n-1}^2) v_1^2 + a_{2,2,n} v_2^2 + \\
& (a_{1,2,n} - 2 a_{2,2,n} \alpha_{n-1}) v_1 v_2 + (b_{1,n} - b_{2,n} \alpha_{n-1}) v_1 + \\
& b_{2,n} v_2 + c_n.
\end{aligned}
$$

This expression plus the linear cofactor $t_n = h_n(x_n, x_{n-1})$ give the stated factorization. ∎

Equations (8.99)–(8.102) are reminiscent of the Riccati relation (7.36) in Section 7.4. Whereas one relation suffices for linear equations, *four* relations are required for the occurrence of the linear form symmetry in a quadratic difference equation. Of course, *nonlinear* form symmetries may occur in the quadratic case if one or more of the equations (8.99)–(8.102) fail; see the discussion before Example 8.10 above. In particular, there are rational recursive equations that do not possess the linear form symmetry; see the Problems section below.

If all coefficients in $E_n(u_0, u_1, u_2)$ are constants except possibly the free term c_n then a simpler version of Corollary 8.6 is easily obtained as follows.

COROLLARY 8.7
The quadratic difference equation

$$a_{0,0}x_n^2 + a_{0,1}x_nx_{n-1} + a_{0,2}x_nx_{n-2} + a_{1,1}x_{n-1}^2 + a_{1,2}x_{n-1}x_{n-2}$$
$$+a_{2,2}x_{n-2}^2 + b_0x_n + b_1x_{n-1} + b_2x_{n-2} + c_n = 0 \qquad (8.103)$$

in a nontrivial field $\mathcal{F}$ has the linear form symmetry with components $h(u, v) = \alpha u + v$ if and only if the following polynomials have a common nonzero root α in $\mathcal{F}$:

$$a_{0,0} - a_{0,1}\alpha + (a_{1,1} + a_{0,2})\alpha^2 - a_{1,2}\alpha^3 + a_{2,2}\alpha^4 = 0, \qquad (8.104)$$
$$a_{0,1} - (a_{0,2} + 2a_{1,1})\alpha + 2a_{1,2}\alpha^2 - 2a_{2,2}\alpha^3 = 0, \qquad (8.105)$$
$$a_{0,2} - a_{1,2}\alpha + 2a_{2,2}\alpha^2 = 0, \qquad (8.106)$$
$$b_0 - b_1\alpha + b_2\alpha^2 = 0. \qquad (8.107)$$

If such a root $\alpha \neq 0$ exists then Eq. (8.103) has the factorization

$$(a_{1,1} - a_{1,2}\alpha + a_{2,2}\alpha^2)t_n^2 + (a_{1,2} - 2a_{2,2}\alpha)t_nt_{n-1} + a_{2,2}t_{n-1}^2$$
$$+(b_1 - b_2\alpha)t_n + b_2t_{n-1} + c_n = 0,$$
$$x_n = -\frac{1}{\alpha}x_{n-1} + \frac{t_n}{\alpha}.$$

The following example gives a straightforward application of Corollary 8.7.

Example 8.11
Consider the difference equation

$$ax_nx_{n-1} + bx_{n-1}x_{n-2} + cx_nx_{n-2} + dx_{n-2}^2 = \delta_n \qquad (8.108)$$

with $a, b, c, d, \delta_n \in \mathbb{R}$. To check whether Eq. (8.108) has a factorization into first-order equations we use Corollary 8.7. In this case, (8.104)–(8.106) take the following forms

$$da^4 - ba^3 + ca^2 - a\alpha = 0 \tag{8.109}$$

$$-2da^3 + 2ba^2 - c\alpha + a = 0 \tag{8.110}$$

$$2da^2 - ba + c = 0 \tag{8.111}$$

while (8.107) holds trivially. If α is a nonzero solution of (8.109)–(8.111) then by dividing (8.109) by α and adding the result to (8.110) we find that

$$-da^3 + ba^2 = 0 \Rightarrow b = da. \tag{8.112}$$

Similarly, multiplying (8.111) by α and adding the result to (8.110) gives

$$ba^2 + a = 0 \Rightarrow a = -ba^2 = -da^3. \tag{8.113}$$

And from (8.111) and (8.112) we obtain

$$da^2 + \alpha(da - b) + c = 0 \Rightarrow c = -da^2. \tag{8.114}$$

From (8.112)–(8.114) it follows that

$$\alpha = \frac{b}{d}, \quad a = -\frac{b^3}{d^2}, \quad c = -\frac{b^2}{d}. \tag{8.115}$$

From (8.115) we conclude that if $b, d \neq 0$ then there is a nonzero $\alpha \in \mathbb{R}$ that makes Corollary 8.7 applicable, provided that the coefficients a, c are given as in (8.115). If (8.115) holds then Eq. (8.108) has a factor equation

$$(-ba + da^2)t_n^2 + (b - 2da)t_n t_{n-1} + dt_{n-1}^2 - \delta_n = 0, \quad \text{or:}$$
$$-bt_n t_{n-1} + dt_{n-1}^2 - \delta_n = 0.$$

Thus, subject to (8.115), Eq. (8.108) has the factorization

$$bt_n t_{n-1} - dt_{n-1}^2 + \delta_n = 0,$$
$$\frac{b}{d}x_n + x_{n-1} = t_n$$

which can also be written in a recursive form as

$$t_n = \frac{d}{b}t_{n-1} - \frac{\delta_n}{bt_{n-1}},$$
$$x_n = -\frac{d}{b}x_{n-1} + t_n.$$

☐

We note that Eq. (8.108) also has a unique recursive form and that an SC factorization of the recursive form is the same as that obtained in Example 8.11. See the Problems below.

In the next example we use Corollary 8.7 to study the solutions of a quadratic difference equation that does not possess a unique recursive form.

Example 8.12

Let a, b, c, d be nonzero real numbers and consider the equation

$$x_n^2 + ax_n x_{n-1} + bx_n x_{n-2} + cx_{n-1}x_{n-2} = d. \tag{8.116}$$

If a linear form symmetry exists under some conditions on the real parameters in Eq. (8.116) then we obtain a factorization of this equation into two first-order equations. To find the linear form symmetry we use Corollary 8.7. In the absence of linear terms in (8.116), Eq. (8.107) holds trivially; the other three equations (8.104)–(8.106) take the following forms

$$1 - a\alpha + b\alpha^2 - c\alpha^3 = 0 \tag{8.117}$$

$$a - b\alpha + 2c\alpha^2 = 0 \tag{8.118}$$

$$b - c\alpha = 0. \tag{8.119}$$

From (8.119) it follows that $\alpha = b/c$. This nonzero value of α must satisfy the other two equations in the above system so from (8.118) we obtain

$$a + \frac{b^2}{c} = 0$$

while (8.117) yields

$$1 - \frac{ab}{c} = 0 \quad \text{or} \quad c = ab.$$

Eliminating b and c from the last two equations gives

$$b = -a^2, \ c = -a^3 \text{ and } \alpha = \frac{1}{a}.$$

These calculations show that the equation

$$x_n^2 + ax_n x_{n-1} - a^2 x_n x_{n-2} - a^3 x_{n-1}x_{n-2} - d = 0 \tag{8.120}$$

has the linear form symmetry $h(u, v) = (1/a)u + v$ and the corresponding factorization

$$a^2 t_n^2 - a^3 t_n t_{n-1} - d = 0, \tag{8.121}$$

$$x_n + ax_{n-1} = at_n. \tag{8.122}$$

Let us use this factorization to obtain information about the real solutions of (8.120). Although the factor equation (8.121) has order one, it is nonrecursive. Thus we start by completing the square in (8.121) to obtain

$$\left(t_n - \frac{a}{2}t_{n-1}\right)^2 = \frac{a^2}{4}t_{n-1}^2 + \frac{d}{a^2}$$

which yields the recursive class of Eq. (8.121)

$$t_n = \frac{a}{2}t_{n-1} + \beta_n \sqrt{\frac{a^2}{4}t_{n-1}^2 + \frac{d}{a^2}} \tag{8.123}$$

where as usual, $\{\beta_n\}$ is a binary sequence in $\{-1, 1\}$. Depending on the values of a and d, different types of behaviors are possible for the solutions of (8.123). We consider a particular case here for illustration; let

$$-1 < a < 0, \ d > 0. \tag{8.124}$$

Under conditions (8.124) every solution of (8.123) with real initial values is real, no matter what binary sequence $\{\beta_n\}$ is chosen. To simplify the exposition, we consider the constant case where $\beta_n = 1$ for all n, i.e., the autonomous equation

$$t_n = \frac{a}{2}t_{n-1} + \sqrt{\frac{a^2}{4}t_{n-1}^2 + \frac{d}{a^2}}.$$

In this case, it is easy to see that the function

$$\psi(t) = \frac{a}{2}t + \sqrt{\frac{a^2}{4}t^2 + \frac{d}{a^2}}$$

is positive and has a unique fixed point

$$\bar{t} = \frac{1}{|a|}\sqrt{\frac{d}{1-a}}.$$

Further, under conditions (8.124) routine calculation shows that the iterate

$$\psi(\psi(t)) = \frac{a}{2}\psi(t) + \sqrt{\frac{a^2}{4}\psi(t)^2 + \frac{d}{a^2}}$$

$$= \frac{a^2}{4}t + \frac{a}{2}\sqrt{\frac{a^2}{4}t^2 + \frac{d}{a^2}} + \sqrt{\frac{a^2}{4}\left[\frac{a}{2}t + \sqrt{\frac{a^2}{4}t^2 + \frac{d}{a^2}}\right]^2 + \frac{d}{a^2}}$$

$$= \frac{a^2}{4}t + \frac{a}{2}\sqrt{\frac{a^2}{4}t^2 + \frac{d}{a^2}} + \sqrt{\frac{a^4}{8}t^2 + \frac{a^3}{4}t\sqrt{\frac{a^2}{4}t^2 + \frac{d}{a^2}} + \frac{d}{4} + \frac{d}{a^2}}.$$

has the property

$$\psi(\psi(t)) > t \text{ if } t < \bar{t} \text{ and } \psi(\psi(t)) < t \text{ if } t > \bar{t}.$$

This fact implies that $\bar{t}$ is stable and globally attracting; see the Appendix. With $t_n \to \bar{t}$ as $n \to \infty$ it can be shown that under conditions (8.124) the corresponding solution $\{x_n\}$ of the cofactor equation (8.122) approaches the fixed point $\bar{x}$ of (8.120) where

$$\bar{x} = \frac{a\bar{t}}{1+a} = \frac{a}{-a(1+a)}\sqrt{\frac{d}{1-a}} = -\frac{1}{1+a}\sqrt{\frac{d}{1-a}}.$$

Of course, Eq. (8.120) possesses a greater variety of solutions if different binary sequences $\{\beta_n\}$ are chosen or if different ranges are considered for parameters such as a. ☐

Although a form symmetry is not known for Eq. (8.116) with arbitrary nonzero real parameters a, b, c, d we may determine the recursive class of (8.116) by completing the squares; see the Problems for this chapter.

Corollary 8.7 may also imply the *non-existence* of a linear form symmetry for Eq. (8.103) as seen in the next example.

Example 8.13

(a) The following quadratic difference equation on $\mathbb{R}$ *cannot* have a linear form symmetry since Eq. (8.106) does not hold for a nonzero α:

$$x_n^2 + x_{n-1}^2 + x_{n-2}^2 = c_n, \quad c_n \in \mathbb{R} \text{ for all } n. \tag{8.125}$$

(b) Eq. (8.4) in Example 8.1 does not have a linear form symmetry. In fact, the more general equation

$$x_n^2 - ax_{n-1}^2 - bx_{n-1} - dx_{n-2} = c_n$$

cannot have a linear form symmetry since Eq. (8.105) does not hold if $a \neq 0$ and Eq. (8.104) fails if $a = 0$. ☐

8.5 An order-preserving form symmetry

In this closing section we formalize an observation whose special cases were encountered earlier.

REMARK 8.5 (**An order-preserving form symmetry**) Note that if $y_n = x_n^2$ then (8.125) is transformed into a linear equation

$$y_n + y_{n-1} + y_{n-2} = c_n$$

which *does* have a linear form symmetry and can thus be factored in the usual way into two first-order difference equations. Then substituting x_n^2 back in the cofactor gives a factorization of (8.125). The substitution $y_n = x_n^2$ is an order-preserving form symmetry since it does not reduce the order of (8.125). This useful idea can be extended to the more general difference equation

$$f(x_n) + a_{1,n}f(x_{n-1}) + \cdots + a_{k,n}f(x_{n-k}) = c_n \tag{8.126}$$

where $\{a_{j,n}\}$ and $\{c_n\}$ are given sequences in a field $\mathcal{F}$ for $j = 1, 2, \ldots, k$ and $f : \mathcal{F} \to \mathcal{F}$ a given function. The substitution $y_n = f(x_n)$ transforms (8.126) into a linear equation which can be factored in the usual way and in which the factor equation is again linear of order $k - 1$. ∎

Example 8.14

Consider once again the equation

$$x_n^2 + x_{n-1}^2 + x_{n-2}^2 = 1 \qquad (8.127)$$

from Example 8.6 and use the order preserving form symmetry $y_n = x_n^2$ to tranform (8.127) into the nonhomogeneous linear equation

$$y_n + y_{n-1} + y_{n-2} = 1. \qquad (8.128)$$

We now use the solutions of (8.128) to study the solutions of (8.127). The eigenvalues of the homogeneous part of (8.128) are roots of the polynomial $\lambda^2 + \lambda + 1$. These roots are

$$\lambda_{\pm} = -\frac{1}{2} \pm \frac{\sqrt{3}}{2} i$$

which are both complex with modulus 1. The argument $\theta = 2\pi/3$ is obtained from $\cos\theta = -1/2$, so the general solution of (8.128) can be written as

$$y_n = C_1 \cos\frac{2\pi n}{3} + C_2 \sin\frac{2\pi n}{3} + \frac{1}{3}$$

since $1/3$ is the unique fixed point of (8.128). C_1, C_2 are constants that depend on initial values $y_0 = x_0^2$, $y_{-1} = x_{-1}^2$ and are calculated from the system

$$-\frac{1}{2}C_1 - \frac{\sqrt{3}}{2}C_2 + \frac{1}{3} = x_{-1}^2$$

$$C_1 + \frac{1}{3} = x_0^2.$$

Thus,

$$C_1 = x_0^2 - \frac{1}{3}, \quad C_2 = \frac{1 - 2x_{-1}^2 - x_0^2}{\sqrt{3}}$$

and the general solution of (8.128) is

$$y_n = \left(x_0^2 - \frac{1}{3}\right)\cos\frac{2\pi n}{3} + \left(\frac{1 - 2x_{-1}^2 - x_0^2}{\sqrt{3}}\right)\sin\frac{2\pi n}{3} + \frac{1}{3}. \qquad (8.129)$$

Only those solutions $\{y_n\}$ of (8.129) that are nonnegative can generate real solutions of (8.127). In particular, since it is necessary that

$$y_1 = 1 - x_0^2 - x_{-1}^2 \geq 0$$

it follows that $x_0^2 + x_{-1}^2 \leq 1$, as expected. If $\{y_n\}$ as defined by (8.129) is non-negative then taking square roots we find that a formula for the corresponding real solutions of (8.127) is

$$x_n = \beta_n \sqrt{\left(x_0^2 - \frac{1}{3}\right) \cos \frac{2\pi n}{3} + \left(\frac{1 - 2x_{-1}^2 - x_0^2}{\sqrt{3}}\right) \sin \frac{2\pi n}{3} + \frac{1}{3}} \qquad (8.130)$$

where $\{\beta_n\}$ is an arbitrary binary sequence in $\{-1, 1\}$. In particular, if

$$x_0, x_{-1} = \pm \frac{1}{\sqrt{3}}$$

then (8.130) reduces to (8.60) in Example 8.6. $\Box$

A comparison of the procedure used in the preceding example with that in Example 8.6 indicates that for Eq. (8.127) the direct approach in Example 8.6 is faster and more informative than the approach in Example 8.14 based on solving a linear equation by the classical method; also see Example 8.4. However, the next example shows that transformation to linear equations can be beneficial in other cases.

Example 8.15
Consider the following difference equation of order three

$$x_n^2 + x_{n-1}^2 + x_{n-3}^2 = 1. \qquad (8.131)$$

with the recursive class

$$s_n = \beta_n \sqrt{1 - s_{n-1}^2 - s_{n-3}^2}. \qquad (8.132)$$

A simple substitution verifies that for each binary sequence $\{\beta_n\}$ with $\beta_n \in \{-1, 1\}$ for all n the following sequence is a real solution of (8.131):

$$\left\{ \frac{\beta_n}{\sqrt{3}} \right\}_{n=-2}^{\infty} . \qquad (8.133)$$

Though simple in nature, these real solutions include sequences of all possible periods as well as bounded, nonperiodic ones. We ask if they are the *only* real solutions of (8.131). Numerical simulations do not reveal any real solutions of (8.132) beyond those listed in (8.133). We *claim* that (8.133) gives all real solutions of (8.131).

To prove this claim, we first transform (8.131) into a nonhomogeneous linear equation using the substitution (8.61) to obtain

$$y_n + y_{n-1} + y_{n-3} = 1$$

which we can also write in the more familiar form

$$y_{n+1} = -y_n - y_{n-2} + 1. \tag{8.134}$$

The eigenvalues of the homogeneous part of (8.134) are roots of the characteristic equation

$$\lambda^3 + \lambda^2 + 1 = 0. \tag{8.135}$$

This equation has one real root λ_0 such that

$$-\frac{3}{2} < \lambda_0 < -1 \tag{8.136}$$

since

$$\left(-\frac{3}{2}\right)^3 + \left(-\frac{3}{2}\right)^2 + 1 = -\frac{1}{8} < 0$$

$$(-1)^3 + (-1)^2 + 1 = 1 > 0.$$

The other roots of (8.135) are found by first dividing

$$\frac{\lambda^3 + \lambda^2 + 1}{\lambda - \lambda_0} = \lambda^2 + (\lambda_0 + 1)\lambda + \lambda_0(\lambda_0 + 1)$$

and then finding the roots of the quadratic equation on the right-hand side as

$$\lambda_\pm = \frac{-(\lambda_0 + 1) \pm \sqrt{(\lambda_0 + 1)^2 - 4\lambda_0(\lambda_0 + 1)}}{2}.$$

By (8.136) $\lambda_\pm$ are complex roots with modulus

$$|\lambda_\pm| = \sqrt{\lambda_0(\lambda_0 + 1)} < 1. \tag{8.137}$$

Now the general solution of (8.134) can be stated as

$$y_n = C_1 \lambda_0^n + \left(\sqrt{\lambda_0(\lambda_0 + 1)}\right)^n (C_2 \cos \theta n + C_3 \sin \theta n) + \frac{1}{3} \tag{8.138}$$

where θ is the argument for complex roots $\lambda_\pm$ and

$$C_i = C_i(y_0, y_{-1}, y_{-2}), \quad i = 1, 2, 3$$

are constants depending on the initial values. By (8.136) the first term λ_0^n in (8.138) is unbounded and eventually oscillating with both positive and negative values. By (8.137) the remaining terms approach $1/3$, the fixed point of (8.134), as $n \to \infty$. Thus if $C_1 \neq 0$ then (8.131) cannot have real solutions. The exceptional solutions corresponding to $C_1 = 0$ are the only ones for which real solutions may occur for (8.131).

The constants C_1, C_2, C_3 are calculated by solving the following system of equations that is obtained from (8.138)

$$C_1 + C_2 + \frac{1}{3} = y_0 \quad (n = 0)$$

$$\frac{C_1}{\lambda_0} + \frac{C_2 \cos \theta - C_3 \sin \theta}{\sqrt{\lambda_0(\lambda_0 + 1)}} + \frac{1}{3} = y_{-1} \quad (n = -1)$$

$$\frac{C_1}{\lambda_0^2} + \frac{C_2 \cos 2\theta - C_3 \sin 2\theta}{\lambda_0(\lambda_0 + 1)} + \frac{1}{3} = y_{-2} \quad (n = -2).$$

We may determine C_1 by eliminating the other two constants. By routine calculation we obtain

$$C_1 = A\left(y_0 - \frac{1}{3}\right) + B\left(y_{-1} - \frac{1}{3}\right) + C\left(y_{-2} - \frac{1}{3}\right) \qquad (8.139)$$

where

$$A = \frac{\lambda_0}{\lambda_0 + 1}, \quad B = -2\sqrt{\lambda_0(\lambda_0 + 1)}\cos\theta, \quad C = \lambda_0^2. \qquad (8.140)$$

It is clear from (8.139) and (8.140) that

$$C_1 = 0 \quad \text{if and only if} \quad y_0 = y_{-1} = y_{-2} = \frac{1}{3}.$$

This proves our claim above that the only real solutions of (8.131) are those given by (8.133). $\quad\square$

Variations of (8.131) are discussed in the Problems for this section. In particular, the third-order equation

$$x_n^2 + x_{n-1}^2 + x_{n-2}^2 + x_{n-3}^2 = 1$$

is seen to have a large variety of real solutions, in contrast to (8.131). The effect of the term x_{n-2}^2 that is missing from (8.131) is most readily explained by the changes that occur in the sizes of eigenvalues of the linear equation that we obtain upon the substitution $y_n = x_n^2$. A direct approach to this problem (showing non-existence of invariant sets in the domain, namely, the unit ball centered at the origin of $\mathbb{R}^3$ rather than using linear equations) is less straightforward.

Example 8.16

Let $a, b \in \mathbb{R}$ with $b > 0$ and consider the difference equation

$$x_n^2 + x_{n-1}^2 + x_{n-2}^2 = a(x_n + x_{n-1} + x_{n-2}) + b. \qquad (8.141)$$

This equation can be written as

$$x_n^2 - ax_n + x_{n-1}^2 - ax_{n-1} + x_{n-2}^2 - ax_{n-2} = b. \qquad (8.142)$$

Substituting

$$y_n = f(x_n) = x_n^2 - ax_n \qquad (8.143)$$

in (8.142) yields the nonhomogeneous linear equation (8.128) whose general solution (8.129) was calculated in Example 8.14. Now completing the square in (8.143) gives

$$\left(x_n - \frac{a}{2}\right)^2 = y_n + \frac{a^2}{4}$$

and this leads to the recursive class

$$x_n = \frac{a}{2} + \beta_n \sqrt{y_n + \frac{a^2}{4}}$$

if $y_n + a^2/4 \geq 0$.   ☐

8.6   Notes

Nonrecursive scalar difference equations have historically been studied in ways that are similar to the studies of scalar differential equations, which do not have natural recursive forms. The approach in this chapter is new and different in that solutions of nonrecursive equations are found either by deriving recursive classes or through reduction of order and factorization.

Quadratic difference equations have not been previously studied as a class. This is surprising in the sense that a large number of other difference equations belong to this remarkable class. This scarcity of results is also understandable since existing methods of study yield relatively little of value that is common to all quadratic difference equations. It is hoped that the new approach presented in this chapter will motivate future studies of this interesting class of difference equations.

The Henon difference equation, more commonly known as the Henon map, is discussed in most introductory texts on difference equations; see, e.g., Devaney (1989) or Strogatz (1994). The logistic equation with delay has been studied in Morimoto (1988, 1989), and Gu and Ruan (2005).

8.7 Problems

8.1 Furnish the details of calculation for Items 1–4 in Example 8.1. The behaviors of all real solutions may be described in terms of the line $v = u$, the three parabolas $v = u^2/b$, $v = u^2/b + b/4$, $v = u^2/b + u$ and the parabolas that are parallel to $v = u^2/b$.

8.2 Following the line of reasoning in Example 8.4, show that the difference equation

$$\sum_{i=0}^{k} \binom{k}{i} x_{n-i} = c_n \tag{8.144}$$

where $\binom{k}{i} = k!/[i!(k-i)!]$ are the binomial coefficients and c_n are positive real numbers for all n has a factorization

$$\sum_{j=0}^{k-1} \binom{k-1}{j} t_{n-j} = c_n,$$

$$x_n + x_{n-1} = t_n$$

in the additive semigroup $(0, \infty)$ (as well as in its subsemigroups $\mathbb{N}$ or $\mathbb{Q} \cap (0, \infty)$ if these contain c_n for all n). Conclude that Eq. (8.144) admits repeated reductions in order until a set of k first-order equations is obtained.

8.3 Use an analysis similar to that in Example 8.6 to find the recursive class and study the solution set of the following equation

$$x_n^2 + x_{n-1}^2 + x_{n-2}^2 + \cdots + x_{n-k}^2 = 1 \tag{8.145}$$

with $k \geq 1$ and

$$\sum_{i=1}^{k} x_{-i}^2 \leq 1. \tag{8.146}$$

In particular, establish the following:
(a) Obtain the analog of the sequence (8.59).
(b) For each binary sequence $\{\beta_n\}$ with $\beta_n \in \{-1, 1\}$ for all n the following sequence is a solution of (8.145):

$$\left\{ \frac{\beta_n}{\sqrt{k+1}} \right\}_{n=-k+1}^{\infty}.$$

Thus (8.145) has periodic solutions of all possible periods as well as bounded nonperiodic solutions.
(c) Every solution of the following recursive equation

$$x_n = \sqrt{1 - x_{n-1}^2 - x_{n-2}^2 - \cdots - x_{n-k}^2}$$

that satisfies (8.146) and is not eventually constant has period $k + 1$. Note that these solutions do not appear in (a).

(d) The substitution $y_n = x_n^2$ converts (8.145) to the nonhomogeneous linear equation

$$y_n + y_{n-1} + \cdots + y_{n-k} = 1.$$

The eigenvalues of the homogeneous part are the $(k + 1)$-th roots of unity. For $k = 3$ use the general solution of the linear equation to determine, using the classical linear theory, a formula for the solutions of the third-order quadratic equation similarly to the one obtained in Example 8.14 for the case $k = 2$.

Note. Recall from Example 8.4 that the state-space of all real solutions of (8.145) is the closed unit ball in $\mathbb{R}^k$.

8.4 Using the ideas in Example 8.15 determine all real solutions of the following third-order difference equation

$$x_n^2 + x_{n-2}^2 + x_{n-3}^2 = 1.$$

8.5 Let $\mathcal{S}$ be invariant so that $\mathcal{S} = \mathcal{M}$ where $\mathcal{M}$ is the largest invariant subset of $\mathcal{S}$ as in Definition 8.5. Does this imply that every solution of the quadratic equation (8.65) is real? Note that invariance of $\mathcal{S}$ is relative to a *particular* binary sequence $\{\beta_n\}$.

8.6 By completing the square, determine the recursive class of Eq. (8.116) in Example 8.12 for arbitrary nonzero real parameters a, b, c, d.

8.7 Assume that $2a + b < 1$ in the quadratic equation (8.63) of Example 8.7.
  (a) Show that (8.63) has a positive fixed point

$$\bar{x} = \sqrt{\frac{c}{1 - 2a - b}}.$$

  (b) Use the fixed point in (a) to show that the nonrecursive equation (8.63) has periodic solutions of all possible periods as well as bounded nonperiodic solutions.

  (c) Use Corollary 8.7 to verify that Eq. (8.63) does *not* have a linear form symmetry.

8.8 Show that for all integers $k \geq 2$ the following "delay" version of equation (8.63) has the same oscillation properties that are mentioned in Example 8.7:

$$x_n^2 = a x_n x_{n-1} + b x_{n-k}^2 + c.$$

8.9 (a) Find the recursive class and discuss the behavior of solutions of the difference equation

$$x_n^2 = a x_n x_{n-1} + b x_{n-2}^2 + c x_{n-1} + d x_{n-2} + e \tag{8.147}$$

which generalizes Eq. (8.63) in Example 8.7.

(b) Show that every solution of every equation in the recursive class (i.e., for every binary sequence $\{\beta_n\}$) with real initial values is a real solution of (8.147) if

$$a \neq 0, \ b > 0, \ e \geq \frac{c^2}{a^2} + \frac{d^2}{4b}.$$

(c) Show that the real solutions in (b) have the same oscillatory properties as those in Example 8.7.

8.10 Prove that inequality (8.77) is equivalent to (8.72).

8.10 (a) Use Corollary 8.5 to determine a factorization for the following variation of the equation in Example 8.1:

$$x_n^2 + ax_{n-1}^2 + bx_n - a^2bx_{n-2} = c$$

where a, b, c are real constants.

(b) Use the result in (a) to explore the existence of real solutions.

8.11 In Example 8.9 let $m = 3$.

(a) Find a periodic solution of (8.84) that is never zero.

(b) Does (8.84) have a period-2 solution? If so what is it? If not then why not?

(c) Find a nonperiodic solution of (8.84).

8.12 Let $\mathcal{F}$ be a nontrivial field and define functions $\phi_n, h_n : \mathcal{F}^2 \to \mathcal{F}$ as follows:

$$\phi_n(u, v) = a_n + \frac{c}{uv}, \quad h_n(u, v) = \frac{1}{u + b_n v}.$$

where $c, a_n, b_n \in \mathcal{F}$ with $a_n, b_n \neq 0$ for all n. Show that

$$\phi_n(h_n(u_0, u_1), h_{n-1}(u_1, u_2))$$

is a quadratic polynomial for every n and determine the difference equation of order 2 that satisfies (8.32). ϕ_n and h_n are examples of non-polynomial factors and form symmetries for a quadratic difference equation. By solving for x_n show also that this equation has a unique recursive expression as a rational equation.

8.13 (a) Use Corollary 8.7 to prove that the quadratic difference equation

$$a_{0,1}x_n x_{n-1} + a_{0,2}x_n x_{n-2} + b_0 x_n + b_1 x_{n-1} + b_2 x_{n-2} = c_n \qquad (8.148)$$

does *not* have a linear form symmetry unless it is linear, i.e., $a_{0,1} = a_{0,2} = 0$. Conclude that the Ladas rational difference equations

$$x_n = \frac{ax_{n-1} + bx_{n-2} + c}{Ax_{n-1} + Bx_{n-2} + C} \qquad (8.149)$$

do not have a linear form symmetry in the nonlinear case $A, B \neq 0$.

(b) Suppose that $A = a_{0,1} \neq 0$ or $B = a_{0,2} \neq 0$. If x_0, x_{-1} are initial values for which

$$Ax_{m-1} + Bx_{m-2} + C = 0$$

in (8.149) for a least $m \geq 1$ then show that (8.148) with $c_n = C$ for all n and $b_0 = a$, $b_1 = b$, $b_2 = c$ has no solutions whose orbit contain the point (x_0, x_{-1}).

(c) State and prove a generalization of (b) for more general rational recursive difference equations and their quadratic versions.

8.14 For the quadratic difference equation

$$a_{0,0}x_n^2 + a_{0,1}x_n x_{n-1} + a_{1,1}x_{n-1}^2 + b_0 x_n + b_1 x_{n-1} + b_2 x_{n-2} = c_n \quad (8.150)$$

assume that all c_n and all the coefficients on the left-hand side are real numbers with $b_2 \neq 0$.

(a) Use Corollary 8.7 to prove that Eq. (8.150) has a factorization into two first-order difference equations as

$$a_{0,1}a_{1,1}t_n^2 + 2b_0 a_{1,1}t_n + a_{0,1}b_2 t_{n-1} = a_{0,1}c_n, \quad (8.151)$$

$$x_n = -\frac{2a_{1,1}}{a_{0,1}}x_{n-1} + \frac{2a_{1,1}}{a_{0,1}}t_n$$

provided that

$$a_{0,1}, a_{1,1} \neq 0, \quad a_{0,0} = \frac{a_{0,1}^2}{4a_{1,1}}, \quad \text{and} \quad (8.152)$$

$$4a_{1,1}^2 b_0 - 2a_{0,1}a_{1,1}b_1 + a_{0,1}^2 b_2 = 0.$$

A discussion similar to that in Example 8.11 may be used here. Note that with the value of $a_{0,0}$ in (8.152) Eq. (8.150) can be written as

$$(a_{0,1}x_n + 2a_{1,1}x_{n-1})^2 + \beta_0 x_n + \beta_1 x_{n-1} + \beta_2 x_{n-2} = 4a_{1,1}c_n$$

where $\beta_j = 4a_{1,1}b_j$ for $j = 0, 1, 2$.

(b) In addition to the restrictions (8.152) on parameters, let $b_2 a_{1,1} < 0$ and $c_n \geq -a_{1,1}(b_0/a_{0,1})^2$ for all n. By completing the square, show that the factor equation (8.151) has a positive solution $\{t_n\}$ that is generated by the recursive first-order equation

$$t_n = -\frac{b_0}{a_{0,1}} + \sqrt{\left(\frac{b_0}{a_{0,1}}\right)^2 + \frac{c_n}{a_{1,1}} - \frac{b_2}{a_{1,1}}t_{n-1}}$$

if $t_0 = (a_{0,1}/2a_{1,1})x_0 + x_{-1} > 0$. Conclude that Eq. (8.150) has real solutions through any initial point (x_0, x_{-1}) in the region of the state-space $\mathbb{R}^2$ defined by the inequality $(a_{0,1}/2a_{1,1})x_0 + x_{-1} > 0$.

(c) If in addition to the hypotheses in Part (b), $c_n = c$ is constant for all n then determine the asymptotic behaviors of solutions $\{t_n\}$ and $\{x_n\}$ in Part (b) for all possible parameter values. Use this information to comment on the class of all solutions of Eq. (8.150).

8.15 (a) Use Theorem 8.2 to derive the factorization in Example 8.10 using the semi-invertible h_n defined by (8.95).

(b) Use this factorization to prove that Eq. (8.97) with $\sigma_n \geq 0$ for all n has a positive solution for each initial point (x_1, x_0) where

$$x_1 = \frac{1}{2}x_0 + t_0 \text{ and } x_0, t_0 > 0.$$

(c) Use the factorization in Example 8.10 to obtain a factorization for the recursive equation

$$x_n = \sqrt{\frac{1}{n}x_n x_{n-1} - \frac{1}{4n^2}x_{n-1}^2 + nx_{n-1} - \frac{1}{2}x_{n-2}}.$$

8.16 Write Eq. (8.108) in recursive form and determine the SC factorization of the resulting recursive equation.

8.17 (a) Show that the difference equation

$$x_n^p - x_{n-1}^p - x_{n-2}^p = c, \quad p > 0, \ c \in \mathbb{R}$$

has a factorization over $\mathbb{R}$ with constant coefficients (based on the Golden Ratio) in both the factor and cofactor equations. Determine this factorization. *Hint:* Let $y_n = x_n^p$.

(b) If p is a positive integer and $c \in \mathbb{Q}$ then show that the equation in Part (a) has a factorization over $\mathbb{Q}$ and find this factorization (coefficients may not be constants).

A

Appendix: Asymptotic Stability on the Real Line

A general necessary and sufficient condition that is known for first-order difference equations defined on the set of all real numbers $\mathbb{R}$ was referred to in the preceding chapters. This condition is not yet as widely known as some other results mentioned earlier (e.g., the Li-Yorke theorem) so for convenience we present it, along with supporting notions and results in this Appendix. These results are taken from Sedaghat (1999); also see Block and Coppel (1992) for related results and Sedaghat (2003) for additional details and applications of these results.

A.1 An inverse-map characterization

Let $f : I \rightarrow I$ be continuous (relative to the usual topology) on a nontrivial interval I which may be unbounded. Let $\bar{x}$ be an isolated fixed point of f, i.e., $f(\bar{x}) = \bar{x}$, and let $U = (\bar{x} - \varepsilon, \bar{x} + \varepsilon)$ be an open interval containing $\bar{x}$ and no other fixed points of f. Assume that the following condition holds:

$$|f(x) - \bar{x}| < |x - \bar{x}| \qquad x \neq \bar{x}, \; x \in U. \tag{A.1}$$

Note that the linearization inequality $|f'(\bar{x})| < 1$ implies (A.1) because of the mean value theorem. In fact, through (A.1) that linear stability is proved.

LEMMA A.1
If K is a nontrivial compact interval such that $f(K) = K$, then K contains either at least two fixed points or a fixed point and a period-2 point.

PROOF   Let $K = [a, b]$ where $a < b$. Then $f(a) \geq a$ and $f(b) \leq b$, so there is at least one fixed point $\bar{x}$ in K. If $\bar{x}$ is unique, then by the hypothesis $f(K) = K$ we conclude that $\bar{x} \neq a, b$, $f(x) > x$ for $x < \bar{x}$, and $f(x) < x$ for $x > \bar{x}$; in fact,

$$\max_{a \leq x \leq \bar{x}} f(x) = b \quad \text{and} \quad \min_{\bar{x} \leq x \leq b} f(x) = a.$$

Now consider the iterate f^2 on $[a, \bar{x}]$, and note that $f^2(K) = K$. If $f^2(a) = a$, then we are done; otherwise, $f^2(a) > a$. Since $a \in f((\bar{x}, b])$ and $(\bar{x}, b] \subset f([a, \bar{x}))$, it follows that there exists $c \in (a, \bar{x})$ such that $f^2(c) = a < c$. Thus, there exists a number $x^* \in (a, c)$ such that $f^2(x^*) = x^*$, i.e., x^* is period-2 point.   ∎

DEFINITION A.1   *(a) Let $\bar{x}$ be a fixed point of f and for each subset $A \subset I$, define the right and left parts of A as*

$$A_r \doteq A \cap [\bar{x}, \infty) \qquad A_l \doteq A \cap (-\infty, \bar{x}].$$

(b) We denote by f_r and f_l the restrictions of f to I_r and I_l, respectively. Since $f_r I_r \subset I$, the inverse map f_r^{-1} may be generally defined on I if we allow the empty set as a possible value of f_r^{-1}. With this convention, we conclude that $f_r^{-1}(x) \subset I_r$ for all $x \in I$, with a similar conclusion holding for f_l and its inverse.

LEMMA A.2
Let $a_i \in I_l$, $b_i \in f_r^{-1}(a_i)$, $i = 1, 2$. If $f(a_1) \geq b_1$ and $f(a_2) \leq b_2$ then there is c between b_1 and b_2 such that $f^2(c) = c$; i.e., the graphs of f_l and f_r^{-1} intersect at c and $\{c, f_r(c)\}$ is a 2-cycle.

PROOF   Note that $f_l \circ f_r(b_1) = f_l(a_1) \geq b_1$ while $f_l \circ f_r(b_2) = f_l(a_2) \leq b_2$. Since $f_l \circ f_r$ is continuous, there is c between b_1 and b_2 such that $f^2(c) = f_l \circ f_r(c) = c$. Note further that $b_1, b_2, c \in I_r$, while $f_r(c) \in I_l$. So $(f_r(c), c) \in f_r^{-1} \cap f_l$.   ∎

LEMMA A.3
If $\bar{x}$ is an isolated fixed point, then a bounded interval $U \subset I$ is a proper I-neighborhood of $\bar{x}$ if:
(i) U is open in I and contains $\bar{x}$;
(ii) $\bar{x}$ is the only fixed point of f that is contained in the closure $\bar{U}$;
(iii) If a is an endpoint of I, then $a \in \bar{U}$ if and only if $a = \bar{x}$.
Note in particular that both U_r and U_l contain $\bar{x}$ and are nonempty; also, every interval neighborhood of $\bar{x}$ contains a proper I-neighborhood.

DEFINITION A.2   *Let $\bar{x}$ be an isolated fixed point of f and let U be a proper I-neighborhood of $\bar{x}$.*
(a) For each $x \in U$ define the lower envelope function of f_r^{-1} on U as

$$\phi(x) \doteq \inf f_r^{-1}(x) = \inf\{u \in U_r : f_r(u) = x\}.$$

Note that $\phi(x) \geq \bar{x} = \inf U_r$ for all $x \in U$ with equality holding if and only if $x = \bar{x}$. By usual convention, $\phi(x) = \infty$ if $f_r^{-1}(x)$ is empty.

(b) For each $x \in U_l$ define the upper envelope function of f_l as

$$\mu(x) \doteq \sup_{x \le u \le \bar{x}} f_l(u).$$

Note that μ is bounded on U_l (because U is proper) and $\mu(x) \ge f(\bar{x}) = \bar{x}$ for all $x \in U_l$.

The next lemma establishes some of the essential properties of ϕ and μ.

LEMMA A.4
Let U be a proper I-neighborhood of an isolated fixed point $\bar{x}$ of f.
(a) μ is a continuous and nonincreasing function on U_l with $\mu(x) \ge f(x)$ for all $x \in U_l$.
(b) If ϕ is real valued on U, then ϕ is a decreasing function on U_l and an increasing function on U_r.
(c) $\phi(x) > f(x)$ for all $x \in U_l$, $x \ne \bar{x}$, if and only if $\phi(x) > \mu(x)$ for all $x \in U_l$, $x \ne \bar{x}$.

PROOF (a) Assume, for nontriviality, that U_l contains points other than $\bar{x}$. It is clear from the definition that μ is nonincreasing and dominates f on U_l. To prove μ is continuous, let $a \in U_l$, $a \ne \bar{x}$ and consider two cases:

Case I. $\mu(a) > f(a)$, so there is a least $b \in (a, \bar{x}]$ such that $\mu(a) = f(b)$. Choose $\delta > 0$ such that $a + \delta < b$, $V = (a - \delta, a + \delta) \subset U_l$ and $f(x) < \mu(a)$ for all $x \in V$. Now, let $x \in V$ and note that if $x > a$ then

$$f(b) \le \mu(x) \le \mu(a) = f(b)$$

while if $x < a$ then

$$\mu(x) = \sup_{a \le u \le \bar{x}} f(u) = \mu(a).$$

Therefore, μ is constant, hence continuous on V.

Case II. $\mu(a) = f(a)$; if μ is not continuous at a, let $x_n \to a$ as $n \to \infty$ and first assume (by taking a subsequence if necessary) that there is $\varepsilon > 0$ such that $\mu(x_n) - \mu(a) \le -\varepsilon$ for all n; but then

$$f(x_n) - f(a) \le \mu(x_n) - \mu(a) \le -\varepsilon$$

for every n, contradicting the continuity of f. So assume (by taking a subsequence if necessary) that $\mu(x_n) \ge \mu(a) + \varepsilon$ for all n. Since μ is nonincreasing, it follows that $x_n < a$ for all n. For each n define $y_n \in [x_n, \bar{x}]$ by the equality $f(y_n) = \mu(x_n)$, and note that

$$f(y_n) > \mu(a) > f(x)$$

for $x \in [a, \bar{x}]$ and all n. Therefore, $x_n \le y_n \le a$ for all n, implying that $y_n \to a$ as $n \to \infty$; however, by the definition of y_n, $f(y_n)$ is not converging to

$\mu(a) = f(a)$ which once again contradicts the continuity of f. This completes the proof of assertion (a).

To prove (b), note that since the sets $f_r^{-1}(x)$ are closed, $\phi(x) \in f_r^{-1}(x)$ for all $x \in U$. Therefore, for each $x \in U$, $\phi(x)$ is the smallest number in I_r with the property that $f_r(\phi(x)) = x$. Since f_r is continuous and $f_r(\bar{x}) = \bar{x}$, the minimality of $\phi(x)$ implies that for $x \in U_l$,

$$f_r(y) \geq x \quad \text{for} \quad y \in [\bar{x}, \phi(x)] \tag{A.2}$$

with the inequality reversed for $x \in U_r$. Now, if (b) is false, and there are $u, v \in U_l$, $u < v$ such that $\phi(u) \leq \phi(v)$, then $\phi(u) \in [\bar{x}, \phi(v)]$ with $f_r(\phi(u)) = u < v$ which contradicts (A.2). The argument for $u, v \in U_r$ is similar.

With regard to (c), necessity being clear from the definition of μ, we proceed to prove the sufficiency; i.e., if there is $u \in U_l$ such that $\phi(u) \leq \mu(u)$, then for some $v \in U_l$, $\phi(v) \leq f(v)$. Choose $v \in [u, \bar{x}]$ so that $\mu(u) = f(v)$. Then by Part (b) and our assumption on u,

$$\phi(v) \leq \phi(u) \leq \mu(u) = f(v)$$

which is the desired inequality for v.   ∎

The next theorem is the key result of this section.

THEOREM A.1

(The Inverse Map Characterization) A fixed point $\bar{x}$ of f is asymptotically stable if and only if there is a proper I-neighborhood U of $\bar{x}$ such that:

$$\begin{cases} \phi(x) > f(x) > x & \text{if} \quad x \in U_l, \ x \neq \bar{x} \\ f(x) < x & \text{if} \quad x \in U_r, \ x \neq \bar{x}. \end{cases} \tag{A.3}$$

PROOF   First assume that (A.3) holds. For convenience, we denote $U_l - \{\bar{x}\}$ by U_l^o, and similarly for U_r. First assume that $f_r^{-1}(x)$ is empty for all $x \in U_l^o$ (or that U_l^o is empty) so $f_r(x) \geq \bar{x}$ for all $x \in U_r$. Now if $x_0 \in U_r^o$ then by (A.3), $\bar{x} \leq f_r(x_0) < x_0$ so we conclude by induction that $f^n(x_0) = f_r^n(x_0)$ decreases to $\bar{x}$ from the right. If $x_0 \in U_l^o$ then either $f^k(x_0) \geq \bar{x}$ for some $k \geq 1$ or $f^n(x_0) < \bar{x}$ for all $n \geq 1$. In the former case, assuming without loss of generality that $f^k(x_0) \in U_r^o$, the sequence $\{f^{k+n}(x_0)\}$ decreases as before to $\bar{x}$. In the second case, condition (A.3) shows that

$$x_0 < f^{n-1}(x_0) < f^n(x_0) < \bar{x}$$

for all n so that the terms $f^n(x_0)$ increase to $\bar{x}$ from the left. Next, assume that $f_r^{-1}(u)$ is nonempty for some $u \in U_l^o$, in which case $f_r^{-1}(x)$ is nonempty for all $x \in [u, \bar{x}]$ by Lemma A.4(b). So we may choose $a \in U_l^o$ sufficiently close to $\bar{x}$ such that $f_r^{-1}(a)$ is nonemtpy, $\mu(a) \in U_r$ and thus $J = [a, \mu(a)] \subset U$. We now show that $fJ \subset J$. If $x \in [a, \bar{x}]$, then by (A.3)

$$a \leq x < f(x) \leq \mu(x) \leq \mu(a)$$

so that

$$f[a, \bar{x}] \subset (a, \mu(a)] \subset J. \tag{A.4}$$

Next, suppose that $x \in [\bar{x}, \mu(a)]$. If $f(x) \geq \bar{x}$ then by (A.3) $f(x) \in [\bar{x}, x) \subset [\bar{x}, \mu(a)]$, while if $f(x) < \bar{x}$, then (A.3) and Lemma 2.1.3(c) imply that

$$\mu(f(x)) < \phi(f(x)) = \inf f_r^{-1}(f_r(x)) \leq x \leq \mu(a)$$

which because of the nonincreasing nature of μ implies that $f(x) > a$. Thus

$$f[\bar{x}, \mu(a)] \subset (a, \mu(a)] \subset J. \tag{A.5}$$

Inequalities (A.4) and (A.5) imply that $fJ \subset J$. Now successive applications of f to J yield a decreasing sequence $J \supset fJ \supset f^2J \supset \cdots$ whose limit $K = \bigcap_{n=0}^{\infty} f^n J$ contains $\bar{x}$ and is thus nonempty. Since $f^n J$ is a compact interval for every n, it follows that K is a compact interval and $fK = K$. Given that $\bar{x}$ is the only fixed point of f in $K \subset J \subset U$, Lemma A.2 implies that $K = \{\bar{x}\}$. Hence, $\bar{x}$ is asymptotically stable.

To prove the converse, suppose that every proper I-neighborhood U of $\bar{x}$ contains a point x_U such that (A.3) fails at x_U. Thus either (I) $x_U \in U_r^o$ and $f_r(x_U) \geq x_U$, or (II) $x_U \in U_l^o$ and $\phi(x_U) \leq f_l(x_U)$ or $f_l(x_U) \leq x_U$ (here ϕ defined with respect to some U works for all smaller neighborhoods contained in U). In case (I), the uniqueness of $\bar{x}$ in U implies that $f_r(x) > x$ for all $x \in (\bar{x}, x_U)$. But then for every $x_0 \in (\bar{x}, x_U)$, no matter how close to $\bar{x}$, the increasing sequence

$$x_0 < f(x_0) < \cdots < f^{n+1}(x_0), \quad \text{if } f^k(x_0) < x_U, \text{ for } 1 \leq k \leq n$$

eventually exceeds x_U; it follows that $\bar{x}$ is not stable. In case (II) the inequality $f_l(x_U) \leq x_U$ implies that $\bar{x}$ is not asymptotically stable in a manner similar to that just described for (I). It remains to show that the other inequality in (II) also implies a lack of asymptotic stability. The first inequality in (II) applied over a sequence $\{U_n\}$ of neighborhoods of $\bar{x}$ whose diameters approach zero, implies that there is a sequence $u_n \to \bar{x}$ from the left such that $\phi(u_n) \leq f_l(u_n)$. Since for each n, $\phi(u_n) \geq \bar{x}$ and also $f_l(u_n) \to \bar{x}$ as $n \to \infty$, we conclude that $\phi(u_n) \to \bar{x}$. Since for each $x \in I_l^o$, $f_r^{-1}(x) \cap I_l^o$ is not empty, two possible cases arise:

Case 1. There is $\delta > 0$ sufficiently small that $\sup \left[f_r^{-1}(x) \cap (\bar{x}, \bar{x} + \delta) \right] < f_l(x)$ for all $x \in (\bar{x} - \delta, \bar{x})$; i.e., the graph of f_r^{-1} near and to the left of $\bar{x}$ lies below the graph of f_l. Let $x_0 \in (\bar{x}, \bar{x} + \delta)$ and note that $f_r(x_0) < \bar{x}$. If $f_r(x_0) > \bar{x} - \delta$, then since $x_0 \in f_r^{-1}(f_r(x_0))$, we see that $f^2(x_0) = f_l(f_r(x_0)) > x_0$. If

$f^2(x_0) < \bar{x} + \delta$, then the preceding argument may be repeated; inductively, the sequence

$$x_0 < f^2(x_0) < f^4(x_0) < \cdots$$

is obtained that moves away from $\bar{x}$ until it exceeds $\bar{x}+\delta$, no matter how close x_0 is to $\bar{x}$. Therefore, $\bar{x}$ is not stable. Now let $x_0 \in (\bar{x} - \delta, \bar{x})$ and note that $f_l(x_0) > \bar{x}$. Repeating the above argument, the sequence $\{f^{2k+1}(x_0)\}$ is seen to increase away from $\bar{x}$, and once again $\bar{x}$ cannot be asymptotically stable.

Case 2. There is a sequence $v_n \to \bar{x}$, $v_n < \bar{x}$, such that

$$\sup\left[f_r^{-1}(v_n) \cap [\bar{x}, \bar{x} + 1/(n+1)]\right] \geq f_l(v_n), \qquad n = 1, 2, 3, \ldots$$

i.e., for each n, there is $w_n \in f_r^{-1}(v_n) \cap [\bar{x}, \bar{x}+1/(n+1)]$ such that $w_n \geq f_l(v_n)$ and $w_n \to \bar{x}$ as $n \to \infty$. The conditions of Lemma 2.1.2 are met with $a_1 = v_n$, $b_1 = w_n$, $a_2 = u_n$ and $b_2 = \phi(u_n)$. It follows that there is c_n between w_n and $\phi(u_n)$ such that $f^2(c_n) = c_n$; i.e., there is a sequence of period-2 points $c_n \to \bar{x}$ as $n \to \infty$, and therefore, $\bar{x}$ is again not asymptotically stable. This concludes the proof. ∎

Condition (A.3) may be thought of as a "right" condition because of f_r^{-1}. There is also a "left" analog of (A.3) which compares f_l^{-1} with f_r and can be useful when the left part f_l is simpler than the right part f_r for the purpose of inversion. These left versions are obtained from the right versions here by making a few minor modifications.

A.2 Equivalent necessary and sufficient conditions

The following result presents a few conditions that are equivalent to that in Theorem A.1.

THEOREM A.2

(Asymptotic Stability) Let $\bar{x}$ be a fixed point of f. The following statements are equivalent:

(a) $\bar{x}$ is asymptotically stable;

(b) There is a proper I-neighborhood U of $\bar{x}$ on which the following inequality holds:

$$[f^2(x) - x](x - \bar{x}) < 0, \qquad x \neq \bar{x}, \; x \in U \subset I \tag{A.6}$$

(c) There is a proper I-neighborhood U of $\bar{x}$ on which (A.3) holds;

(d) There is a proper I-neighborhood U of $\bar{x}$ such that:

$$[f(x) - x](x - \bar{x}) < 0, \qquad x \neq \bar{x}, \; x \in U \subset I \tag{A.7}$$

and over $U_l - \{\bar{x}\}$, the graph of f_r^{-1} lies above the graph of f;

(e) Inequality (A.7) holds on U, and for the parameterizations of f_l and f_r^{-1} given by $(x_l(t), y_l(t))$ and $(y_r(s), x_r(s))$ respectively, it is ture that:

$$x_l(t) = y_r(s) \quad \text{implies} \quad x_r(s) > y_l(t).$$

(f) There is a proper I-neighborhood U of $\bar{x}$ such that for $x \in U - \{\bar{x}\}$,

$$(\phi(x) - f(x))(\phi(x) - x)(f(x) - x)(x - \bar{x}) < 0. \tag{A.8}$$

PROOF We show that (a)$\Rightarrow$(b)$\Rightarrow$(c)$\Rightarrow$(a) and (c)$\Rightarrow$(d)$\Rightarrow$(e)$\Rightarrow$(f)$\Rightarrow$(c). First, if (a) is true, then there is a sufficiently small I-neighborhood U of $\bar{x}$ on which the equality $f^2(x) = x$ holds only at $\bar{x}$ and every point in U is attracted to $\bar{x}$. Hence, the continuous function $f^2(x) - x$ either does not change its sign over U, or if it does, then the sign change can occur only at $\bar{x}$. If $f^2(x) > x$ for all $x \in U_r^o$, and $x_0 \in U_r^o$, then $f^2(x_0) > x_0$; if $f^2(x_0) \in U_r$ also, then another application of f^2 leads further away from $\bar{x}$ and the process continues until the trajectory $\{f^{2n}(x_0)\}$ exits U_r, no matter how close x_0 is to $\bar{x}$. Thus $\bar{x}$ cannot be stable, contradicting (a). Similarly, (a) is contradicted if $f^2(x) < x$ for all $x \in U_l^o$. Now (A.6) follows and (b) is established.

Next, suppose that (b) is true. Then (A.7) must hold, since otherwise there is either a fixed point other than $\bar{x}$ in U at which (b) would be false, or else, $f(x) < x$ (respectively, $f(x) > x$) for all $x < \bar{x}$ (respectively, $x > \bar{x}$) in U, in which case choosing x_0 sufficiently close to $\bar{x}$ so that $f(x_0) \in U$ implies that $f^2(x_0) < x_0$ (respectively, $f^2(x_0) > x_0$) also, again contradicting (b).

To establish (c), it remains to show that $\phi(x) > f_l(x)$ for $x \in U_l$. This is clear if $\phi(x) \geq a > 0$ for all $x \in U_l^o$; otherwise, arguing as in the last two cases in the proof of Theorem A.1, we conclude that there is either a sequence of period-2 points converging to $\bar{x}$ from the left, or else, there is $x' \in U_l^o$ close to $\bar{x}$ such that $x_0 = f(x') \in U_r$ and $f^2(x_0) \in U_r$ with $f^2(x_0) > x_0$. Since in either case (b) is contradicted, we must assume that (c) holds. Finally, in Theorem A.1 it was established that (c) implies (a).

Next, note that (d) follows easily from (c) because conditions (A.3) imply (A.7), and by Lemma A.4(c) ϕ (hence also the graph of f_r^{-1}) dominates f on U_l^o if and only if ϕ dominates μ, hence also f_l. In light of Lemma A.4(c), (e) is just a rephrasing of (d), hence equivalent to it. Statement (f) is an immediate consequence of (e), or equivalently (d), which implies that $\phi(x) > f(x)$ for all $x \in U - \{\bar{x}\}$ (for $x > \bar{x}$, the graph of f_r^{-1} lies above the identity line if and only if f_r lies below that line).

Finally, assume (f) holds. For $x < \bar{x}$, $\phi(x) > \bar{x} > x$, so if $\phi(u) < f(u)$ for some $u < \bar{x}$, then $f(u) > \bar{x} > u$ and (A.8) fails. Hence $\phi(x) > f(x)$ for all x, and so by (A.8) $f(x) > x$. For $x > \bar{x}$, the product $(\phi(x) - f(x))(\phi(x) - x)$ is always positive, since both f and the identity line always lie on the same side of f_r^{-1}. Therefore, by (A.8) $f(x) - x < 0$ and condition (A.3) is established.

For continuous maps of the real line, Theorem A.2 settles the question of asymptotic stability conclusively, although it does not explicitly address instability, neutral stability, etc. For a detailed discussion of these issues we refer to Sedaghat (2003).

References

Agarwal, R.P. (2000) *Difference Equations and Inequalities*, 2nd ed., Dekker, New York.

Akin, E. (1993) *The General Topology of Dynamical Systems*, Amer. Math. Society, Providence.

Alseda, L. and Llibre, J. (1993) "Periods for triangular maps," *Bull. Austral. Math. Soc.*, **47,** 41–53.

Andreassian, A. (1974) "Fibonacci sequences modulo m," *Fibonacci Quart.*, **12,** 51–54.

Bertram, W. (2007) "Difference problems and differential problems," in *Contemp. Geom. Topol. and Related Topics*, Proceedings of Eighth Int. Workshop on Differential Geometry and its Applications, Cluj-Napoca, 73–87.

Block, L. and Coppel, W.A. (1992) *Dynamics in One Dimension*, Springer, New York.

Byers, B. (1983) "Topological semiconjugacy of piecewise monotone maps of the interval," *Trans. Amer. Math. Soc.*, **276,** 489–495.

Camouzis, E. and Ladas, G. (2008) *Dynamics of third-order Rational Difference Equations with Open Problems and Conjectures*, Chapman and Hall/CRC Press, Boca Raton.

Ceccherini-Silberstein, T. and Coornaert, M. (2010) *Cellular Automata and Groups*, Springer, Berlin.

Chan, D.M., Chang, E-R., Dehghan, M., Kent, C.M., Mazrooei-Sebdani, R. and Sedaghat, H. (2006) "Asymptotic stability for difference equations with decreasing arguments," *J. Difference Eqs. and Appl.*, **12,** 109–123.

Cheng, S.S. (2003) *Partial Difference Equations*, Taylor and Francis, London.

Dehghan, M., Kent, C.M., Mazrooei-Sebdani, R., Ortiz, N.L., and Sedaghat, H. (2008a) "Dynamics of rational difference equations containing quadratic terms," *J. Difference Eq. Appl.*, **14,** 191–208.

Dehghan, M., Kent, C.M., Mazrooei-Sebdani, R., Ortiz, N.L., and Sedaghat, H. (2008b) "Monotone and oscillatory solutions of a rational difference equation containing quadratic terms," *J. Difference Eq. Appl.*, **14,** 1045–1058.

Dehghan, M., Mazrooei-Sebdani, R., and Sedaghat, H. (2011) "Global behavior of the Riccati difference equation of order two," *J. Difference Eq. Appl.*, in press.

Devaney, R.L. (1989) *An Introduction to Chaotic Dynamical Systems*, 2nd ed., Addison-Wesley, Redwood City.

DeVault, R., Ladas, G. and Schultz, S.W. (1998) "On the recursive sequence $x_{n+1} = A/x_n + 1/xn - 2$," *Proc. Amer. Math. Soc.*, **126**, 3257–3261.

Elaydi, S. (1999) *An Introduction to Difference Equations*, 2nd ed., Springer, New York.

El-Metwally, H.A., Grove, E.A., Ladas, G., and Voulov, H.D. (2001) "On the global attractivity and the periodic character of some difference equations," *J. Difference Eqs. and Appl.*, **7**, 837–850.

Fisher, M.E. and Goh, B.S. (1984) "Stability results for delayed-recruitment in population dynamics," *J. Math. Biol.*, **19**, 147–156.

Fraleigh, J.B. (1976) *A First Course in Abstract Algebra*, 2nd ed, Addison-Wesley, Reading.

Franke, J.E., Hoag, J.T., and Ladas, G. (1999) "Global attractivity and convergence to a two-cycle in a difference equation," *J. Difference Eqs. and Appl.*, **5**, 203–209.

Gil, M.I. (2007) *Difference Equations in Normed Spaces: Stability and Oscillations*, Elsevier, Amsterdam.

Grove, E.A., Ladas, G., McGrath, L.C., and Teixeira, C.T. (2001) "Existence and behavior of solutions of a rational system," *Comm. Appl. Nonlin. Anal.*, **8**, 1–25.

Grove, E.A. and Ladas, G. (2005) *Periodicities in Nonlinear Difference Equations*, CRC Press, Boca Raton.

Gu, E.G, and Ruan, J. (2005) "On some global bifurcations of the domains of feasible trajectories: An analysis of recurrence equations," *Int'l J. Bifurc. and Chaos*, **15**, 1625–1639.

Hamaya, Y. (2005) "On the asymptotic behavior of solutions of neuronic difference equations," in *Proc. Int'l. Conf. on Difference Equations, Special Functions and Appl.*, Elaydi et al. (editors), World Scientific, 258–265.

Hicks, J.R. (1965) *A Contribution to the Theory of the Trade Cycle*, 2nd ed., Clarendon Press, Oxford (First ed. 1950, Oxford University Press, Oxford).

Hungerford, T.W. (1974) *Algebra*, Springer, New York.

Jordan, C. (1965) *Calculus of Finite Differences*, 3rd ed., Chelsea, New York.

Kent, C.M. and Sedaghat, H. (2004a) "Convergence, periodicity and bifurcations for the 2-parameter absolute-difference equation," *J. Difference Eqs. and Appl.*, **10**, 817–841.

Kent, C.M. and Sedaghat, H. (2004b) "Global stability and boundedness in $x_{n+1} = cx_n + f(x_n - x_{n-1})$," *J. Difference Eqs. and Appl.*, **10**, 1215–1227.

Kocic, V.L. and Ladas, G. (1993) *Global Behavior of Nonlinear Difference Equations of higher order with Applications*, Kluwer, Dordrecht.

Kulenovic, M.R.S. and Ladas, G. (2002) *Dynamics of second-order Rational Difference Equations with Open Problems and Conjectures*, Chapman and Hall/CRC, Boca Raton.

Mickens, R. (1991) *Difference Equations, Theory and Applications*, 2nd ed., CRC Press, Boca Raton.

Milnor, J. and Thurston, W. (1977, preprint) "On iterated maps of the interval," I, II in: *Lecture Notes in Math.*, **1342**, Springer, New York, 1988.

Morimoto, Y. (1988) "Hopf bifurcation in the simple nonlinear recurrence equation $x(t + 1) = Ax(t)(1 - x(t - 1))$," *Phys. Lett.*, **A134**, 179–182.

Morimoto, Y. (1989) "Variation and bifurcation diagram in difference equation $x(t + 1) = ax(t)(1 - x(t) - bx(t - 1))$," *Trans. IEEIC*, **72**, 1–3.

Philos, Ch.G., Purnaras, I.K. and Sficas, Y.G. (1994) "Global attractivity in a nonlinear difference equation," *Appl. Math and Comp.*, **62**, 249–258.

Puu, T. (1993) *Nonlinear Economic Dynamics*, 3rd. ed., Springer, New York.

Renault, M. (1996) "Properties of the Fibonacci sequence under various moduli," Master's Thesis, Wake Forest University.

Rothe, F. (1992) "Snapback repellers and semiconjugacy for iterated entire mappings," *Manuscripta Math.*, **74**, 299–319.

Samuelson, P.A. (1939) "Interaction between the multiplier analysis and the principle of acceleration," *Rev. Econ. Stat.*, **21**, 75–78.

Sedaghat, H. (1997) "A class of nonlinear, second-order difference equations from macroeconomics," *Nonlin. Anal. TMA*, **29**, 593–603.

Sedaghat, H. (1998) "Bounded oscillations in the Hicks business cycle model and other delay equations," *J. Difference Eqs. and Appl.*, **4**, 325–341.

Sedaghat, H. (1999) "An inverse map characterization of asymptotic stability on the line," *Rocky Mt. J. Math*, **29**, 1505–1519.

Sedaghat, H. (2003) *Nonlinear Difference Equations: Theory with Applications to Social Science Models*, Kluwer Academic, Dordrecht.

Sedaghat, H. (2004a) "Periodicity and convergence for $x_{n+1} = |x_n - x_{n-1}|$," *J. Math. Anal. Appl.*, **291**, 31–39.

Sedaghat, H. (2004b) "On the equation $x_{n+1} = cx_n + f(x_n - x_{n-1})$," Proc. 7th Int'l. Conf. on Difference Equations and Appl., *Fields Inst. Communications*, **42**, 323–326.

Sedaghat, H. (2008) "Reduction of order of separable second-order difference equations with form symmetries," *Int'l. J. Pure and Applied Math.*, **47**, 155–163.

Sedaghat, H. (2009a) "Every homogeneous difference equation of degree one admits a reduction in order," *J. Difference Eqs. and Appl.*, **15**, 621–624.

Sedaghat, H. (2009b) "Global behaviors of rational difference equations of orders two and three with quadratic terms," *J. Difference Eqs. and Appl.*, **15**, 215–224.

Sedaghat, H. (2009c) "Periodic and chaotic behavior in a class of second-order difference equations,", *Adv. Stud. Pure Math.*, **53**, 321–328.

Sedaghat, H. (2009d) "Reduction of order in difference equations by semiconjugate factorizations," *Int'l. J. Pure and Applied Math.*, **53**, 377–384.

Sedaghat, H. (2009e) "Semiconjugate factorizations and reduction of order in difference equations," http://arxiv.org/abs/0907.3951

Sedaghat, H. (2010a) "Factorizations of difference equations by semiconjugacy with application to nonautonomous linear equations," http://arxiv.org/abs/1005.2428

Sedaghat, H. (2010b) "Semiconjugate factorization of nonautonomous higher-order difference equations," *Int'l. J. Pure and Applied Math.*, **62**, 233–245.

Sedaghat, H. (2010c) "Reductions of order in difference equations defined as products of exponential and power functions", *J. Difference Eqs. and Appl.*, 1563-5120, in press.

Smital, J. (2008) "Why it is important to understand the dynamics of triangular maps," *J. Difference Eqs. and Appl.*, **14**, 597–606.

Strogatz, S.H. (1994) *Nonlinear Dynamics and Chaos*, Addison-Wesley, Reading.

Wall, D.D. (1960) "Fibonacci series modulo m," *Amer. Math. Monthly*, **67**, 525–532.

Wolfram, S. (1984) *Cellular Automata and Complexity: Collected Papers*, Addison-Wesley, Reading.

Index

algebraically closed, 56, 181
almost periodic, 201
aperiodic, 8, 14, 179, 248
asymptotic stability
    inverse-map characterization, 300
    necessity and sufficiency, 302
asymptotically stable, 103
    cycle or periodic solution, 156
    global, 95, 96, 122
    local, 121
autonomous, 9, 28, 188, 284
    absolute value equation, 79
    difference equation, 39
    Riccati equation, 231

bifurcation, 96, 178, 203, 206
    in state-space, 244
    period-doubling, 157, 179
    tangent, 157, 179
bijection, 18, 36, 108, 256
binary sequence, 62, 241, 245, 262, 266, 272
binomial coefficients, 291
business cycle, 143
    boom-bust cycle, 144

cellular automata (CA), 25, 28
chain
    cofactor, 54, 169, 179
    factor, 54, 169, 179
    reduction, 54
change of variables, 58
chaotic, 94, 159, 179, 208
characteristic polynomial, 117, 126, 180
cofactor
    chain, 54, 149, 179

equation, 46, 107, 110, 128, 132, 146, 169, 253, 258, 276, 279
complementary solution, 181
complete SC factorization, 56, 74, 179
completing the square method, 265
complex numbers
    field, 171
    multiplicative subgroup, 184
conjugate equivalents, 58
conjugate maps, 36, 38
coordinate transformation, 36

degenerate quadratic difference equation, 260
dense
    orbit, 125, 187
    set, 126
    subgroup, 183
Descartes (rule of signs), 117
difference equation
    absolute value, 79, 134, 245
    autonomous, 6, 188, 203
    expow (power/exponential), 189
    factorization, not SC, 253
    functional, 17
    Henon, 260
    homogeneous of degree one, or HD1, 66
    linear, 123, 179
    linear argument, 139, 170
    linear, nonhomogeneous, 22
    Lyness, 23
    nonautonomous, 6, 206
    nonrecursive or implicit, 6, 239
    partial, 25
    quadratic, 259
    rational, 23, 50, 54, 67

T - #0487 - 101024 - C0 - 234/156/18 - PB - 9781138374126 - Gloss Lamination